Wireless Communications
Signal Processing Perspectives

ISBN 0-13-620345-0

90000

9 780136 203452

PRENTICE HALL SIGNAL PROCESSING SERIES

Alan V. Oppenheim, *Series Editor*

Wireless Communications
Signal Processing Perspectives

H. Vincent Poor

Princeton University

Gregory W. Wornell

Massachusetts Institute of Technology

EDITORS

Prentice Hall PTR
Upper Saddle River, New Jersey 07458
http://www.phptr.com

Library of Congress Cataloging-in-Publication Data

Wireless communications: signal processing perspectives / [edited by]
  H. Vincent Poor, Gregory W. Wornell.
        p.   cm.—(Prentice Hall signal processing series)
    Includes bibliographical references and index.
    ISBN 0–13–620345–0
    1. Wireless communication systems.   2. Signal processing.
  I. Poor, H. Vincent.   II. Wornell, Gregory W.   III. Series.
  TK5103.2.W5718   1998   98–9676
  621.382—dc21   CIP

Editorial/production supervision: *Jane Bonnell*
Cover design director: *Jerry Votta*
Cover design: *Anthony Gemmellaro*
Copyeditor: *Mary Lou Nohr*
Manufacturing manager: *Alan Fischer*
Acquisitions editor: *Bernard M. Goodwin*
Editorial assistant: *Diane Spina*
Marketing manager: *Miles Williams*

 © 1998 by Prentice Hall PTR
Prentice-Hall, Inc.
A Simon & Schuster Company
Upper Saddle River, New Jersey 07458

Prentice Hall books are widely used by corporations and government agencies
for training, marketing, and resale.
The publisher offers discounts on this book when ordered in bulk quantities.
For more information, contact Corporate Sales Department, Phone: 800–382–3419;
FAX: 201–236–7141; E-mail: corpsales@prenhall.com
Or write: Prentice Hall PTR, Corporate Sales Dept., One Lake Street, Upper Saddle River,
NJ 07458.

Printed in the United States of America
10 9 8 7 6 5 4 3 2 1

ISBN 0-13-620345-0

Prentice-Hall International (UK) Limited, *London*
Prentice-Hall of Australia Pty. Limited, *Sydney*
Prentice-Hall Canada Inc., *Toronto*
Prentice-Hall Hispanoamericana, S. A., *Mexico*
Prentice-Hall of India Private Limited, *New Delhi*
Prentice-Hall of Japan, Inc., *Tokyo*
Simon & Schuster Asia Pte. Ltd., *Singapore*
Editora Prentice-Hall do Brasil, Ltda., *Rio de Janeiro*

Contents

Preface

These are exciting times in which to be involved in wireless communications research. The field is growing at an explosive rate, stimulated by a host of important emerging applications ranging from third-generation mobile telephony [5, 10, 17, 18, 20], wireless personal communications [1, 7, 8, 9, 13, 14, 16], and wireless subscriber loops, to radio and infrared indoor communications [2, 4, 6, 11, 15], nomadic computing [3, 12], and wireless tactical military communications [10]. These and other newly envisioned networks have both profound social implications and enormous commercial potential. For system planners and designers, the projections of rapidly escalating demand for such wireless services present major challenges, and meeting these challenges will require sustained technical innovation on many fronts.

CHARACTERISTICS OF WIRELESS SERVICES

Many of the main technical challenges stem from three key characteristics inherent in existing and envisioned wireless services. First, the communication channels over which radio-frequency (RF), infrared, underwater acoustic, and other wireless systems must operate are all complex and highly dynamic. These channels suffer from numerous physical impairments that severely impact system performance, important examples of which include fading due to multipath propagation and interference from extra-network sources. Moreover, channel characteristics often change at rates that can be significant relative to system time scales, particularly in mobile communications applications.

Second, many emerging wireless applications are aimed toward providing universal access at relatively high data rates. Providing these capabilities is increasingly accomplished through the use of random multiple-access protocols, which while natural, lead to a still more complicated wireless channel. In particular, in such cases multiple-access/cochannel interference is at least as significant an impairment as noise and other forms of interference in limiting system performance. Another aspect of the universal access paradigm is that network demands more generally are highly dynamic, with the users entering and leaving the network having diverse quality-of-service requirements. At the same time, the network itself is often reconfigurable and typically part of a larger heterogeneous system of networks, meaning that the associated network resources are also highly dynamic as well.

Finally, driving much of the development of wireless technology is the need for truly portable communications. Many applications demand a system infrastructure that can be rapidly and flexibly deployed. And the end users themselves require a lightweight, compact interface to the network in the form of a pocket-sized, battery-powered transceiver or terminal. As such, complexity and power consumption are critical issues in the design of mobile systems and give rise to important practical constraints.

What all these diverse challenges have in common is the very central role that signal processing ultimately has to play in meeting them. Indeed, increasingly we are seeing many key problems in wireless communication system design being approached from signal processing perspectives, yielding solutions in the form of advanced signal processing algorithms and having implementations on flexible digital signal processing architectures. In turn, this evolution is stimulating both significantly heightened interest in signal processing methodologies within the wireless communications community and considerable recent growth of interest within the signal processing community in wireless communications applications.

The idea of this book evolved from these basic observations. As such, its aim is to provide a signal processing perspective on the field of wireless communications by describing the state of the art and recent research developments in this area, and also by identifying key directions in which further research is needed. The treatment comprises eight contributed chapters and an epilogue, spanning some of the main focus areas of signal processing research at both the physical and network layers of the wireless system hierarchy.

OVERVIEW OF CHAPTERS

A brief outline of the constituent chapters is as follows.

Chapter 1 is the first of four that focus primarily on issues at the physical layer. In this chapter, Wornell focuses on the role of signal processing in creating and exploiting diversity for counteracting the effects of multipath-induced signal fading. Multipath propagation effects generally lead to a channel signal-to-noise ratio characteristic that varies as a function of frequency, time, and space; diversity techniques take advantage of the fact that typically not all parts of such channels fade simultaneously. This chapter describes the key ways in which linear signal processing algorithms can be used at the transmitters and receivers of multiuser communication systems to realize effective forms of diversity with very low computational complexity. Such diversity is achieved by, in effect, spreading the transmission of symbols spectrally, temporally, and/or spatially to within the limits imposed by bandwidth, delay, and other physical system constraints. Spread-spectrum code-division multiple-access (CDMA) protocols and multiple-element antenna arrays at transmitters and receivers are discussed as means for realizing spreading of this type. All such techniques have the characteristic that they improve both average and worst-case performance and can be used in conjunction with—or as an alternative to—coding in such systems. The various techniques in Chapter 1 are described and interrelated within a common multirate/multichannel signal processing framework. This framework not only provides some useful new perspectives on traditional approaches for making use of diversity but also lends itself naturally to the description of several promising, more recently introduced methods.

In Chapter 2, Honig and Poor describe the analogous role that signal processing algorithms have to play in suppressing interference in the receivers of wireless systems. The primary focus of the chapter is on suppression of interference that arises due to the nonorthogonal multiplexing inherent in random-access protocols such as CDMA. While these formats allow systems to optimize their use of bandwidth in channels subject to fading, bursty traffic, and time-varying user populations, overall system performance depends critically on the degree to which the accompanying increase in interference is effectively eliminated at the receiver. Unlike the ambient noise that limits all electronic communications, this multiple-access interference is highly structured. This structure provides both challenges and opportunities for the use of advanced signal processing to mitigate the effects of the interference. Chapter 2 develops the basic signal processing concepts relevant to this and closely related classes of interference suppression problems such as multipath mitigation, narrowband interference suppression, and beamforming. These problems are treated within the basic framework of the CDMA transmission protocol, although many of the techniques described are equally applicable to any system in which structured interference is a major impairment. Moreover, because typically the wireless channel is rapidly time-varying, the emphasis in the chapter is on adaptive methods. This is a very active area of current research, and a number of open issues are discussed.

Wireless channels are ultimately limited by their bandwidths. As data rates become more demanding with respect to the channel bandwidth, dispersion and the attendant intersymbol interference become a critical, performance-limiting issue. The equalization of wireless channels presents major signal processing challenges not present in more traditional wireline equalization, again both because of the rapid time-variation in the channel and because of the additional sources of interference that compound the problem. In Chapter 3, Papadopoulos discusses the principal issues arising in the problem of equalizing wireless channels in multiuser environments and describes several of the key algorithmic structures being explored for their solution. The development of this chapter emphasizes a powerful multiple-input multiple-output linear systems perspective within which equalizers that efficiently and jointly mitigate both intersymbol and multiple-access interference are described. As the development reflects, the resulting equalizers constitute natural and powerful generalizations of equalizers used in many traditional communications applications. Representative examples of both training-based and blind algorithms from this particularly active area of research are discussed.

In Chapter 4, Paulraj, Papadias, Reddy, and van der Veen turn to the problem of space-time processing for wireless systems. This chapter is complementary to the preceding three in that it also considers issues of spatial diversity, interference suppression, and equalization in the wireless channel. However, the focus of this chapter is on explicit space-time signal processing strategies that can be employed in these pursuits. In particular, this chapter develops array signal processing for both single-user and multiuser systems with the underlying assumptions of linear modulation and time-division multiple-access transmission. Key elements of this treatment are the issues of blind channel identifiability and linear channel equalizability in the presence of both intersymbol interference and cochannel interference.

Ensuring that emerging wireless networks operate efficiently while meeting the quality-of-service requirements of end users increasingly requires sophisticated strategies for dynamic allocation of power, bandwidth, and other resources. Moreover, many of the key system design issues are particularly challenging because they span both the physical layer and the network layer within the wireless communication system hierarchy—two layers whose development has traditionally been treated separately and by culturally different subcommunities. In Chapter 5, Tse and Hanly explore some key aspects of the interplay between these two layers in wireless architectures. In particular, they discuss natural and practical notions of capacity for a cellular system in terms of the quality-of-service requirements of the participating users. Central to their development is the powerful concept of a user's effective bandwidth, which summarizes the fraction of a cell's resources that is required to support a user at its desired target signal-to-interference ratio, given the class of interference suppression techniques being implemented at the physical layer. Through this framework, several important insights into the problem of optimal power control are obtained.

Focusing further on the network layer, one of the major thrusts of current research in wireless communications is the development of techniques for introducing multimedia capability into wireless networks. The high bandwidth requirements and quality-of-service expectations of multimedia traffic place major demands on all signal processing functions of the wireless system. In Chapter 6, Haskell, Messerschmitt, and Yun consider the major signal processing issues that arise from the needs of such multimedia transmission. In particular, the chapter examines four primary signal processing functions—data compression, encryption, modulation, and error control—in this context and considers the impact of multimedia traffic on the backbone network. The authors propose a novel architectural framework for the design of both network protocols and signal processing algorithms that allows for the resolution of most issues arising in wireless multimedia networks. In outlining this framework, the authors also identify a number of active research areas, both in networking and signal processing.

In Chapter 7, Ramchandran and Vetterli explore issues of data compression (i.e., source coding) and error control (i.e., channel coding) in wireless multimedia networks in more detail. In particular, this chapter focuses on the interaction between source and channel coding in such contexts. The heterogeneous nature of both the information sources and the physical channels in these applications introduces issues that are not addressed by more traditional information-theoretic approaches to these problems. Joint source and channel coding techniques are featured as an approach to some of the key challenges in these contexts, and promising techniques based on multiresolution signal processing in the form of wavelet- and subband-based source coders together with multilevel error-correcting coders are emphasized. The chapter illustrates how such techniques can be particularly well suited to the demands of speech, image, and video transmission over wireless channels, allowing, for example, rate adaptation and flexible control over noise immunity.

Finally, while much of the discussion in Chapters 1 through 7 applies, in principle, to general wireless communication systems, these chapters are largely developed with RF systems in mind. As an illustration of other kinds of propagation environments and how they have their own unique sets of issues in terms of wireless system design, in Chapter 8, Brady and Preisig explore wireless communication through the underwater acoustic channel. In practice, this channel is an important one in a number of scientific and military applications involving underwater communication, and it presents particularly challenging problems for signal processing design. Indeed, many of the impairments encountered on wireless RF channels are experienced at even more severe levels in the underwater acoustic channel. As such, this latter channel provides a useful context in which to explore and develop some of the most aggressive and signal-processing-intensive emerging techniques.

Finally, in the epilogue, Viterbi provides a philosophical view of the forces driving the development of wireless technology. In particular, he identifies four laws, two each from the natural sciences and the social sciences, that have formed the basis for the development of digital wireless communication networks. This essay describes their interaction, as well as their logical support for spread-spectrum multiple-access techniques.

Audience for This Book

Collectively, these nine contributions represent a sampling of some of the main themes and directions being pursued within this active field of research. While broad in scope, this volume does not attempt to be comprehensive in its coverage. Indeed, given the brisk pace at which developments are currently taking place, thorough coverage would be almost impossible. Instead, the goal of the book is a more modest one. The topics are representative rather than exhaustive, and the treatment is aimed toward developing perspectives and insights that will allow readers to appreciate what some of the fundamental challenges are, what the scope of current activities is, and where some of the major research opportunities lie. For newcomers, this book can be used as a starting point for navigating the rapidly growing body of literature on various aspects of the topic. For those already active in the field, this book can provide an opportunity to reflect on one's work in the context of developments in other aspects of the topic and to explore interconnections between these developments that may lead to fundamentally new research directions.

More broadly, it is our hope that this volume will be a useful resource to the dual audience it is intended to serve: both to the signal processing community as it becomes more active in the wireless communications area, and to the communications community as it increasingly embraces signal processing algorithms and architectures in the development of efficient wireless systems of the future. More generally, there is tremendous opportunity for major advances to come from expanded dialog and interaction between these two communities, and thus it is also our hope that projects such as this can ultimately be vehicles for fostering such collaboration.

H. Vincent Poor
Gregory W. Wornell

REFERENCES

[1] I. Brodsky, *Wireless: The Revolution in Personal Telecommunications.* (Artech: Boston, 1995)

[2] K.-C. Chen, et al., eds., *IEEE J. Select. Areas Commun.*, Issue on Wireless Local Communication, vol. 14, Nos. 3–4, Apr.–May, 1996.

[3] P. Cochran, ed., *Proc. IEEE*, Special Issue on Communications in the Twenty-first Century, vol. 85, No. 10, Oct. 1997.

[4] P. T. Davis and G. R. McGuffin, *Wireless Local Area Networks.* (McGraw-Hill: New York, 1995)

[5] K. Fehrer, *Wireless Digital Communications: Modulation and Spread Spectrum Applications.* (Prentice-Hall: Upper Saddle River, NJ, 1995)

[6] J. J. Fernandes, P. A. Watson, and J. C. Neves, "Wireless LANs: Physical Properties of Infrared Systems vs. Mmw Systems," *IEEE Commun. Mag.*, vol. 32, no. 8, pp. 68–73, Aug. 1994.

[7] V. K. Garg and J. E. Wilkes, *Wireless and Personal Communications Systems.* (Prentice-Hall: Upper Saddle River, NJ, 1996)

[8] V. K. Garg, K. Smolik, and J. E. Wilkes, *Applications of CDMA in Wireless/Personal Communications.* (Prentice-Hall: Upper Saddle River, NJ, 1997)

[9] D. Grillo, A. Sasaki, R. A. Skoog, and B. Warfield, eds., *IEEE J. Select. Areas Commun.*, Issue on Personal Communications—Services, Architecture, and Performance Issues, vol. 15, no. 8, Oct. 1997.

[10] J. M. Holtzman and D. J. Goodman, *Wireless and Mobile Communications.* (Kluwer: Boston, 1994)

[11] J. H. Kahn and J. R. Barry, "Wireless Infrared Communications," *Proc. IEEE*, vol. 85, no. 2, pp. 265–298, Feb. 1997.

[12] L. Kleinrock, "Nomadic Computing—An Opportunity," *ACM SIGCOMM Comput. Commun. Rev.*, vol. 25, pp. 36–40, Jan. 1995.

[13] K. I. Park, *Personal and Wireless Communications: Digital Technology and Standards.* (Kluwer: Boston, 1996)

[14] K. Pahlavan and A. H. Levesque, *Wireless Information Networks.* (Wiley: New York, 1995)

[15] T. S. Rappaport, *Wireless Communications: Principles and Practice.* (Prentice-Hall: Upper Saddle River, NJ, 1996)

[16] T. S. Rappaport, B. E. Woerner, and J. H. Reed, *Wireless Personal Communications: The Evolution of Personal Communications Systems.* (Kluwer: Boston, 1996)

[17] S. Sampei, *Applications of Digital Wireless Technologies to Global Wireless Communications.* (Prentice-Hall: Upper Saddle River, NJ, 1997)

[18] G. L. Stüber, *Principles of Mobile Communications.* (Kluwer: Boston, 1996)

[19] *Universal Communications—Proc. Military Commun. Conf. (MILCOM)*, San Diego, Calif., Nov. 6–8, 1995.

[20] A. J. Viterbi, *CDMA: Principles of Spread Spectrum Communications.* (Addison-Wesley: Reading, Mass., 1995)

List of Contributors

David Brady
Northeastern University

Stephen V. Hanly
University of Melbourne, Australia

Paul Haskell
DiviComm, Inc.

Michael L. Honig
Northwestern University

David G. Messerschmitt
University of California, Berkeley

Constantinos B. Papadias
Lucent Technologies (Bell Labs Research)

Haralabos C. Papadopoulos
Massachusetts Institute of Technology

Arogyaswami J. Paulraj
Stanford University

H. Vincent Poor
Princeton University

James C. Preisig
Woods Hole Oceanographic Institution

Kannan Ramchandran
University of Illinois, Urbana-Champaign

Vellenki U. Reddy
Indian Institute of Science, Bangalore, India

David N. C. Tse
University of California, Berkeley

Alle-Jan van der Veen
Delft University of Technology, The Netherlands

Martin Vetterli
Ecole Polytechnique Fédérale de Lausanne, Switzerland

Andrew J. Viterbi
Qualcomm, Inc.

Gregory W. Wornell
Massachusetts Institute of Technology

Louis Yun
ArrayComm, Inc.

1

Linear Diversity Techniques for Fading Channels

Gregory W. Wornell

Signal fading due to multipath propagation is a dominant source of impairment in wireless communication systems, often severely impacting performance. However, the effects of fading can be substantially mitigated through the use of diversity techniques via appropriately designed signal processing algorithms at both the transmitters and receivers. Practical, high-performance systems require that such diversity techniques be efficient in their use of resources such as power, bandwidth, and hardware, and that they meet often tight computational and delay constraints. This chapter develops a common, multirate-signal-processing-oriented framework within which the relative benefits of various types of diversity are discussed. Several examples of resource-efficient linear diversity methods—both traditional and more recently proposed—are highlighted.

Fading in signal strength arises primarily from multipath propagation of a transmitted signal due to reflections off physical objects, which gives rise to spatially distributed standing wave patterns of constructive and destructive interference.[1]

---

[1]We will be focusing primarily on radio-frequency (RF) systems, although much of what we discuss also applies to acoustic-frequency systems used in underwater environments. Many of the underlying principles also apply in infrared (IR) systems, although technological and other considerations often preclude direct implementations of the kinds of processing discussed here.

These standing wave patterns depend not only on the geometry of the constituent propagation paths from transmitter to receiver but on the carrier frequency of the transmitted signal as well. As a result, signal strength varies both with spatial location and with frequency. Moreover, when the receiver is in motion through the standing wave pattern, time variation in signal strength is experienced.

Figure 1.1 illustrates an example of a multipath-induced standing wave pattern in two dimensions. In this example, a single-frequency tone is transmitted from a particular source location, and a single, perfectly reflecting, linear boundary creates one indirect propagation path in addition to the direct path to a given receiver location. We emphasize that increasing the frequency of the transmitted signal would lead to more closely spaced contours in Figure 1.1.

The scenario represented by Figure 1.1 is obviously highly simplified. For example, the effects of path loss, i.e., attenuation in signal strength that occurs with

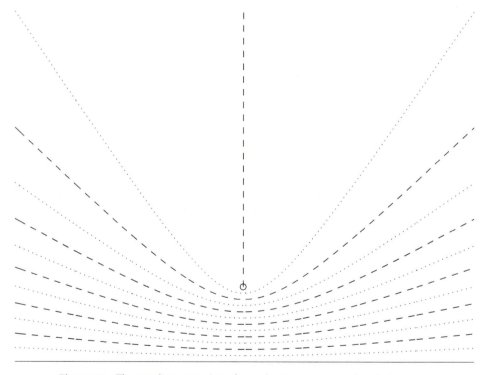

Figure 1.1 The standing wave (interference) pattern associated with the transmission of a single tone in a two-dimensional propagation environment with a single perfect reflecting boundary, indicated by the solid line, and no path losses. The symbol ○ denotes the transmitter location, and the dashed curves indicate the contours of points where maximum constructive interference would be experienced by a receiver. Between are dotted curves indicating corresponding contours of points where maximum destructive interference would be experienced. A similar interpretation is obtained when the roles of transmitter and receiver are reversed.

propagation distance, have been neglected; these would otherwise affect the interference pattern. More generally, in more realistic physical environments involving many reflectors and therefore many propagation paths, the standing wave pattern is considerably more complicated. In such cases, deterministic characterization of the particular interference patterns may be difficult to obtain. However, useful statistical characterizations of the effects of such fading can often be exploited.

To compensate for the effects of signal fading due to multipath propagation, three main forms of diversity are traditionally exploited to varying degrees in wireless communication systems: spectral diversity, temporal diversity, and spatial diversity. Each of these forms is discussed in detail in this chapter. We begin by summarizing their salient features.

Spectral diversity is effective when the fading is frequency-selective, i.e., varies as a function of frequency. This form of diversity can be practically exploited when the available bandwidth for transmission is large enough that individual multipath components can begin to be resolved, or equivalently when it is large enough that different subbands of the transmission bandwidth experience effectively independent fading. As we discuss, examples of systems that take advantage of frequency diversity are direct-sequence or frequency-hopped spread-spectrum communication systems, which are designed to utilize wideband transmission formats.

Temporal diversity is effective when the fading is time-selective, i.e., fluctuates with time.[2] This form of diversity is exploited by introducing memory into the transmitted symbol stream in the form of coding or precoding to effect temporal spreading of symbols. Since temporal diversity schemes introduce delay, the degree to which this form of diversity can be exploited depends on delay constraints in the system relative to the coherence time of the channel,[3] which, in turn, is a function of both the vehicle speeds in mobile applications and the carrier frequency in the system. Delay constraints are often quite stringent for two-way voice communication but can, in principle, be significantly milder for broadcast and other applications.

Finally, spatial diversity—which involves the use of sufficiently spaced, multiple-element antenna arrays at either the receiver, the transmitter, or both—exploits the spatial variation in fading corresponding to the standing wave interference patterns. Among other advantages, spatial diversity can be exploited even in situations where the fading channel is neither frequency-selective nor time-selective—or when system constraints preclude the use of spectral or temporal

---

[2]The term "fading" is sometimes reserved exclusively for describing this phenomenon of temporal variation specifically. Throughout this chapter, however, the term fading is used in its broader sense to describe effects induced by multipath propagation more generally, whether or not they are accompanied by significant time variation.

[3]The coherence time of the channel is the correlation time of the fading process: it is a measure of how far apart two time samples have to be to experience effectively independent fades.

diversity. The extent to which this form of diversity can be exploited depends on issues such as physical size constraints and cost.

A detailed outline of the chapter is as follows. We begin with a discussion in Section 1.1 of a basic system and channel model that will be used throughout. For reference, Section 1.2 summarizes the performance characteristics of systems that do not exploit diversity. In the remainder of this chapter, we focus on several inter-related classes of low-complexity, linear signal processing algorithms for exploiting each of the three forms of diversity, both individually and jointly. In particular, spectral and temporal diversity techniques are discussed in Sections 1.3 and 1.4, respectively. The emphasis in these sections is on effectively single-user (i.e., point-to-point) systems. Natural generalizations of these diversity techniques are developed for use in multiuser scenarios in Section 1.5; these are of particular interest in many contemporary applications. Section 1.6 describes and contrasts techniques for exploiting diversity through the use of both receiver and transmitter antenna arrays in wireless systems, and finally Section 1.7 contains some concluding remarks.

1.1 SYSTEM AND FADING CHANNEL MODELS

The model for the channel between a particular transmitter and receiver pair in a wireless system generally consists of two components: a linear, time-varying filter that captures the effects of multipath fading in the transmission medium, and an additive noise term representing both receiver noise and, often more significantly, sources of co-channel interference. When pulse amplitude modulation (PAM) is employed in the system, it is convenient to work with an equivalent discrete-time baseband model for the passband system.[4] In particular, as depicted in Figure 1.2, the response of the channel to an input sequence $y[n]$ is given by

$$r[n] = \sum_k a[n; k]y[n - k] + w[n], \qquad (1.1)$$

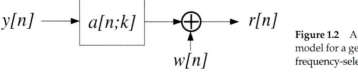

Figure 1.2   A discrete-time baseband model for a generally time- and frequency-selective fading channel.

[4]It is worth commenting at the outset that while such discrete-time models sometimes constitute an oversimplification of the physical systems they represent, they are generally adequate for capturing the key features of such systems and the associated channels, and at the same time lead rather naturally to algorithms having efficient digital signal processor (DSP) based implementations.

where the sequence $w[n]$ represents the additive noise and interference, and where the kernel $a[n; k]$ is the response of the channel at time n to a unit-sample input at time $n - k$.

It is often convenient to model the noise $w[n]$ as a zero-mean, complex-valued, circularly symmetric stationary white Gaussian process with variance

$$E\left[|w[n]|^2\right] = \mathcal{N}_0 \mathcal{W}_0,$$

where $\mathcal{W}_0$ is the underlying system bandwidth. While this is a reasonably realistic model for the receiver noise, it is generally an overly simplistic (and rather pessimistic) model for cochannel interference; more realistic models for cochannel interference that take into account its inherent structure lead to more effective interference suppression algorithms, several of which are discussed in Chapters 2 and 3 as well as other chapters of this volume.

The time-varying frequency response of the channel is related to the kernel $a[n; k]$ via[5]

$$A(\omega; n) = \sum_k a[n; k]e^{-j\omega k}. \tag{1.2}$$

When $A(\omega; n)$ does not vary with n, the channel is said to be time-nonselective, and $a[n; k]$ specializes to the kernel of a linear time-invariant system. This fading model is generally applicable to systems in which any transmitter or receiver motion is sufficiently slow that time variations in the channel are on scales much longer than the symbol duration. In this case, as depicted in Figure 1.3 the input-output relation involves simple convolution, i.e.,

$$r[n] = \sum_k a[k]y[n - k] + w[n], \tag{1.3}$$

where the unit-sample response $a[k]$ is given by $a[k] = a[0; k]$ and has the associated frequency response $A(\omega) = A(\omega; 0)$. In such cases, temporal diversity cannot be exploited.

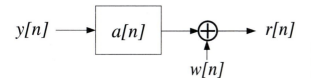

Figure 1.3   A discrete-time baseband model for a frequency-selective, time-nonselective fading channel.

[5]We adopt the useful convention of using parentheses ($\cdot$) to denote continuous-valued arguments and brackets [$\cdot$] to denote discrete-valued arguments. For functions of two arguments where the first is continuous and the second is discrete (as in the case of time-varying frequency responses), we use the convenient mixed notation ($\cdot$; $\cdot$].

By contrast, when $A(\omega; n)$ in (1.2) does not vary with ω, we have $a[n; k] = a[n]\delta[k]$, with $\delta[n]$ denoting the unit-sample, i.e.,

$$\delta[n] \triangleq \begin{cases} 1 & n = 0 \\ 0 & \text{otherwise,} \end{cases} \tag{1.4}$$

in which case the channel is said to be frequency-nonselective. This fading model is appropriate for use with systems in which any frequency variations are on scales much larger than the system bandwidth. As such, this model is applicable to narrowband channels where the delay spread—i.e., the effective length of the sequence $a[n; k]$ with n fixed—is smaller than the symbol duration. Such is the case, for example, in indoor applications when the bandwidth is less than about 200 kHz, and outdoor applications when the bandwidth is less than about 20 kHz. In this scenario, as depicted in Figure 1.4, the input-output relation involves modulation, i.e.,

$$r[n] = a[n]y[n] + w[n]. \tag{1.5}$$

On such channels, spectral diversity cannot be exploited.

Finally, channels that are both time- and frequency-nonselective can be expressed in the form

$$r[n] = ay[n] + w[n], \tag{1.6}$$

where $a = A$ is a random variable. On such channels, neither temporal nor spectral diversity can be exploited, though spatial diversity often can. Spatial diversity exploits the fact that different gains a are encountered (experienced) by different elements of the antenna array.

As we will generally assume throughout this chapter, the transmitter typically does not have any knowledge of the fading channel kernel $a[n; k]$. Exceptions are when a feedback path exists from receiver to transmitter through which at least partial channel state information can be passed. In contrast, the receiver is generally able to infer information about the fading channel kernel from the received waveform, through the use of either training data sent by the transmitter or a "blind" algorithm (examples of which are discussed in Chapters 3 and 4). In this chapter, we assume that the receiver has perfect knowledge of the fading channel

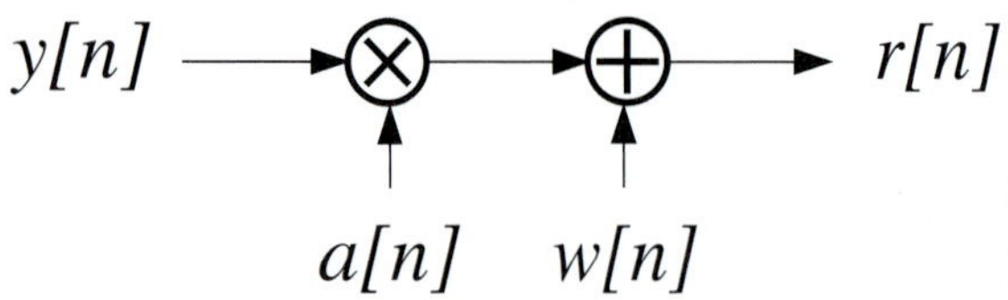

Figure 1.4   A discrete-time baseband model for a time-selective but frequency-nonselective fading channel.

through which the information was transmitted; this assumption allows us to develop useful bounds on the performance attainable in practice. In turn, useful refinements of these bounds can be obtained by separately assessing and subsequently incorporating the effects of channel identification limitations on diversity methods. However, such refinements are not explored in this chapter.

A variety of statistical characterizations of fading behavior have proven useful in designing receivers and transmitters for wireless systems and in evaluating the resulting performance. One of the most widely used, and the one we use in this chapter, is the stationary Rayleigh fading channel model, which is appropriate when there are a large number of superimposed propagation paths.[6] In this case, for fixed values of k, the kernel $a[n; k]$ is well modeled as a zero-mean, stationary, complex-valued, circularly symmetric Gaussian sequence based on a Central Limit Theorem argument. Furthermore, uncorrelated scattering is generally assumed— i.e., sequences corresponding to distinct values of k are statistically independent, so that[7]

$$E\left[a[n; k]a^*[n - m; l]\right] = R_a[m; k]\,\delta[k - l].$$

With uncorrelated scattering, the time-varying channel frequency response $A(\omega; n]$ is then stationary in both n and ω and satisfies

$$E\left[A(\omega; n]\right] = 0 \qquad\qquad (1.7\text{a})$$

$$E\left[\left|A(\omega; n]\right|^2\right] = \sigma_a^2. \qquad\qquad (1.7\text{b})$$

The noise $w[n]$, fading kernel $a[n; k]$, and channel input $y[n]$ are reasonably modeled as mutually independent. Other aspects of the characterization of discrete-time Rayleigh fading channels based on Bello's continuous-time framework [2] are discussed in [3].

The remainder of the chapter considers methods for transmitting a continuous- or discrete-valued sequence $x[n]$ of complex-valued symbols having average energy $\mathcal{E}$ per symbol over such channels at a prescribed rate of $\mathcal{R}$ complex symbols per second. From traditional Nyquist sampling theory, we know that such rates can be achieved over a passband channel having a bandwidth of at least $W_0 = \mathcal{R}$ Hz, corresponding to a spectral efficiency of at most 1 (complex) symbol/s/Hz [4]. In practice, at least a small amount of excess bandwidth is generally required to facilitate timing recovery. However, when available, larger amounts of excess bandwidth can also be used to dramatically improve performance of wireless systems in particular, as we will discuss.

---

[6]When a direct propagation path is also included, a Ricean model [1] is more appropriate.

[7]We use the operator $*$ to denote convolution, and the superscript * to denote complex conjugation.

1.2 TRANSMISSION WITHOUT DIVERSITY

To illustrate the impact of fading on system performance before exploiting diversity, we consider a representative uncoded system in which a stream of quadrature phase-shift keying (QPSK) symbols $x[n]$ is transmitted over a nonselective fading channel (1.6). A typical signal set for these systems would be of the form

$$x[n] \in \{+\sqrt{\mathcal{E}}, -\sqrt{\mathcal{E}}, +j\sqrt{\mathcal{E}}, -j\sqrt{\mathcal{E}}\}.$$

First, note that the signal-to-noise ratio (SNR) associated with the channel takes the form

$$\alpha = \frac{\mathcal{E}|a|^2}{\mathcal{N}_0 \mathcal{W}_0}.\tag{1.8}$$

For a particular realization (i.e., fixed a), the channel is Gaussian and the associated QPSK bit-error rate is given by

$$\mathcal{P}_a = \mathcal{Q}(\sqrt{\alpha}),\tag{1.9}$$

where

$$\mathcal{Q}(v) = \frac{1}{\sqrt{2\pi}} \int_v^\infty e^{-t^2/2}\, dt.\tag{1.10}$$

Thus, the bit-error rate achieved depends strongly on the channel gain a. If the fading is sufficiently severe that α drops below some minimum SNR level necessary for adequate performance, then the system is said to experience an "outage."[8] Since with a complex-valued Gaussian amplitude the SNR (1.8) is an exponentially distributed random variable with mean

$$\frac{1}{\zeta_0} \triangleq E[\alpha] = \frac{\sigma_a^2 \mathcal{E}}{\mathcal{N}_0 \mathcal{W}_0},\tag{1.11}$$

then for a given threshold the outage probability can be readily calculated as the integral under the tail of the exponential density. This outage probability characterizes, in an appropriate sense, worst-case performance on the fading channel.

Average performance on the fading channel is also of interest in practice. Taking the expectation of (1.9) with respect to the exponential density for α yields an average bit-error rate of

$$\mathcal{P} = \frac{1}{2}\left(1 - \frac{1}{\sqrt{2\zeta_0 + 1}}\right),\tag{1.12}$$

---

[8]This threshold effect is often pronounced for coded systems as error-correcting codes typically exhibit rapid performance degradation when SNR falls below a prescribed level.

with ζ_0 as given by (1.11). By contrast, a Gaussian channel with a deterministic gain $\bar{a}$ that is the root-mean-square (RMS) value of a in the fading channel, i.e.,

$$\bar{a} = \sqrt{E\left[\,|a|^2\right]}$$

has the same *average* SNR as the fading channel but supports a dramatically better average bit-error rate of

$$\mathcal{P}_0 = \mathcal{Q}\left(\sqrt{E[\alpha]}\right) = \mathcal{Q}\left(1/\sqrt{\zeta_0}\right). \tag{1.13}$$

These two bit-error rate curves are depicted in Figure 1.5, where we see that to achieve a modest bit-error rate of 10^{-3}, in excess of 15 dB more transmission power is required on the fading channel than on the corresponding Gaussian channel. At lower error rates, the differences are even more striking and arise because at high SNR (low ζ_0), (1.12) decays only as the reciprocal of SNR, i.e.,

$$\mathcal{P} \sim \zeta_0, \tag{1.14}$$

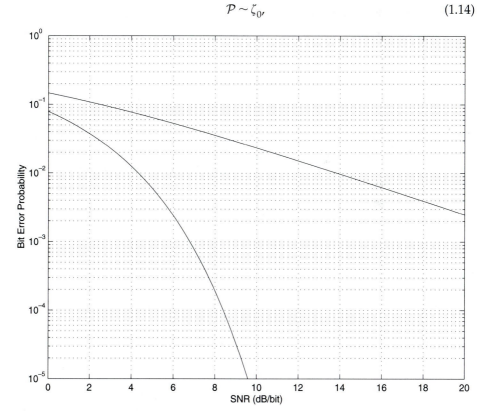

Figure 1.5 Bit-error rate behavior for QPSK transmission. The top curve corresponds to the performance on the Rayleigh fading channel, and the bottom corresponds to the performance on a Gaussian channel with the same average SNR.

while (1.13) decays exponentially with SNR, i.e.,

$$\mathcal{P}_0 \sim \sqrt{\zeta_0} e^{-1/(2\zeta_0)}. \tag{1.15}$$

With these observations in mind, the goal of diversity is to improve both the worst-case and average performance of systems operating in the presence of fading. Ultimately, the performance of the corresponding Gaussian channel discussed above provides a useful performance benchmark for various linear diversity strategies. For example, as we will see first, the Gaussian channel performance can be approached arbitrarily closely through the use of spectral diversity.

1.3 SPECTRAL DIVERSITY

Spectral diversity is obtained by means of a spread-spectrum system, which exploits a bandwidth significantly greater than the $\mathcal{W}_0$ Hz that would otherwise be required for transmission of the symbol stream $x[n]$ at rate $\mathcal{R} = \mathcal{W}_0$ symbols/s. The most effective spectral diversity is obtained by making use of the excess bandwidth through an appropriately designed error-correction code. However, we restrict our attention to linear methods for generating and exploiting spectral diversity, a class of which are the simple spread-spectrum systems that take the form depicted in Figure 1.6 and which effectively correspond to a repetition coding strategy. As this figure reflects, the symbol stream $x[n]$ is upsampled by an integer factor ρ,[9] the result of which is processed by a linear time-invariant (LTI) filter (the code) whose unit-sample response $h[n]$ has unit energy, i.e.,

$$\sum_n \left| h[n] \right|^2 = 1,$$

yielding

$$y[n] = \sum_k x[k] h[n - k\rho]. \tag{1.16}$$

---

[9]Upsampling by a factor ρ involves inserting $\rho - 1$ zeros between each of the symbols, i.e., if $z[n]$ denotes the output of a rate-ρ upsampler with input $x[n]$, then

$$z[n] = \begin{cases} x[n/\rho] & n = \dots, -\rho, 0, \rho, 2\rho, \dots \\ 0 & \text{otherwise.} \end{cases}$$

As a result, the average power in $z[n]$ is reduced from that in $x[n]$ by a factor of ρ. This reduction ensures that SNR degrades in the proper manner with bandwidth expansion due to increased noise at the front end of the receiver. As we will see, this noise enhancement is more than compensated for by the diversity benefit inherent in the bandwidth expansion.

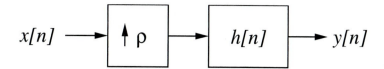

Figure 1.6 A multirate model for direct-sequence, spread-spectrum modulation.

The samples constituting $y[n]$ are referred to as "chips," and thus for each symbol in the original stream, ρ chips are generated in the new stream.

When the chips $y[n]$ are pulse-amplitude modulated at the correspondingly higher rate $\rho\mathcal{R} = \rho\mathcal{W}_0$ (the chip rate), the symbols $x[n]$ continue to be transmitted at rate $\mathcal{R} = \mathcal{W}_0$ (the symbol rate), but now using a bandwidth of $\rho\mathcal{W}_0$. Thus, the spectral efficiency is reduced to ρ^{-1} symbols/s/Hz.

The detailed characteristics of the transmitted spectrum—and how much spectral diversity can be ultimately be exploited—are controlled by the choice of $h[n]$ or, equivalently, its frequency response

$$H(\omega) = \sum_n h[n]e^{-j\omega n}.$$

In particular, as will become apparent, $H(\omega)$ should be broadband, with its energy spread as uniformly as possible in frequency.

The associated whitened matched filter for this system when used in conjunction with the frequency-selective fading channel of Figure 1.3 takes the following form: with

$$\tilde{a}[n] = h[n] * a[n]$$

denoting the equivalent channel, the received signal $r[n]$ is first processed with a matched filter, whose unit-sample response is $\tilde{a}^*[-n]$, and downsampled by a factor of ρ. This first stage of processing, which takes place at the chip rate, is referred to as a RAKE receiver, after Price and Green [5]. The second stage of processing takes place at the symbol rate and performs the whitening. In particular, the output of the RAKE receiver is further processed by an anticausal system with frequency response $1/G^*(\omega)$, where $G(\omega)$ is the minimum-phase spectral factor of the associated folded spectrum

$$S_{\tilde{a}}(\omega) = \frac{1}{\rho}\sum_{l=0}^{\rho-1}\left|H\left(\frac{\omega + 2\pi l}{\rho}\right)\right|^2\left|A\left(\frac{\omega + 2\pi l}{\rho}\right)\right|^2,$$

i.e.,

$$G(\omega)G^*(\omega) = S_{\tilde{a}}(\omega).$$

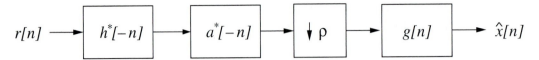

$$r[n] \longrightarrow \boxed{h^*[-n]} \longrightarrow \boxed{a^*[-n]} \longrightarrow \boxed{\downarrow \rho} \longrightarrow \boxed{g[n]} \longrightarrow \hat{x}[n]$$

Figure 1.7 Whitened matched filter for direct-sequence, spread-spectrum system in frequency-selective fading.

The result $\hat{x}[n]$ is then related to $x[n]$ according to

$$\hat{x}[n] = g[n] * x[n] + v[n], \tag{1.17}$$

where $g[n]$ is the causal unit-sample response of the system with frequency response $G(\omega)$, and the noise $v[n]$ is white with the same variance as $w[n]$. It can be shown, as in the continuous-time case [4], that the collection of output samples $\hat{x}[n]$ of this whitened matched filter, which is depicted in Figure 1.7, is a sufficient statistic for making decisions about $x[n]$.

The whitened matched filter, as well as the subsequent processing, often simplify in practice. For example, when $h[n]$ is the unit-sample response of an allpass filter [6], i.e.,

$$|H(\omega)|^2 = 1, \qquad \text{whence} \qquad h[n] * h^*[-n] = \delta[n], \tag{1.18}$$

and when the coherence bandwidth θ_a of the channel is such that $a[n]$ has a delay spread (effective length) L satisfying[10]

$$L = \rho W_0 / \theta_a \le \rho, \tag{1.19}$$

then

$$S_{\tilde{a}}(\omega) = \sum_{l=0}^{L-1} |a[l]|^2 \triangleq \mu^2. \tag{1.20}$$

Thus, the whitening filter becomes a simple normalizing gain $1/\mu$, and the equivalent channel is a simple additive white Gaussian noise channel that is free of inter-symbol interference (ISI), i.e.,

$$\hat{x}[n] = \mu x[n] + v[n], \tag{1.21}$$

where, as before, $\operatorname{var} v[n] = \operatorname{var} w[n] = \mathcal{N}_0 W_0$. Thus, simple symbol-by-symbol detection will suffice in recovering the original stream.

To evaluate the diversity benefit obtained through this spread-spectrum strategy, we first note that under Rayleigh fading with uncorrelated scattering, the $a[n]$ are independent, identically distributed zero-mean Gaussian random vari-

[10]The coherence bandwidth of the channel is a measure of how far apart two frequencies have to be in order to experience effectively independent fades, so L is a measure of the number of available degrees of spectral diversity.

ables. Furthermore, consistent with (1.7b), the frequency-dependent SNR in the original channel, i.e.,

$$\alpha(\omega) = \frac{\mathcal{E}|A(\omega)|^2}{\mathcal{N}_0 \mathcal{W}_0} \tag{1.22}$$

has mean

$$\frac{1}{\zeta_0} \triangleq E[\alpha] = \frac{\sigma_a^2 \mathcal{E}}{\mathcal{N}_0 \mathcal{W}_0} \tag{1.23}$$

and the same variance. By contrast, the SNR in the equivalent channel (1.21), i.e.,

$$\gamma = \frac{\mathcal{E}\mu^2}{\mathcal{N}_0 \mathcal{W}_0} \tag{1.24}$$

is an Lth-order Erlang random variable with the same mean but reduced variance; specifically, using (1.20),

$$E[\gamma] = \frac{1}{\zeta_0} \tag{1.25}$$

$$\mathrm{var}\,\gamma = \frac{1}{L\zeta_0^2}\, . \tag{1.26}$$

This reduction in variance has a substantial impact on average bit-error-rate performance. For example, for a QPSK symbol stream $x[n]$, the bit-error probability is [1]

$$\mathcal{P} = \frac{1}{2}\left[1 - \frac{1}{\sqrt{2\zeta_0 L + 1}} \sum_{k=0}^{L-1} \binom{2k}{k}\left(\frac{\zeta_0 L}{2(2\zeta_0 L + 1)}\right)^k\right], \tag{1.27}$$

where ζ_0 is as defined in (1.23). At high SNR, (1.27) has the asymptotic form

$$\mathcal{P} \sim \zeta_0^L. \tag{1.28}$$

As we would expect, (1.27) and (1.28) specialize to (1.12) and (1.14), respectively, when there is no spectral diversity. More significantly, in the limit of infinite spectral diversity ($L \to \infty$), (1.27) and (1.28) approach the equivalent Gaussian channel performance (1.13) and (1.15), respectively, i.e.,

$$\lim_{L \to \infty} \mathcal{P} = \mathcal{Q}\left(1/\sqrt{\zeta_0}\right). \tag{1.29}$$

This result can be obtained by exploiting the fact that $\mu^2 \to \sigma_a^2$ as $L \to \infty$, using the (strong) law of large numbers. Figure 1.8 depicts the bit-error rates (1.27) as a function of SNR for different numbers of degrees of spectral diversity (and, therefore, different amounts of bandwidth expansion).

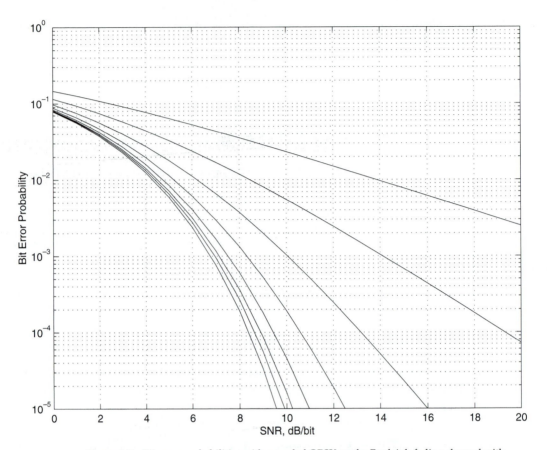

Figure 1.8 Bit-error probabilities with uncoded QPSK on the Rayleigh fading channel with spectral diversity exploited by a RAKE receiver. The top curve corresponds to the performance without spectral diversity ($L = 1$), and the bottom curve indicates the performance potential with infinite spectral diversity ($L \rightarrow \infty$) obtained via infinite bandwidth expansion. The successively lower curves between these two extremes represent the performance obtained with $L = 2, 4, 18, 16, 32$, and 64 degrees of spectral diversity, respectively.

We conclude this section with some remarks on the design of the filter $h[n]$ and on the implementation of the system. First, consistent with the preceding analysis, selecting $h[n]$ to be the unit-sample response of an allpass system allows the maximum number of degrees of spectral diversity to be exploited. From this perspective, even the simple choice $h[n] = \delta[n]$ is sufficient and allows the individual multipath components to be directly resolved. In practice, this choice is typically not used, primarily because the associated transmitted signal $y[n]$ has a large peak-power requirement, which is often undesirable from the perspective of transmitter amplifier design.[11] As a result, to achieve more uniform power characteristics, $h[n]$ is more

---

[11]This choice can also be undesirable when there are transmission security considerations.

typically chosen to be a binary-valued sequence $(h[n] \in \{\pm 1/\sqrt{\rho}\})$ of length ρ. With this constraint, (1.18) cannot be satisfied exactly but can be approximated by appropriate design of the binary taps of $h[n]$. A variety of pseudorandom shift-register sequences have proven popular for this purpose—see, e.g., [7, 8] and the references therein.

When $h[n]$ and the length L of $a[n]$ satisfy—at least approximately—the conditions (1.18) and (1.19), respectively, the RAKE receiver is often most easily implemented in the manner reflected in Figure 1.7. First, the received sequence $r[n]$ is correlated with $h[n]$, which is implemented by filtering with $h^*[-n]$. Exploiting (1.18), we see that this operation eliminates the effect of $h[n]$ at the transmitter without affecting the noise statistics. As a result, from the point of view of subsequent processing, it is as if the filter $h[n] = \delta[n]$ were used to implement the spread-spectrum system. In particular, provided that the bandwidth expansion factor ρ is chosen so that there is no ISI after this correlation operation, then symbol-by-symbol decisions can be made after subsequent processing in the form of matched filtering and downsampling, where the filter is matched to the channel and therefore has unit-sample response $a^*[-n]$.

It is also important to note that the choice of $h[n]$ also determines the spectral characteristics of the transmitted signal, which is an important consideration in system design for both public and secure communication. From this perspective it is useful to recognize that the allpass condition (1.18) also ensures that if the original symbols $x[n]$ are uncorrelated, then the transmit spectrum will be white, i.e., $S_y(\omega)$ will be constant.

Structure in the code $h[n]$ can also facilitate implementation. For example, when $h[n]$ is of length ρ and binary-valued, i.e., $h[n] \propto \pm 1$, then the spread-spectrum modulation with a binary phase-shift keying (BPSK) symbol stream, i.e., $x[n] \propto \pm 1$, can be implemented very efficiently with high-speed digital logic circuitry. In particular, a sequence is created in which each symbol in the original stream $x[n]$ is replicated ρ times in succession, and the result is bitwise exclusive-or'd with a periodically replicated version of the length-ρ sequence to generate the transmission $y[n]$. Most practical implementations of direct-sequence spread-spectrum employ this approach.[12]

A final comment: the general spread-spectrum approach described in this section extends readily to multiuser systems, either as time-division or code-division multiple access systems. In such systems, many users share the total expanded bandwidth. The transmissions of the users are distinguished from one another by assigning different filters $h[n]$ to the different users; these filters serve as distinct

---

[12]In practice, this exclusive-or process is often used with a periodically replicated binary code sequence having length greater than ρ. The resulting modulation effectively corresponds to using different codes for successive symbols. While this generalization does not allow any additional diversity benefit to be realized, it sometimes has implementational and other advantages.

"signatures" for each of the users. A framework for such systems is explored later in the context of Section 1.5.

1.4 TEMPORAL DIVERSITY

In systems where the transmitters, receivers, and/or reflectors are in motion so that time-varying multipath propagation is experienced, temporal diversity can be exploited either as an alternative to, or in conjunction with, spectral diversity. This section discusses natural techniques for exploiting temporal diversity in such scenarios. Our focus in this section is on single-user scenarios; multiuser extensions are subsequently developed in Section 1.5. We first consider temporal diversity methods that are bandwidth-preserving and then turn our attention to extensions that involve bandwidth expansion.

In general, to obtain the benefits of temporal diversity, the transmission of each symbol is effectively spread in time, thus avoiding an entire symbol encountering a deep fade. Indeed, since SNR in a time-selective fading channel fluctuates from time sample to time sample, we can improve at least worst-case performance (outage probability) in the transmission of individual symbols by appropriately spreading the transmission of each symbol over a large number of time samples. As we will see, this generally also leads to an improvement in average performance.

In practice, such spreading can be accomplished in a variety of ways such as through the generation of memory in the transmitted stream by applying error-correction coding to the original symbols. Moreover, if necessary, this spreading can be accomplished without bandwidth expansion through the use of coded modulation [9]. When used on fading channels, coding is generally combined with interleaving, the purpose of which is to scramble the coded data stream so that symbols within a codeword experience effectively independent fading. This strategy reduces the coding complexity required to achieve a given level of fidelity, allowing shorter lengths in the case of block codes, or fewer states in the case of convolutional codes. However, although such codes can be used to approach capacity, they can be impractical in attempts to exploit large numbers of degrees of time diversity because such nonlinear methods often require high computational complexity at the receiver for decoding in such regimes.

A computationally efficient alternative for obtaining a temporal diversity benefit is to use linear processing in the form of what is referred to as spread-response precoding [10] (see also [11]). This strategy requires signal processing algorithms at the transmitter and receiver whose complexity grows linearly with the number of degrees of diversity being exploited.

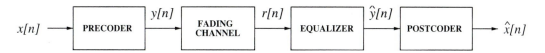

Figure 1.9   A spread-response precoding system.

1.4.1 Spread-Response Precoding

A typical system employing spread-response precoding is depicted in Figure 1.9. In the transmitter of this system, the symbol stream $x[n]$ is transformed into a precoded symbol stream $y[n]$. At the receiver, the data $r[n]$ is first processed by an equalizer to produce $\hat{y}[n]$, then by a postcoder to produce $\hat{x}[n]$, from which decisions about the original symbols $x[n]$ are made. In general, the linear precoding (and postcoding) may be either time-invariant or periodically time-varying. The time-invariant case is conceptually somewhat simpler and is discussed first, but the periodically time-varying pre- and postcoders we discuss later have important advantages for implementations in practice.

In the LTI case, the spread-response precoding filter that constitutes the transmitter portion of the system of Figure 1.9 is characterized by its (possibly complex-valued) unit-sample response, which we denote by $h[n]$. With this notation, the transmitted sequence is

$$y[n] = x[n] * h[n] = \sum_{k} x[k]h[n - k]. \tag{1.30}$$

We now focus on the design of $h[n]$, which can also be viewed as a discrete-time pulse shape, so as to achieve the desired symbol spreading.

For this application, it is natural to require $h[n]$ to be the unit-sample response of an (at least approximately) allpass filter, which as we recall satisfies the condition (1.18). The condition (1.18) implies that allpass filters are lossless systems, producing outputs $y[n]$ that are orthogonal (i.e., distance- and energy-preserving) transformations on their inputs $x[n]$. Such transformations have a number of key properties; as just one example, (1.18) implies that $x[n]$ can be conveniently and stably recomputed from $y[n]$ via

$$x[n] = y[n] * h^*[-n] = \sum_{k} y[k]h^*[k - n]. \tag{1.31}$$

From a practical standpoint, it is generally necessary to restrict our attention to finite impulse response (FIR) precoders. However, the only lossless FIR filters are the shifted unit-samples, i.e., $h[n] = \delta[n - n_0]$ for arbitrary n_0. Nevertheless, many infinite impulse response (IIR) lossless filters have sufficiently localized temporal support that they can be truncated without significantly altering their characteristics. More generally, a wide variety of FIR filters closely approximate the losslessness (allpass) condition (1.18), where a suitable measure of loss is based on the energy in the approximation error, as developed in [10]. More importantly, we will see that perfect losslessness can be achieved with FIR systems when we relax the strict time-invariance constraint and allow periodically time-varying filters.

Among FIR precoders that meet or approximate the losslessness condition (1.18), those of interest have the energy in their unit-sample response spread or dispersed as uniformly as possible over the filter length, so that the energy allocated to each symbol in the stream is spread most efficiently in time. As discussed in [10], a variety of useful measures of the spreading or dispersion capabilities of a precoder can be developed; for real-valued FIR systems, maximum spreading occurs when the associated impulse response $h[n]$ is binary-valued, i.e., $h[n] = \pm N^{-1/2}$, where N is the filter length. In this case, the length N of the precoder impulse response determines the number of degrees of temporal diversity L that can be exploited. In particular, L is determined by the length of the precoding filter relative to the coherence time (in samples) τ_a of the fading; specifically,

$$L = N/\tau_a. \tag{1.32}$$

The transmission characteristics of spread-response precoding are straightforward to develop. First, with lossless precoding, the transmitted stream $y[n]$ has the same power spectrum as the original coded data $x[n]$, i.e.,

$$S_y(\omega) = S_x(\omega).$$

Hence, when $x[n]$ is a sequence of independent symbols, each with energy $\mathcal{E}$, then $y[n]$ is a complex-valued, wide-sense stationary white sequence with variance $\mathcal{E}$. Moreover, with sufficient spreading (large N), the transmitted samples $y[n]$ have each an effectively Gaussian distribution regardless of the symbol constellation associated with $x[n]$—this is suggested by a Central Limit Theorem argument since via (1.30) each $y[n]$ is the balanced sum of a large number of independent random variables.[13] The associated peak-to-average power characteristics and receiver synchronization requirements are therefore different from some conventional wireless systems and require correspondingly different implementations.

The temporal diversity inherent in spread-response precoding transmissions can be exploited via an appropriately designed equalizer and postcoder, as illustrated in Figure 1.9. Consistent with the emphasis in the chapter, we focus on linear receiver structures. It is important to remark, however, that nonlinear receivers can be designed to extract even more substantial diversity benefits than the linear receivers described here. For example, maximum likelihood sequence detection can be implemented with the Viterbi algorithm, although this is generally imprac-

---

[13]It is worth emphasizing however, that $y[n]$ is not, even asymptotically, a Gaussian process unless the independent symbols constituting $x[n]$ are Gaussian.

tical when the number of degrees of temporal diversity is large. Potentially more practical nonlinear receivers exploit the efficient iterated-decision algorithms described in [12, 35] and [13].

When we use a linear time-varying equalizer with kernel $b[n; k]$ to compensate for the fading, the equalized stream takes the form

$$\hat{y}[n] = \sum_k b[n; k] r[n - k].$$

Typically, the kernel $b[n; k]$, or equivalently, its time-varying frequency response

$$B(\omega; n] = \sum_k b[n; k] e^{-j\omega k}$$

is chosen based on the specific fading channel coefficients $a[n; k]$ and the noise statistics.

In turn, the postcoder inverts the transformation of input symbols that takes place during precoding and is simply a linear filter whose unit-sample response is [cf. (1.31)] a time-reversed version of the lossless precoding filter $h[n]$, i.e.,

$$\hat{x}[n] = h^*[-n] * \hat{y}[n] = \sum_k \hat{y}[k] h^*[k - n]. \tag{1.33}$$

A natural equalizer choice for such systems arises from a useful second-order characterization for the equivalent channel consisting of the fading channel together with precoding, equalization, and postcoding. As developed in [10], provided that the physical channel is sufficiently ergodic and that the lossless precoder effects spreading that is large compared to the coherence time, the equivalent channel can be well modeled as a simple additive white noise channel that is free of fading when any of a broad class of equalizers is used. Specifically, for a white input stream $x[n]$,

$$\hat{x}[n] \approx \mu x[n] + v[n], \tag{1.34}$$

where μ is a nonrandom gain and $v[n]$ is an additive distortion that is uncorrelated with $x[n]$. Moreover, the quality of the approximation in (1.34) improves with the number of degrees of temporal diversity involved and is asymptotically exact. As (1.34) implies, when used in conjunction with a linear equalizer and postcoder, spread-response precoding transforms the effects of fading into a comparatively more benign form of additive white interference that is uncorrelated with the symbol stream. This behavior allows each symbol in the original stream $x[n]$ to effectively experience the average characteristics of the fading channel during transmission.

It is important to emphasize that this equivalent channel has some characteristics that distinguish it from the usual additive white noise channel. In particular, the equivalent distortion $v[n]$ in (1.34) can be expressed as the sum of two uncorrelated components, one due to the receiver noise and one due to the ISI induced by fading, i.e.,

$$v[n] = v_{\text{NOISE}}[n] + v_{\text{ISI}}[n].$$ (1.35)

Thus, because the ISI term has the property that its variance is proportional to the transmitted power, the overall "equivalent noise" power in the equivalent model has a dependence on symbol energy. Ultimately, this property is taken into account in designing a suitable equalizer for this system.

The gain μ in (1.34) as well as the variances σ_{NOISE}^2 and σ_{ISI}^2 of the components of (1.35) all depend on the choice of equalizer kernel $b[n; k]$. Remarkably, when the equalizer is selected so as to maximize the associated signal-to-interference+noise ratio (SINR)

$$\gamma = \frac{|\mu|^2}{\sigma_{\text{NOISE}}^2 + \sigma_{\text{ISI}}^2},$$ (1.36)

a minimum mean-square error (MMSE) type equalizer is obtained [10]. In particular, for sufficiently slowly varying channels, the optimum equalizer has time-varying frequency response

$$B(\omega; n] \propto \frac{A^*(\omega; n]}{1 + \alpha(\omega; n]},$$ (1.37a)

where

$$\alpha(\omega; n] = \frac{\mathcal{E}|A(\omega; n]|^2}{\mathcal{N}_0 \mathcal{W}_0}$$ (1.37b)

denotes the SNR at time n and frequency ω in the original channel. When the arbitrary scale factor is appropriately chosen for this equalizer, the resulting $\hat{x}[n]$ is an MMSE linear estimate of $x[n]$.

For frequency-nonselective channels of the form (1.5), this equalization process specializes to

$$\hat{y}[n] = b[n]r[n],$$

where

$$b[n] \propto \frac{a^*[n]}{1 + \alpha[n]},$$ (1.38a)

with

$$\alpha[n] = \frac{\mathcal{E}|a[n]|^2}{\mathcal{N}_0 \mathcal{W}_0}$$ (1.38b)

analogously denoting the SNR at time n in the original channel.

In both frequency-selective and frequency-nonselective cases, the associated receivers can be implemented via computationally efficient recursive linear filtering by recognizing the MMSE property of the equalizer and exploiting a state-space description of the received data. Algorithms of this type are developed in [14, 35], and more generally, this state-space framework also leads to equalizers that can be used even with rapidly time-varying channels; these issues are discussed more generally in Chapter 3.

With the optimized equalizer (1.37) (or (1.38) when applicable), the resulting SINR (1.36) takes the form [10]

$$\gamma = \cfrac{1}{E\left[\cfrac{1}{\alpha(\omega; n] + 1} \right]} - 1 = \frac{1}{\zeta_0 e^{\zeta_0} E_1(\zeta_0)} - 1, \tag{1.39}$$

with

$$\frac{1}{\zeta_0} \triangleq E\left[\alpha(\omega; n]\right] = \frac{\sigma_a^2 \mathcal{E}}{\mathcal{N}_0 \mathcal{W}_0} \tag{1.40}$$

denoting the mean SNR in the original channel, and $E_1(\cdot)$ denoting the exponential integral, i.e.,

$$E_1(v) = \int_v^\infty \frac{e^{-t}}{t} \, dt. \tag{1.41}$$

As a result, in this regime where the effective number of degrees of temporal diversity is high, the associated QPSK bit-error rate is well approximated by

$$\mathcal{P} = \mathcal{Q}\left(\sqrt{\gamma}\right). \tag{1.42}$$

At high SNR ($\zeta_0 \ll 1$), the associated SINR takes the form

$$\gamma \sim \frac{1/\zeta_0}{\ln(1/\zeta_0)}, \tag{1.43}$$

which, when combined with a corresponding $\mathcal{Q}$-function approximation, leads in turn to an associated bit-error probability of the form

$$\mathcal{P} \sim \sqrt{\frac{\ln(1/\zeta_0)}{1/\zeta_0}} \, \exp\left(-\frac{1}{2} \frac{1/\zeta_0}{\ln(1/\zeta_0)}\right). \tag{1.44}$$

The asymptotic bit-error probability (1.42) bounds the performance obtained in practice with FIR precoders that exploit finitely many degrees (L) of temporal diversity. This can be seen in Figure 1.10, where achievable QPSK bit-error rate is plotted as a function of the average received SNR per bit $\mathcal{E}\sigma_a^2/(2\mathcal{N}_0\mathcal{W}_0)$ as the

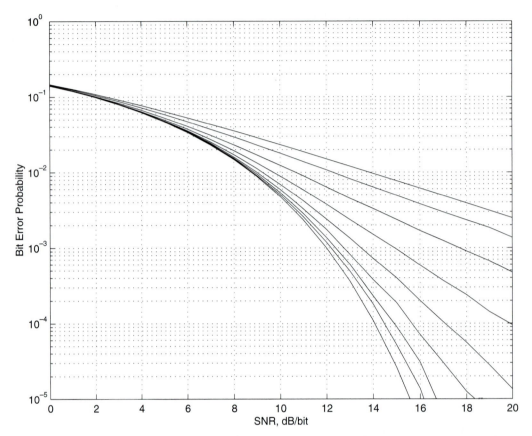

Figure 1.10 QPSK bit-error probabilities on the Rayleigh fading channel when spread-response precoding is employed. The top solid curve corresponds to the performance without precoding ($L = 1$), and the bottom solid curve indicates the performance bound corresponding to the infinite temporal-diversity case ($L \rightarrow \infty$). The successively lower solid curves between these two extremes represent the performance corresponding to $L = 2, 4, 8, 16, 32, 64$, and 128 degrees of temporal diversity, respectively.

number of degrees of temporal diversity being exploited is varied. In the simulations, optimized periodically time-varying lossless precoders were used; their detailed construction arises naturally in the context of Section 1.5, where multiuser generalizations of spread-response precoding are developed.

As Figure 1.10 reflects, there are diminishing returns as L becomes large. In particular, for a given target bit-error rate, there is corresponding value of L that is adequate for capturing most of the diversity benefit that can be attained. In turn, this choice of L effectively bounds the delay that is needed to exact this maximum possible diversity benefit. As the curves reflect, this value of L (and associated delay) are inversely related to the target bit-error rate.

1.4.2 Incorporating Bandwidth Expansion

Infinite temporal diversity in the form of spread-response precoding without bandwidth expansion cannot achieve the performance of the equivalent Gaussian channel, as is apparent from a comparison of (1.15) and (1.44). Equivalently, comparing Figures 1.8 and 1.10, we see that among linear strategies, large numbers of degrees of spectral diversity have a more substantial impact on performance than do the corresponding number of degrees of temporal diversity. From this we conclude that using large bandwidths leads to improvements that generally cannot be achieved with arbitrarily large delays on jointly time- and frequency-selective channels (when restricting attention to linear processing). However, it is possible to close the gap between the infinite temporal diversity performance and the infinite spectral diversity performance by combining spread-response precoding with bandwidth expansion. And it is important to emphasize that this potential exists even when the underlying channel is frequency-nonselective so that no spectral diversity is to be gained.[14]

To incorporate bandwidth expansion into the spread-response precoding system, it suffices to upsample the symbol stream $x[n]$ by an (integer) bandwidth expansion factor of ρ prior to precoding, downsample the output of the postcoder by the same factor, and replace the equalizer (1.37a) with

$$B(\omega; n] \propto \frac{A^*(\omega; n]}{1 + \alpha(\omega; n]/\rho} \, ,$$

where $\alpha(\omega; n]$ is still given by (1.37b).

Exploiting the analysis developed in [3], we can determine the additional diversity benefit of bandwidth expansion when the number of degrees of temporal diversity is large. In particular, the corresponding optimized SINR is

$$\gamma = \rho \left[\left(E\left[\frac{1}{1 + \alpha(\omega; n]/\rho} \right] \right)^{-1} - 1 \right] = \rho \left[\frac{1}{\rho\zeta_0 e^{\rho\zeta_0} E_1(\rho\zeta_0)} - 1 \right], \quad (1.45)$$

which is larger than (1.39) for $\rho > 1$ at every ζ_0, and reflects a reduction in ISI effects. With ζ_0 fixed, it is straightforward to verify that by choosing L to be sufficiently large, the performance gap to the infinite spectral diversity case can be made arbitrarily small. Indeed, a large ρ approximation to (1.45) yields $\gamma \rightarrow 1/\zeta_0$ as

---

[14]As an aside, it is straightforward to verify that one simple way to achieve the infinite spectral diversity performance on an infinite-bandwidth, frequency-nonselective fading channel is to use a simple symbol-repetition strategy. Using L repetitions while preserving the original symbol rate gives this result for large L, assuming sufficient symbol interleaving is used.

$\rho \to \infty$, and hence $\mathcal{Q}\left(\sqrt{\gamma}\right)$ approaches (1.29). Also, for a fixed ρ and high SNR $(1/\zeta_0 \gg \rho)$, (1.45) takes the form [cf. (1.43)]

$$\gamma \approx \rho \left[\frac{1/(\rho\zeta_0)}{\ln\left(1/(\rho\zeta_0)\right)} \right]. \tag{1.46}$$

These features are apparent in Figure 1.11, where QPSK bit-error rate is plotted as a function of SNR for several different values of ρ, again assuming the maximum possible temporal diversity benefit has been obtained.

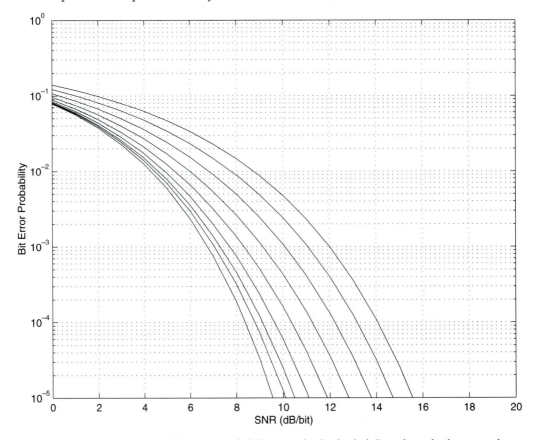

Figure 1.11 QPSK bit-error probabilities on the Rayleigh fading channel when spread-response precoding with a large number of degrees of temporal diversity is combined with bandwidth expansion. The top curve corresponds to the performance without bandwidth expansion ($\rho = 1$), and the bottom curve indicates the performance bound corresponding to the infinite bandwidth expansion ($\rho \to \infty$). The successively lower curves in between represent the performance corresponding to bandwidth expansion factors of $\rho = 2, 4, 8, 16, 32, 64$, and 128, respectively.

1.4.3 Coherence-Time Scaling

We conclude this section with a brief discussion of how spread-response precoding system performance varies with delay and computational complexity.

First, from (1.32) we see that to achieve the performance corresponding to a fixed number of degrees of temporal diversity requires a delay (precoder length) that grows in proportion to the coherence time of the channel. Conversely, given a fixed delay constraint, larger diversity benefits can be achieved on channels with shorter coherence times (i.e., higher vehicle speeds in mobile applications), at least in principle.

The computational requirements of the transmitter and receiver processing in spread-response precoding systems are rather different, however. In particular, efficient implementations of these systems require a constant number of computations per symbol per degree of temporal diversity being exploited, *independent* of the channel coherence time.

Such implementations arise by observing that when the coherence time in the channel (in samples) is τ_a, only samples of the channel response spaced at least τ_a apart in time are effectively independently faded. As a result, with $J = \lceil \tau_a \rceil$, only every Jth sample of the precoder unit-sample response need be nonzero, and the number of nonzero coefficients corresponds to the number of degrees of temporal diversity to be exploited. Hence, if $h[n]$ is a prototype precoder of length L designed for memoryless fading ($\tau_a = 1$), then an appropriate precoder for fading channels with a coherence time τ_a is obtained by simply upsampling $h[n]$ by a factor τ_a, i.e.,

$$h^{(\tau_a)}[n] = \begin{cases} h[n/J] & n = \cdots, -J, 0, J, 2J, \cdots \\ 0 & \text{otherwise.} \end{cases} \tag{1.47}$$

1.5 DIVERSITY METHODS FOR MULTIUSER SYSTEMS

Temporal and spectral diversity can also be efficiently exploited in multiuser systems and involve rich generalizations of the methods discussed in Sections 1.3 and 1.4. In this section, we develop such generalizations within a common framework. We begin with a broader discussion of multiuser communication.

Wireless communication systems for coordinating communication among multiple users typically employ a cellular architecture, which we will consider for the purposes of illustration. With such a system, the coverage area is partitioned into contiguous cells, each of which contains a base station and multiple users, which are referred to as "mobiles." There is no direct ("peer-to-peer") communication between users. Rather, each mobile sends its message directly to the base unit in its cell,[15] and this local base unit routes the message to the base unit in the cell of the intended recipient. In turn, this remote base unit broadcasts the message within its cell for the receiving mobile to pick up.

Our discussion focuses on the intracell communication involved in such scenarios—both base-to-mobile (referred to as the "forward link") and mobile-to-base (referred to as the "reverse link") transmission. We consider a typical scenario in which forward- and reverse-link communication within each cell takes place on separate (i.e., noninterfering) channels, and some total fixed bandwidth is shared for each of the two links within a cell. In particular, for each link a total of $\rho M W_0$ is available for the M mobiles, each of which is transmitting or receiving a message at a rate of W_0 symbols/s. Initially, we restrict our attention to the case for which $\rho = 1$, corresponding to the nominal spectral efficiency of 1 symbol/s/Hz.

1.5.1 Multiuser Fading Channels

The forward and reverse links have rather different characteristics. For the reverse link, the base unit receives the superposition of uncoordinated and individually faded transmissions from each of the M mobiles in the cell, from which it must extract and separate the individual messages. For the forward link, a particular mobile receives the superposition of M coordinated and identically faded transmissions broadcast by the base, from which it must extract the message for which it is the intended recipient. Nevertheless, the multiuser channel models for both links share a common form. In particular, as depicted in Figure 1.12, in a useful equivalent discrete-time baseband model for the passband communication channel, the received signal $r[n]$ takes the form

$$r[n] = \sum_{m=1}^{M} \sum_{k} a_m[n; k] y_m[n - k] + w[n]. \qquad (1.48)$$

In general, the randomly time-varying kernels $a_m[n; k]$ capture the effects of multipath fading due both to fluctuations in the media and to the relative motions of mobiles in the system, as well as the effects of asynchronism among the mobiles' transmissions in the case of the reverse link. Meanwhile, $w[n]$ captures both

---

[15]In more sophisticated cellular systems, base units in adjacent cells also receive and process the message, particularly when the user is near a cell boundary. However, we consider a simplified scenario without such "soft handoff" processing.

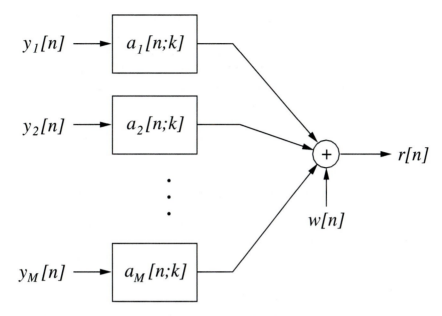

Figure 1.12   General multiuser fading channel model, where $a_m[n; k]$ denotes the randomly time-varying linear kernel corresponding to the mth user.

receiver noise and any sources of interference not otherwise taken into account, including interference from other cells.

The difference between the models for the forward and reverse links is in the relationships between the channel kernels. For the forward link, the $a_m[n; k]$ are all identical in the model, i.e.,

$$a_1[n; k] = a_2[n; k] = \cdots = a_M[n; k] \triangleq a[n; k].$$

However, for the reverse link, the mobiles are usually sufficiently well separated from one another that they experience independent fading, so the kernels $a_m[n; k]$ for different m are mutually independent. As in the single-user scenarios discussed earlier, we continue to assume in both cases that the receiver is able to obtain reliable estimates of the channel and noise coefficients.

1.5.2 Multiple-Access and Multiplexing Formats

We begin with a preliminary discussion of frequency-division, time-division, and code-division transmission formats. On the reverse link, these correspond to what are referred to as multiple-access protocols; on the forward link, they constitute multiplexing strategies. For convenience, we summarize their salient features in the reverse link context.

There are a variety of ways for a collection of users to share some common available bandwidth in reverse-link transmission. One example is frequency-division multiple-access (FDMA), whereby the total bandwidth is partitioned into subbands, a separate one of which is allocated to each user for transmission of its message. From a diversity perspective, FDMA systems are inefficient because with each mobile using a narrowband subchannel of the full wideband channel, an important source of potential spectral diversity is sacrificed. Moreover, such systems can be inefficient when the user transmissions are intermittent unless sophisticated dynamic channel reassignment strategies are employed. And intermittency of this type is quite common—in many such systems, there are large numbers of potential users, only a small fraction of whom are actively transmitting at any time.

As one alternative, time-division multiple-access (TDMA) systems partition each signaling interval into a sequence of time slots, one of which is assigned to each user for transmission of the associated segment of its message. In contrast to FDMA systems, the wideband format of TDMA systems means they are able, at least in principle, to obtain a substantial, spectral-diversity benefit. However, TDMA systems face essentially the same channel reassignment challenges as those of FDMA systems in dynamic user environments. Furthermore, TDMA formats do not provide an intrinsic, temporal-diversity benefit, though they are frequently used in conjunction with error-correction coding, which does allow an effective if computationally somewhat expensive temporal diversity benefit to be realized.

More generally, code-division multiple-access (CDMA) protocols can be used, which generalize the spread-spectrum systems discussed in Section 1.3. In a typical direct-sequence CDMA system, for example, each symbol to be transmitted is effectively represented by a finite sequence of ρ chips in the transmission. For example, in a traditional BPSK scheme, a 0-bit is signaled by transmission of a suitably chosen code sequence of length ρ, and a 1-bit by transmission of the binary complement of this sequence. When the code sequences for the different users are appropriately chosen, the result is a multiple-access system with the same potential frequency-diversity benefit of TDMA, but where intermittency issues are often more easily handled.

Although we do not develop such approaches in this chapter, it is worth pointing out that a more sophisticated alternative to these modulation formats is to directly apply nonlinear processing in the form of error-correcting coding to each mobile's stream prior to transmission. If this coding is suitably chosen, the individual symbol streams can be separated at a base unit by one of a number of nonlinear decoding algorithms, examples of which are those based on the concept of successive cancellation (stripping) [15, 16].

1.5.3 Orthogonal Multiuser Modulation

Many FDMA, TDMA, and CDMA systems of the type discussed in Section 1.5.2 are naturally viewed as forms of orthogonal multiuser modulation and can be described within a convenient multirate signal processing framework [3, 17].

The multirate system framework for multiuser communication takes the following form. The coded symbol stream of the mth user, which we denote by $x_m[n]$, is modulated onto a distinct signature sequence (i.e., discrete-time transmit pulse shape) $h_m[n]$ to produce $y_m[n]$, which is transmitted within the total available bandwidth. Figure 1.13 depicts this process, which consists of upsampling $x_m[n]$ by a factor M, the number of users, followed by linear time-invariant filtering with the signature sequence, i.e.,

$$y_m[n] = \sum_k x_m[k] h_m[n - kM].$$ (1.49)

To obtain a useful representation for signature sequences, we first express the signature set as a vector sequence, i.e.,[16]

$$\mathbf{h}[n] = \begin{bmatrix} h_1[n] & h_2[n] & \cdots & h_M[n] \end{bmatrix}^{\mathrm{T}}.$$ (1.50)

The "spread" of a signature set refers to the extent to which the energy in the constituent signatures is temporally dispersed. When each of the component signatures $h_m[n]$ is of finite length, the signature set is said to have finite spread.

Among such systems, it is natural to restrict our attention to those having the property that in the absence of fading, and with perfect synchronism among users, there is no interference between symbols either within a user's stream or among users. This restriction is equivalent to requiring that the signature sets satisfy certain orthogonality conditions—specifically, that the signature sequences together

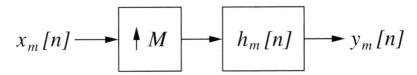

Figure 1.13 Modulation of the mth user's coded symbol stream $x_m[n]$ onto a signature sequence $h_m[n]$ for transmission.

---

[16] The superscript ^{T} denotes transposition.

with all translates by integer multiple of M constitute a complete orthonormal set. The associated conditions on $\mathbf{h}[n]$ are:[17]

$$\sum_k \mathbf{h}[k - nM]\mathbf{h}^\dagger[k - mM] = \delta[n - m]\mathbf{I} \tag{1.51a}$$

$$\sum_k \mathbf{h}^\dagger[n - kM]\mathbf{h}[m - kM] = \sum_{k,i} h_i^*[n - kM]h_i[m - kM] = \delta[n - m], \tag{1.51b}$$

where $\mathbf{I}$ denotes the identity matrix of appropriate size. Orthogonal modulation formats are desirable from a variety of perspectives and simplify receiver design.

For $M \geq 2$, we can infer from multirate filterbank theory [18] that a rich collection of signature sets satisfies (1.51), even when we restrict our attention to finite-length signatures. This property can be conveniently seen in the frequency domain. To this end, we express the set of Fourier transforms corresponding to (1.50) in the form

$$\mathbf{H}(\omega) = \sum_{n=-\infty}^{+\infty} \mathbf{h}[n]e^{-j\omega n} \triangleq \left[H_1(\omega)\ H_2(\omega)\ \cdots\ H_M(\omega)\right]^\mathrm{T}. \tag{1.52}$$

This representation leads to the so-called polyphase factorization

$$\mathbf{H}(\omega) = \mathbf{Q}(M\omega)\,\boldsymbol{\Delta}(\omega), \tag{1.53}$$

where $\mathbf{Q}(\omega)$ is referred to as the polyphase matrix, and $\boldsymbol{\Delta}(\omega)$ is the Fourier transform of the delay chain of order M, i.e.,

$$\boldsymbol{\delta}[n] = \left[\delta[n]\ \delta[n-1]\ \cdots\ \delta[n - M + 1]\right]^\mathrm{T},$$

whence

$$\boldsymbol{\Delta}(\omega) = \left[1\ e^{-j\omega}\ \cdots\ e^{-j\omega(M-1)}\right]^\mathrm{T}.$$

For a signature set to be orthonormal, it is necessary and sufficient that the associated polyphase matrix be paraunitary—i.e., that it satisfy

$$\mathbf{Q}(\omega)\mathbf{Q}^\dagger(\omega) = \mathbf{I} \tag{1.54}$$

for all ω. From this perspective, choosing a signature set is equivalent to choosing a paraunitary matrix.

Several important special cases are immediately apparent. For example, the polyphase matrix corresponding to TDMA systems is

$$\mathbf{Q}(\omega) = \mathbf{I},$$

while that corresponding to (ideal) FDMA systems has (k, l)th element

$$[\mathbf{Q}(\omega)]_{k,l} = e^{j(\omega - 2\pi k)l/M}, \qquad 0 \leq \omega \leq \pi. \tag{1.55}$$

---

[17]The superscript † denotes the conjugate-transpose operator.

In contrast, for discrete Fourier transform (DFT) based multiplexing, for which the signatures are complex-valued, $\mathbf{Q}(\omega)$ is the inverse of the DFT matrix, i.e.,

$$[\mathbf{Q}(\omega)]_{k,l} = e^{j2\pi kl/M}. \tag{1.56}$$

Finally, for direct-sequence CDMA systems that use Hadamard sequences as signatures, the polyphase matrix is

$$\mathbf{Q}(\omega) = \boldsymbol{\Xi}, \tag{1.57}$$

where $\boldsymbol{\Xi}$ is the Hadamard matrix of appropriate dimension. Recall that the Hadamard matrix of dimension M, viz., $\boldsymbol{\Xi}_M$, where M is a power of two, is defined recursively: for $M = 2, 4, \ldots,$

$$\boldsymbol{\Xi}_M = \frac{1}{\sqrt{2}} \begin{bmatrix} \boldsymbol{\Xi}_{M/2} & \boldsymbol{\Xi}_{M/2} \\ \boldsymbol{\Xi}_{M/2} & -\boldsymbol{\Xi}_{M/2} \end{bmatrix},$$

where $\boldsymbol{\Xi}_1 = 1$.

The multirate signal processing framework for orthogonal multiuser modulation is also a convenient one for describing several more recently developed classes of multiple-access and multiplexing formats that inherently provide both spectral and temporal diversity benefits. We discuss these systems next.

1.5.4 Spread-Signature CDMA Systems

In traditional CDMA systems such as those corresponding to (1.57), the signature sequences $h_m[n]$ used in the modulation (1.49) have a length N equal to the intersymbol period (upsampling rate) M. In this way, the signatures are used in a nonoverlapping manner for consecutive symbols of any particular user. In this section, however, we focus on the case in which the signature length N is significantly greater than M, so that signatures are used in highly lapped manner. This format is referred to as "spread-signature CDMA" [3]. Lapped and nonlapped formats are readily distinguished in the frequency domain: in lapped systems, the polyphase matrix $\mathbf{Q}(\omega)$ depends explicitly on ω, while for nonlapped systems, $\mathbf{Q}(\omega)$ is a constant independent of ω. Moreover, it is worth stressing that even with the use of extensive overlapping, interference among symbols both within a user's stream and from different users can be entirely avoided on ideal channels through imposition of the orthogonality conditions (1.51).

Using longer signature sequence lengths for a given symbol rate has the advantage of allowing a temporal diversity benefit to be realized in time-selective fading. Indeed, it is natural to interpret the signature as playing the role of a spread-response precoder for a particular user. Thus, the spread of a signature determines the temporal extent over which a symbol is transmitted, which can be controlled independently of the symbol rate. As in the case of spread-response pre-coding, the longer this symbol duration, the better the immunity to fades within the symbol interval. Moreover, analogously the best signatures in these applications have their energy spread as uniformly as possible over the length to which they are limited by delay constraints while meeting the orthogonality constraints (1.51). When length-N signature sequences have perfectly uniform energy distribution, i.e., $|h_m[n]| = N^{-1/2}$, the corresponding signature sets are said to be "maximally spread" [3]. Orthogonal signature sets of this type that are also real-valued are particularly attractive computationally since they require no multiplications—only additions and sign changes.

A class of real-valued, maximally spread signature sets is developed in [3] for arbitrary large values of M and N and tabulated for several particular values of M and N. As an example, for $M = 2$ and $N = 8$, the non-zero taps of $h_0[n]$ and $h_1[n]$ are given in Table 1.1. As discussed in [3], these signature sets are, rather interestingly, closely related to a number of orthogonal systems developed independently in a variety of other fields for wide ranging applications. For example, they are closely related to sequences constructed by Golay [19] [20] and Turyn [21], and later Taki, et. al. [22] and Tseng and Liu [23]. Similar constructions appear in the work of both Shapiro [24] and, later, Rudin [25].

It is important to emphasize that these multirate systems have characteristics that are markedly different from those typically used in traditional signal processing applications: maximally spread signature sets are localized in neither time nor frequency. In fact, the constituent signatures are fully broadband. As an illustration of this, Figure 1.14 depicts the energy density $|H_m(\omega)|^2$ for the signature corresponding to $m = 0$ for the case $M = 2$ and $N = 1024$. It is this spreading in both time and frequency that provides an effective form of combined temporal and spectral diversity for combating fading in wireless systems.

Figure 1.14 also implies that when the users' symbol streams $x_m[n]$ are white, the corresponding transmitted signals $y_m[n]$ are broadband with power effectively uniformly distributed over the bandwidth. Moreover, as with spread-response precoding systems, regardless of whether the symbol streams $x_m[n]$ are discrete-valued, when these symbols are independent, the transmissions $y_m[n]$ are also mar-

TABLE 1.1   Non-zero taps of the length $N = 8$ maximally spread signature sequences in a two-user ($M = 2$) system

$n =$	0	1	2	3	4	5	6	7
$\sqrt{8}h_0[n]$	+1	+1	+1	-1	+1	+1	-1	+1
$\sqrt{8}h_1[n]$	+1	+1	+1	-1	-1	-1	+1	-1

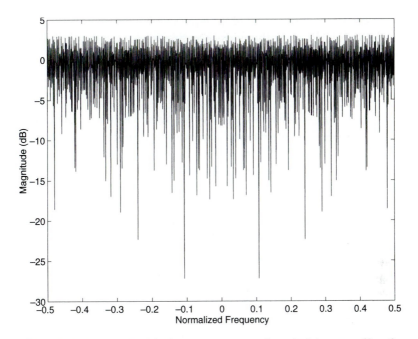

Figure 1.14   Magnitude of the frequency response of a typical signature of length $N = 1024$ in a maximally spread, two-user signature set.

ginally Gaussian due to the lapped manner in which signatures are used.

Some attractive linear equalization and demodulation methods have been developed for use with spread-signature CDMA. These generalize the structure used for spread-response precoding. In particular, a three-stage receiver can be used to recover the mth transmitted message, the front end of which is depicted in Figure 1.15. First, the received data $r[n]$ is processed by a linear equalizer according to

$$\hat{y}_m[n] = \sum_k b_m[n; k] r[n - k], \tag{1.58}$$

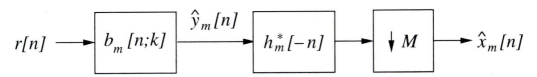

Figure 1.15   Receiver structure for extracting the symbol stream of the mth user. The first stage is equalization, producing $\hat{y}_m[n]$, and the second stage is demodulation, producing $\hat{x}_m[n]$. A final stage (not shown) is decoding.

where $b_m[n; k]$ is the kernel of the equalizer. In the second stage, the equalized data is demodulated from the corresponding signature sequence via a discrete-time matched-filter and downsample operation, viz.,

$$\hat{x}_m[n] = \sum_k \hat{y}_m[k] h_m^*[k - Mn]. \tag{1.59}$$

The final stage of the message recovery, which is not depicted in Figure 1.15, consists of decoding the demodulated stream $\hat{x}_m[n]$ to recover the transmitted symbols. When the symbols are uncoded, a simple slicer can be used at this stage to implement symbol-by-symbol decisions.

As in the case of spread-response precoding, the equivalent system consisting of modulation, propagation, equalization, and demodulation has some appealing asymptotic characteristics that simplify decoding. In particular, if the channel is ergodic and the signature sequences are long relative to the channel coherence time, then for a broad class of equalizers, the original set of coupled fading channels in the multiuser system is effectively transformed into a set of decoupled, simple additive white noise channels. Specifically, when the symbol streams $x_m[n]$ are white, then

$$\hat{x}_m[n] \approx \mu x_m[n] + v_m[n], \tag{1.60}$$

where the accuracy of the approximation increases with the signature length. In (1.60), μ is a (complex-valued) nonrandom constant, and the $v_m[n]$ are mutually uncorrelated, zero-mean, quasi-Gaussian white noise sequences that are uncorrelated with the streams $x_m[n]$. Furthermore, the variance of the noise $v_m[n]$ takes the form

$$\text{var } v_m[n] = \sigma_{\text{NOISE}}^2 + \sigma_{\text{ISI}}^2 + \sigma_{\text{MAI}}^2, \tag{1.61}$$

reflecting the fact that the equivalent noise consists of three components. The first component is due to the original noise $w[n]$ in the system after processing by the equalizer and thus is proportional to the original noise power. The second component is due to ISI in the mth user's stream induced by the fading process. Finally, the third term is due to multiple-access (i.e., interuser) interference (MAI) resulting from the effects of fading in the channel and asynchronism among users; this term is zero in the forward link scenario. The ISI term is again proportional to the user's transmit power, and the MAI term is proportional to a linear combination of the transmit powers of all the other users. Hence, the overall noise power in the equivalent model again has a dependence on signal power, which distinguishes this channel from the usual additive white noise channel.

It is important to emphasize that in the preceding discussion the *average* transmit powers of the various users are the relevant quantities in computing the equivalent noise power. In particular, the smaller the fraction of time a user is actively transmitting symbols at some fixed symbol energy, the smaller the corresponding average transmit power. The characteristic that reduced activity levels directly and dynamically translate into SINR enhancement is an extremely attractive feature of not only spread-signature CDMA systems but of CDMA systems more generally.

As in spread-response precoding systems, the particular choice of equalizer determines the overall normalized SINR

$$\gamma = \frac{|\mu|^2}{\sigma_{\text{NOISE}}^2 + \sigma_{\text{ISI}}^2 + \sigma_{\text{MAI}}^2} . \tag{1.62}$$

For frequency-selective slow fading channels, it has been shown [3] that SINR (1.62) is maximized when the time-variant frequency response of the equalizer is of the MMSE form

$$B_m(\omega; n] \propto \frac{A_m^*(\omega; n]}{1 + \dfrac{1}{M} \displaystyle\sum_{k=1}^{M} \alpha_k(\omega; n]} , \tag{1.63a}$$

where $\alpha_m(\omega; n]$ denotes the SNR at time n and frequency ω of mth user in the original channel, i.e.,

$$\alpha_m(\omega; n] = \frac{\mathcal{E}_m |A_m(\omega; n]|^2}{\mathcal{N}_0 \mathcal{W}_0} , \tag{1.63b}$$

with $\mathcal{E}_m$ the transmission power associated with this user.

The numerator of (1.63a) is a conventional matched filter (i.e., RAKE receiver), so that the denominator can be viewed as an additional compensation stage that takes into account the special characteristics of the equivalent noise in this context. However, it is worth noting that for reverse-link transmission involving a large number of users, it follows from a law of large numbers argument that the denominator of (1.63a) is effectively constant, in which case the RAKE receiver alone suffices. More generally, when the number of users is not large, an efficient recursive implementation of the equalizer (1.63) can be developed by means of the state-space framework developed in [14]; see Chapter 3.

1.5.5 CDMA Performance Characteristics

There are some important differences in the performance characteristics of conventional and spread-signature CDMA systems. In both types of systems, excess bandwidth—i.e., bandwidth beyond what is otherwise needed to support the totality of all users' symbol rates—is generally necessary in practice to achieve typical target bit-error rates. In general, such excess bandwidth is most efficiently exploited through the use of error-correction coding. However, a common and computationally efficient alternative is to use linear processing. Within the orthogonal multiuser modulation framework linear processing is accommodated by simply changing the upsampling rate in the modulation depicted in Figure 1.13 from M to $M' = \rho M$ where $\rho \geq 1$ denotes the desired bandwidth expansion factor, so that (1.49) becomes

$$y_m[n] = \sum_k x_m[k] h_m[n - \rho k M].$$ (1.64)

The amount of excess bandwidth required to achieve a prescribed level of performance generally differs for the two types of systems. We illustrate the key characteristics in a reverse-link scenario in which power control is employed to ensure that the transmissions from all mobiles are received at the same average power level.

Let us first consider spread-signature systems. Given a bandwidth expansion factor of ρ, the optimum equalizer for spread-signature CDMA takes the modified form

$$B(\omega; n] \propto \frac{A_m^*(\omega; n]}{1 + \dfrac{1}{\rho M} \displaystyle\sum_{k=1}^{M} \alpha_k(\omega; n]},$$ (1.65)

where $\alpha_k(\omega; n]$ remains as given by (1.63b). The equivalent SINR associated with each user's transmission then follows as [3]

$$\gamma = \rho M \left[\frac{1}{\beta} - 1 \right],$$ (1.66)

where

$$\beta = \frac{M - 1}{M} + \frac{(\rho M \zeta_0)^M}{M!} \left[(-1)^{M+1} e^{\rho M \zeta_0} E_1(\rho M \zeta_0) + \sum_{k=0}^{M-2} (-1)^{M-k} \frac{k!}{(\rho M \zeta_0)^{k+1}} \right],$$ (1.67)

with, due to power control,

$$1/\zeta_0 = E[\alpha_m(\omega; n]], \qquad \text{all } m.$$ (1.68)

When the resulting system is used in conjunction with, for example, an uncoded QPSK symbol stream, the associated symbol error probability is well approximated by $Q(\sqrt{\gamma})$.

The SINR (1.66) is bounded according to

$$0 < \gamma < \rho \, \frac{M}{M-1} \, , \tag{1.69}$$

where the upper bound is attained at high SNR ($\zeta_0 \to 0$). Moreover, when the number of users M is large, (1.66) is well-approximated by

$$\gamma \approx \frac{1}{\zeta_0 + 1/\rho} \, , \tag{1.70}$$

which at high SNR yields $\gamma \approx \rho$, consistent with (1.69).

With conventional CDMA systems, the bandwidth requirement as a function of the number of users is significantly different because temporal diversity is not involved. In this scenario, bandwidth expansion serves two functions: it increases the available spectral diversity that can be exploited to improve transmission performance as discussed in Section 1.3, and it reduces the multiple access interference between users. In particular, when we extend the analysis in Section 1.3 to the multiuser case, the noise source is augmented with an interference source. In an M user system with bandwidth expansion ρM, there are $M-1$ interferers, the power spectral density for the kth of which is

$$\frac{\mathcal{E}_k |A_k(\omega)|^2}{\rho M}.$$

Hence, the power spectrum for total noise and interference experienced by the mth user is

$$\gamma = \frac{\mathcal{E}_m \mu^2}{\mathcal{N}_0 \mathcal{W}_0 + \sum_{k \neq m} \dfrac{\mathcal{E}_k |A_k(\omega)|^2}{\rho M}} \, ,$$

which via a law of large numbers can be conveniently approximated as

$$\gamma \approx \frac{\mathcal{E}_m \mu^2}{\mathcal{N}_0 \mathcal{W}_0 + (M-1) \dfrac{\mathcal{E}_k \sigma_a^2}{\rho M}} \, . \tag{1.71}$$

Moreover, at high SNR ($\zeta_0 \to 0$), we can further approximate (1.71) via

$$\gamma \approx \rho \, \frac{M}{M-1} \, . \tag{1.72}$$

Hence, replacing ζ_0 in (1.27) with $1/\gamma$, as defined in (1.71) or (1.72), and appropriately redefining L as

$$L = \rho M W_0/\theta_a, \tag{1.73}$$

we obtain the associated bit-error probability behavior. It is worth pointing out that these results also accurately characterize the performance of spread-signature CDMA systems in time-nonselective fading environments.

The approximation (1.72) and upper bound in (1.69) reflect that the performance of both conventional and spread-signature CDMA systems, respectively, is MAI-limited at high SNR. This behavior is characteristic of many CDMA implementations. When power control is not employed, this behavior can give rise to problematic near-far effects, whereby transmissions from nearby mobiles entirely bury those from more distant ones. However, it is important to emphasize that this is not an inherent limitation of either CDMA systems in general or of spread-signature CDMA systems in particular. Rather, it is a consequence of the specific receiver structure imposed, and these effects can be avoided through the use of even linear receivers provided they are appropriately designed. Indeed, a variety of efficient near-far resistant linear receivers for conventional CDMA systems are discussed in detail in Chapter 2. Related receivers for spread-signature systems have also been developed [14, 35], and have computationally efficient recursive implementations; these are special cases of the state-space algorithms discussed in Chapter 3.[18]

Nevertheless, even without near-far resistant receivers, reasonable performance can be achieved with sufficient bandwidth expansion. This is illustrated in Figure 1.16, which depicts the excess bandwidth (i.e., reduction in bits/s/Hz/user) that is required to achieve a target error rate in the high SNR regime in uncoded CDMA systems. Note that a bandwidth expansion factor of ρ corresponds to a spectral efficiency of $2\rho^{-1}$ bits/s/Hz/user since QPSK symbol streams are involved. From this figure, we see that for conventional CDMA systems, the per-user excess bandwidth requirements generally decrease with the number of users; as the number of users increases, so does the overall system bandwidth, which allows a substantial spectral diversity benefit to be realized. This benefit more than offsets the corresponding increase in MAI.

On the other hand, for spread-signature CDMA systems, a substantial temporal diversity benefit is obtained independently of total system bandwidth. For these systems, then, the per-user excess bandwidth requirements increase with the number of users in order to mitigate the increasing effects of MAI. When the num-

---

[18]Furthermore, with only a modest additional increase in complexity, still better interference rejection can be achieved, at least in principle, in spread-signature CDMA systems through the use of the nonlinear iterated-decision equalizers [12, 35].

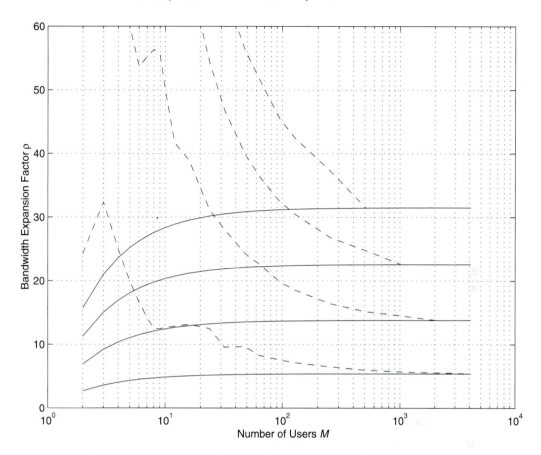

Figure 1.16   Excess bandwidth required to achieve target QPSK bit-error rates in an M transmitter multiple-access system with spread-signature CDMA. The successively higher solid curves correspond to bit-error rates of 10^{-2}, 10^{-4}, 10^{-6}, and 10^{-8}. The corresponding dashed curves describe the performance of a conventional CDMA system, or a spread-signature system used in an environment without time-selective fading.

ber of users is large, corresponding to large system bandwidths, both conventional and spread-signature systems achieve similar efficiencies because maximal use of diversity is made by both systems in this regime. However, when the number of users is smaller, spread-signature systems offer a distinct advantage in providing a more extensive diversity benefit.

It should be pointed out that CDMA systems exhibit dramatically different behavior on the forward link due to the absence of MAI. In fact, forward-link performance is effectively identical to that obtained for the corresponding single-user systems. As such, the forward-link performance of spread-signature CDMA coincides with that of spread-response precoding, as depicted in Figure 1.10 for various degrees of temporal diversity.

1.5.6 Coherence-Time Scaling

As with spread-response precoding systems, the computational complexity required to implement spread-signature CDMA is a function of the number of degrees of temporal diversity to be exploited and is independent of the coherence time characteristics of the channel. In fact, signatures are matched to the coherence-time characteristics of the channel in a similar manner: signatures are obtained by upsampling prototypes so that the non-zero coefficients have a spacing that is on the order of the coherence time. In particular, with memoryless-fading prototype signatures $h_m[n]$, new signatures are derived via

$$h_m^{(\tau_a)}[n] = \begin{cases} h[n/J] & n = \cdots, -J, 0, J, 2J, \cdots \\ 0 & \text{otherwise.} \end{cases} \tag{1.74}$$

In this case, however, the upsampling factor J must satisfy certain constraints so as to ensure that the new signatures also meet the orthogonality conditions (1.51). Specifically, J is chosen to be the smallest integer that is at least as large as τ_a but is not a multiple of a prime factor of M [3].

1.5.7 Efficient Implementations of Spread-Response Precoding

Spread-signature CDMA can be used to obtain an efficient implementation of the spread-response precoding concept described in Section 1.4.1. To develop this perspective, first note that the particular class of spread-response precoding systems described in Section 1.4.1 corresponds to spread-signature CDMA systems in which $M = 1$. However, some difficulties arise in this special case. In particular, recall that FIR LTI filters cannot be simultaneously lossless (allpass) and spread (length greater than one). Thus, if we wish to restrict our attention to LTI filters for spread-response precoding applications, we are forced to sacrifice perfect losslessness.[19] However, if we relax the time-invariance constraint and allow periodically time-varying filters, then both losslessness and maximal spreading can be achieved.

A binary-valued (maximally spread), linear periodically time-varying (LPTV) precoder can be constructed as follows. First, the symbol stream $x[n]$ is transformed into a set of K parallel substreams

$$x_m[n] = x[nK + m] \tag{1.75}$$

via a serial-to-parallel converter, which can be implemented via the multirate structure shown in Figure 1.17. These subsequences are then treated as K virtual

---

[19]However, some approximately lossless solutions are developed in [10] and [11].

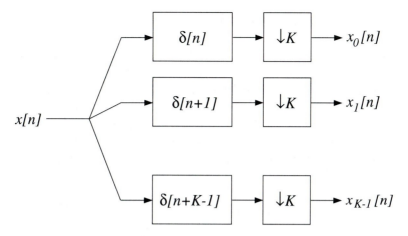

Figure 1.17 Construction of the different phases of an input for linear periodically time-varying precoding: serial-to-parallel conversion.

users and multiplexed in the manner used for forward-link transmission with spread-signature CDMA. In particular, these substreams are upsampled, filtered by the appropriate maximally spread signatures described in Section 1.5.4, and combined as depicted in Figure 1.18. The result is a computationally highly efficient (multiply-free) lossless precoder with optimum spread for a given delay constraint.[20]

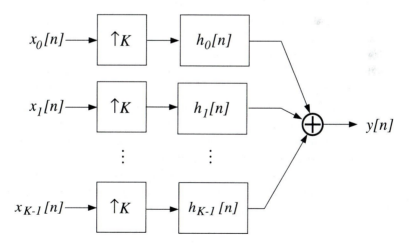

Figure 1.18 Construction of the output of a linear periodically time-varying precoder from the phases of its input.

[20]Note that as an LPTV system, this precoder is characterized by K unit-sample responses: the response to a unit sample $\delta[n - n_0]$ is $h_m[n + m - n_0]$, where $m = n_0 \bmod K$.

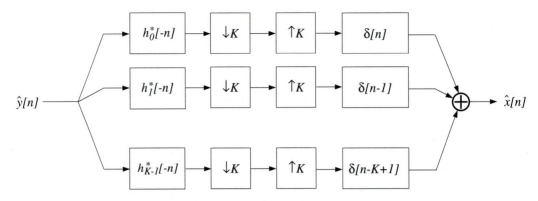

Figure 1.19   Postcoder for use with linear periodically time-varying precoding.

The interpretation of LPTV precoding involving virtual users is also useful in terms of receiver design. Indeed, the equalization and demodulation process follows immediately from the corresponding receiver for spread-signature CDMA in forward-link transmission, and the postcoder in particular takes the form depicted in Figure 1.19.

This precoding system based on maximally spread signatures for the $K = 2$ case was what was used in the simulations of Figure 1.10. Comparisons between the results presented in [10] and [3] reflect that for a given delay constraint, using maximally spread LPTV precoders instead of LTI precoders reduces both the bit-error rate and the required computational complexity.

1.6 SPATIAL DIVERSITY

The final form of diversity we examine in this chapter is spatial diversity in the form of multiple-element antenna arrays. This form of diversity can be used alone or in conjunction with spectral or temporal diversity. Whether such arrays are available at the transmitter or receiver has a major impact on the resulting diversity benefit. To emphasize their differences, we consider the two cases separately, although it is worth keeping in mind that in practice, arrays at both transmitter and receiver can be exploited simultaneously to varying degrees.

Also, to simplify the presentation and to isolate and distinguish the benefits of spatial diversity from those of temporal and spectral diversity, we restrict our attention to time- and frequency-nonselective fading channels where only a spatial diversity benefit can be obtained. However, generalizations of the approaches we discuss can be applied to more general selective fading channels.

As we discuss first, the use of multiple-element antennas at the receiver is fairly easily exploited. In essence, multiple copies of the transmitted stream are received; these copies can be efficiently combined by use of the appropriate spatial matched filter. As the number of antennas increases, the outage probability is driven to zero and the effective channel approaches an infinite-power additive white Gaussian noise channel, which dramatically improves communication even without additional coding. However, receiver antenna diversity can be impractical in a number of applications such as broadcasting or forward-link (base-to-mobile) transmission in cellular systems. In such scenarios, the use of multiple antennas at the transmitter is significantly more attractive.

Transmitter antenna diversity, which we discuss subsequently, is generally less straightforward to exploit, particularly when bandwidth expansion is not feasible and when there is no feedback path to provide the transmitter with knowledge of the channel parameters. To optimally exploit this form of diversity requires the use of suitably designed error-correction coding at each antenna element [26, 27]. As a lower complexity alternative, we will discuss a class of practical and bandwidth-efficient linear techniques that nevertheless yield a substantial improvement in system performance.

1.6.1 Receiver Antenna Diversity

For a system with an L-element receiver antenna array, the associated nonselective fading channel model takes the form depicted in Figure 1.20, where the signal $r_l[n]$ obtained at the lth array element takes the form

$$r_l[n] = a_l x[n] + w_l[n], \tag{1.76}$$

with $x[n]$ denoting the transmitted symbol stream, and with $w_l[n]$ denoting the receiver noise at that element, which is complex-valued, zero-mean, white, circularly symmetric Gaussian noise of variance $\mathcal{N}_0 \mathcal{W}_0$. Given sufficient physical separation among the constituent elements, the fading coefficients $a_0, a_1, \ldots, a_{L-1}$ can be modeled as mutually independent with identical variances σ_a^2. It is not difficult to accommodate the case in which the antennas are close enough that the a_l are correlated, though arrays in this configuration yield a smaller diversity benefit for a given number of array elements. Likewise, we consider a typical scenario in which the receiver noises $w_l[n]$ can be reasonably modeled in practice as being mutually independent; when these noises are correlated, additional performance enhancement is possible, at least in principle. As in earlier parts of the chapter, we continue to assume that the channel parameters a_l have been accurately measured at the receiver.

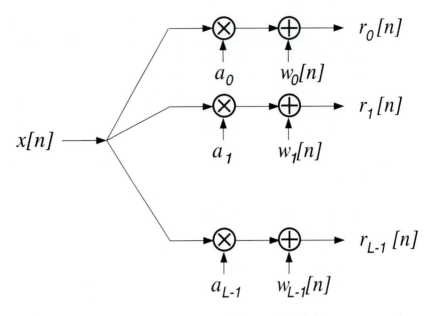

Figure 1.20   Channel model when nonselective Rayleigh fading is experienced with a multiple-element receiver antenna array.

The signals obtained at the different array elements are linearly combined using a weight-and-sum process, i.e.,

$$\hat{x}[n] = \sum_{l=0}^{L-1} b_l r_l[n] \tag{1.77}$$

where b_l are the weights. Substituting (1.76) in (1.77) reveals that the combiner output takes the form

$$\hat{x}[n] = \mu x[n] + v[n], \tag{1.78}$$

where the gain μ and white noise variance $\sigma^2 = \mathrm{var}\, v[n]$ depend on the choice of the weights b_l.

A variety of combining strategies used in practice correspond to different choices for the b_l's. For example, with selection combining, the weights take the form [28]

$$b_l = \begin{cases} a_l^*/|a_l| & \text{if } l = \arg\max_k |a_k| \\ 0 & \text{otherwise,} \end{cases} \tag{1.79}$$

corresponding to a receiver that makes use of only the strongest of the received signals $r_l[n]$. However, it is straightforward to show that the SNR at the output of the combiner, i.e.,

$$\gamma = \frac{|\mu|^2}{\sigma^2} \tag{1.80}$$

is maximized when, in general, all of the received signals are used and are combined by means of weights b_l of the form

$$b_l \propto a_l^*, \tag{1.81}$$

which corresponds to spatial matched filtering. This processing is referred to as maximal ratio combining [28] and results in an output SNR of

$$\alpha = \frac{\mu^2 \mathcal{E}}{\mathcal{N}_0 \mathcal{W}_0} \tag{1.82}$$

with

$$\mu^2 = \sum_{l=0}^{L-1} |a_l|^2. \tag{1.83}$$

This output SNR is therefore an Lth-order Erlang random variable with mean L/ζ_0 and variance L/ζ_0^2, where ζ_0 is as given in (1.11).

Comparing these results with those obtained in Section 1.3, we see that this SNR is a factor of L greater than that obtained with spectral diversity. This gain makes receiver antenna diversity overwhelmingly more effective than other forms of diversity we have discussed thus far. In fact, the preceding implies that we would have to boost the transmitted power by a factor of L in a single-element antenna system with spectral diversity to achieve the performance of a system without spectral diversity but with spatial diversity in the form of an L-element receiver antenna array.

For a QPSK symbol stream, the bit-error probability is obtained through an appropriate renormalization of (1.27) that involves replacing ζ_0 with ζ_0/L, yielding

$$\mathcal{P}[L] = \frac{1}{2}\left[1 - \frac{1}{\sqrt{2\zeta_0 + 1}} \sum_{k=0}^{L-1} \binom{2k}{k}\left(\frac{\zeta_0}{2(2\zeta_0 + 1)}\right)^k\right]. \tag{1.84}$$

The variation with SNR $1/\zeta_0$ is depicted in Figure 1.21. Note that for any particular SNR, L can be chosen to make $\mathcal{P}[L]$ arbitrarily small because $\alpha \to \infty$ as $L \to \infty$. As we develop next, the behavior with transmitter antenna diversity is dramatically different.

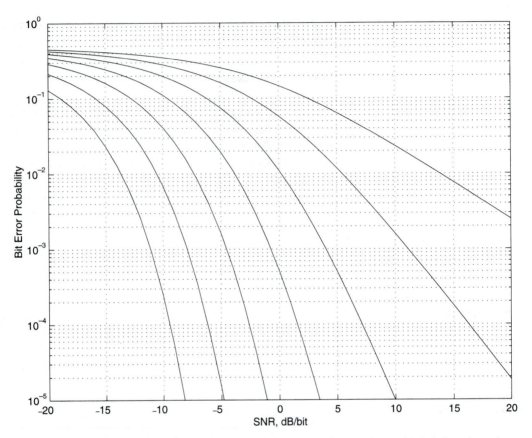

Figure 1.21 Bit-error probabilities, using uncoded QPSK on the Rayleigh fading channel with receiver antenna diversity exploited with maximal ratio combining. The top curve corresponds to the performance without spatial diversity ($L = 1$), and the successively lower curves represent the performance obtained with $L = 2, 4, 8, 16, 32$, and 64 receiver array elements, respectively.

1.6.2 Transmitter Antenna Diversity

For a system with an L-element transmitter antenna array, the associated nonselective fading channel model takes the form depicted in Figure 1.22, where the (generally complex-valued) transmission from the lth array element we denote using $y_l[n]$ for $l = 0, 1, \ldots, L - 1$. At the receiver we obtain

$$r[n] = w[n] + \sum_{l=0}^{L-1} a_l y_l[n], \tag{1.85}$$

where $w[n]$ denotes the receiver noise, which is complex-valued, zero-mean, white circular Gaussian noise with variance $\mathcal{N}_0 \mathcal{W}_0$. As in the case of receiver antenna diversity, given sufficient physical separation among the constituent elements,

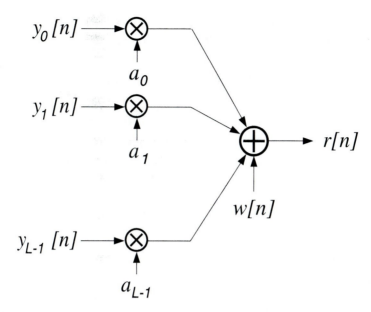

Figure 1.22 Channel model for transmission in nonselective Rayleigh fading via a multiple-element transmitter antenna array.

the fading coefficients $a_0, a_1, \ldots, a_{L-1}$ can be modeled as being mutually independent, complex-valued, zero-mean circular Gaussian random variables with variance σ_a^2.[21] We continue to assume these a_l are known at the receiver.

1.6.2.1 Beamforming: Diversity with Feedback

When a feedback path exists between the receivers and the transmitter through which perfect side information about the channel parameters can be sent, then transmitter antenna diversity can achieve performance identical to that we described for receiver antenna diversity in the previous section.

Consider, for example, a scenario with a single receiver and a transmitter that has exact knowledge of the channel gains a_l to that receiver. In this case, it is sufficient for the sequence emitted from the lth antenna element to be of the form

$$y_l[n] = b_l x[n], \tag{1.86}$$

where the b_l satisfy the constraint

$$\sum_{l=0}^{L-1} |b_l|^2 = 1$$

so that the total transmitted power is independent of L.

---

[21]Again, although not considered here, it is possible to accommodate arrays in which the antenna elements are close enough that the a_l are correlated.

It is straightforward to verify that choosing the b_l so as to optimize the SNR at which the symbol stream $x[n]$ is received yields

$$b_l = \frac{a_l^*}{\sqrt{\sum_k |a_k|^2}} . \qquad (1.87)$$

With these gains, which correspond to a beamforming process in which the individual paths from each antenna element combine coherently, a received signal of the form

$$r[n] = \hat{x}[n] = \mu x[n] + w[n] \qquad (1.88)$$

is produced, where

$$\mu^2 = \sum_{l=0}^{L-1} |a_l|^2. \qquad (1.89)$$

Comparing these results to those of Section 1.6.1 we see that the effective SNR at the receiver is therefore identical to that obtained with receiver antenna diversity: with perfect channel knowledge at the transmitter and a single receiver, the diversity benefit is identical to that obtained through the use of receiver antenna diversity with the same number of antenna elements. Hence, Figure 1.21 reflects the performance of this transmitter antenna diversity system as well.

When more than one receiver is involved, it is generally not possible to achieve this SNR enhancement at every receiver since the transmitter must attempt to beamform to multiple receivers simultaneously, which is an inherently more constrained problem and generally cannot yield the same performance as beamforming to each user individually. However, when there are large numbers of receivers (corresponding, for example, to a broadcast scenario), strategies of the type we explore in the next section are more appropriate. These strategies also have the advantage that channel state information is not required at the transmitter.

1.6.2.2 Linear Antenna Precoding

A rather general framework for describing linear transmitter antenna diversity strategies is that developed in [29]. It encompasses a variety of efficient schemes proposed over the last several years, including the methods of Wittneben [30], Winters [31], Hiroike et. al. [32], and Weerackody [33], as well as more traditional approaches such as those described in Jakes [28]. In this framework, the ability to exploit diversity is created by processing the symbol stream to be transmitted with a different linear filter at each antenna element. This processing is referred to as linear antenna precoding [29] and effects a form of spatial spreading or dispersion that is analogous to the temporal spreading generated in spread-response precoding systems.

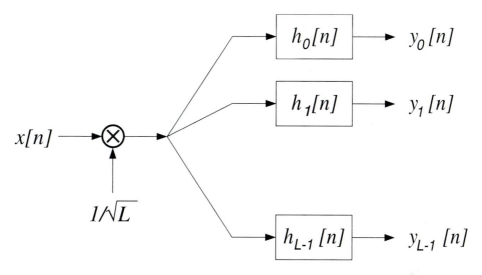

Figure 1.23   Linear antenna precoding (direct-form).

We first restrict our attention to the case in which this linear processing is specifically time-invariant. Important periodically time-varying alternatives are developed in Section 1.6.2.3.

As depicted in Figure 1.23, we use $h_l[n]$ to denote the generally complex-valued, unit-sample response of the filter associated with the lth antenna element and refer to this as the "signature" of the element in this context. The collection of L signatures associated with the array is referred to as the signature set. With this notation, we have

$$y_l[n] = \frac{1}{\sqrt{L}} \sum_{k=-\infty}^{+\infty} h_l[k]x[n-k].\tag{1.90}$$

With $H_l(\omega)$ denoting the associated Fourier transform of each signature, i.e.,

$$H_l(\omega) = \sum_n h_l[n]e^{-j\omega n},\tag{1.91}$$

the total average transmitted power is constrained to be independent of L by imposing the condition

$$\frac{1}{L}\sum_{l=0}^{L-1}\left|H_l(\omega)\right|^2 = 1, \qquad \text{for all } \omega.\tag{1.92}$$

Eq. (1.92) can be verified by noting that the power transmitted from the lth antenna element is

$$\operatorname{var} y_l[n] = \frac{1}{2\pi L}\int_{-\pi}^{\pi}\left|H_l(\omega)\right|^2 S_x(\omega)\,d\omega,$$

where $S_x(\omega)$ is the power spectrum of the symbol stream $x[n]$. Thus, the total transmitted power is

$$\sum_{l=0}^{L-1} \operatorname{var} y_l[n] = \frac{1}{2\pi} \int_{-\pi}^{\pi} \left[\frac{1}{L} \sum_{l=0}^{L-1} |H_l(\omega)|^2 \right] S_x(\omega)\, d\omega,$$

which is equal to $\operatorname{var} x[n]$ when (1.92) is satisfied.

Linear antenna precoding has a powerful interpretation as a channel transformation strategy. To see this, observe that by specializing (1.85), the received signal can be expressed in the form

$$r[n] = a[n] * x[n] + w[n], \tag{1.93}$$

where

$$a[n] = \frac{1}{\sqrt{L}} \sum_{l=0}^{L-1} a_l h_l[n] \tag{1.94}$$

is the unit-sample response of the "effective" channel generated by the antenna precoder. This channel has frequency response

$$A(\omega) = \frac{1}{\sqrt{L}} \sum_{l=0}^{L-1} a_l H_l(\omega), \tag{1.95}$$

which is a zero-mean, 2π-periodic, Gaussian random process in frequency ω, with variance σ_a^2. Hence, we see that the antenna precoding effectively transforms the original nonselective fading channel into a frequency-selective fading channel. Naturally, the specific properties of the resulting equivalent frequency selective channel depend on the characteristics of the antenna signatures $h_l[n]$ through the parameterization (1.94).

This key channel transformation observation has important implications for receiver design. Indeed, it implies, for example, that any of a variety of traditional approaches to detection in the presence of intersymbol interference can be exploited. Examples include maximum likelihood sequence detection, decision-feedback equalization, or linear equalization [4]. When suitably designed, such receivers can exploit this inherent frequency diversity to substantially improve system performance.

As in earlier parts of the chapter, we continue to restrict our attention to the use of a linear equalizer with unit-sample response $b[n]$ and denote the equalized signal by $\hat{x}[n]$, i.e.,

$$\hat{x}[n] = b[n] * r[n] = \sum_k b[k] r[n - k]. \tag{1.96}$$

In general, since the receiver knows the fading coefficients, the equalizer $b[n]$ will depend on the channel response $a[n]$.

With linear equalizers, linear antenna precoding systems have some useful properties that are counterparts to those of spread-response precoding systems in the context of time-diversity. In particular, when the number of antenna elements L is large, then for a broad class of signatures and equalizers, the relationship between the equalizer output $\hat{x}[n]$ and a white symbol stream input $x[n]$ can be approximated by [29]

$$\hat{x}[n] \approx \mu x[n] + v[n], \tag{1.97}$$

where μ is a nonrandom constant and $v[n]$ is an equivalent noise that is white and uncorrelated with $x[n]$. The quality of the approximation (1.97) increases with L and, analogous to earlier results in the chapter, is asymptotically exact.

As was the case with spread-response precoding systems using linear receivers, (1.97) implies that this transmitter antenna diversity system transforms the effects of fading into a form of additive white interference that is uncorrelated with the symbol stream. In particular, $v[n]$ in (1.97) is composed of two uncorrelated components, one due to the receiver noise and one due to the ISI induced by the multiple transmissions, i.e.,

$$v[n] = v_{\text{NOISE}}[n] + v_{\text{ISI}}[n]. \tag{1.98}$$

The class of signatures for which the approximation (1.97) is valid is quite large. It includes, for example, those corresponding to the unrealizable transmit antenna diversity scheme in which each antenna is assigned a distinct portion of the available bandwidth. The associated antenna signatures are the unit-sample responses of ideal bandpass filters, i.e.,

$$H_l(\omega) = \begin{cases} \sqrt{L} & l\pi/L < |\omega| < (l+1)\pi/L \\ 0 & \text{elsewhere in } |\omega| < \pi. \end{cases}$$

This class also includes a large number of realizable systems in general and FIR systems in particular. For example, it can be shown [29] that antenna signatures of length L are within this class whenever the matrix

$$\mathbf{H} = \begin{bmatrix} h_0[0] & h_0[1] & \cdots & h_0[L-1] \\ h_1[0] & h_1[1] & \cdots & h_1[L-1] \\ \vdots & \vdots & \ddots & \vdots \\ h_{L-1}[0] & h_{L-1}[1] & \cdots & h_{L-1}[L-1] \end{bmatrix} \tag{1.99}$$

is unitary.

An example of a signature set in this class is that for which $\mathbf{H} = \mathbf{I}$, so that $h_l[n] = \delta[n-l]$. This corresponds to a scheme explored both by Wittneben [30] for the case $L = 2$ and, more generally, by Winters [31] (see also Seshadri and Winters [34]). In these schemes, each antenna transmits a delayed copy of the sequence

$x[n]$. Other possible choices would include $\mathbf{H} = \mathbf{F}$ or $\mathbf{H} = \mathbf{\Xi}$, which are again the DFT and Hadamard matrices, respectively. The former corresponds to a generally complex-valued signature set and can be viewed as a finite length variant of the frequency band allocation example described earlier. For $L = 2$ the DFT- and Hadamard-based schemes specialize to a common scheme also explored by Wittneben [30].

Although these various choices for $\mathbf{H}$ all lead to a similar diversity benefit, as we will discuss shortly, the detailed transmission characteristics are signature dependent. For example, when $\mathbf{H} = \mathbf{I}$, the transmitted signals $y_l[n]$ have the same amplitude characteristic (i.e., marginal probability density function) as $x[n]$ regardless of L. By contrast, when $\mathbf{H} = \mathbf{\Xi}$, the amplitude distribution for each $y_l[n]$ is effectively Gaussian for large L, corresponding to large peak-to-average transmitted power. Nevertheless, when L is at least moderately large, all such transmitter antenna diversity schemes have the property that the signal component of the received waveform has an effectively Gaussian amplitude distribution due to the superposition of the multiple transmissions.

An important implication of (1.97) is that using a transmitter antenna array reduces variations in system performance as experienced by different receivers. To appreciate what this means in practice, consider a scenario in which there is a collection of suitably separated receivers all at roughly the same radius from the transmitter, as depicted in Figure 1.24. With a single-element transmit antenna, the SNR of the transmitted data as measured at a receiver will vary from receiver to receiver. However, when diversity via a multiple-element antenna cluster is exploited, this variation from receiver to receiver is reduced. In fact, when the

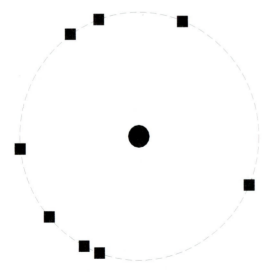

Figure 1.24   A collection of receivers, each denoted by the symbol ■, at a constant radius from a central transmitter, which is denoted by ●.

number of antenna elements is large, (1.97) implies that variation from receiver to receiver is effectively eliminated, in which case all receivers "see" the data at the same effective SNR as the outage probability is driven to zero.

The gain μ in (1.97) as well as the variances σ_{NOISE}^2 and σ_{ISI}^2 of the components of (1.98) all depend on the choice of equalizer unit-sample response $b[n]$. Maximizing the SINR

$$\gamma = \frac{|\mu|^2}{\sigma_{\text{NOISE}}^2 + \sigma_{\text{ISI}}^2} \tag{1.100}$$

with respect to the choice of $b[n]$ to obtain the best possible average performance at the receiver leads to another MMSE type (modified RAKE) equalizer, viz.,

$$B(\omega) \propto \frac{A^*(\omega)}{1 + \alpha(\omega)}, \tag{1.101a}$$

where

$$\alpha(\omega) = \frac{\mathcal{E}|A(\omega)|^2}{\mathcal{N}_0 \mathcal{W}_0} \tag{1.101b}$$

is again the SNR at frequency ω in the corresponding frequency-selective channel, i.e., (1.37b). Computationally efficient implementations of these MMSE equalizers can also be developed, with or without an additional FIR constraint on $b[n]$.

By analogy with the results obtained for spread-response precoding, when this optimum equalizer is used, the resulting SINR can be expressed in the form [29]

$$\gamma = \frac{1}{E\left[\dfrac{1}{\alpha(\omega) + 1}\right]} - 1 = \frac{1}{\zeta_0 e^{\zeta_0} E_1(\zeta_0)} - 1 \tag{1.102}$$

with

$$\frac{1}{\zeta_0} = E[\alpha(\omega)] = \frac{\sigma_a^2 \mathcal{E}}{\mathcal{N}_0 \mathcal{W}_0}. \tag{1.103}$$

In turn, the associated average QPSK bit-error rate is well approximated by

$$\mathcal{P} = \mathcal{Q}\left(\sqrt{\gamma}\right). \tag{1.104}$$

But since (1.104) is identical to the corresponding expression (1.42) for spread-response precoding, the error probability vs. SNR behavior is identical for the system when the number of degrees of diversity is large for each. As a result, the asymptotic expression (1.44), valid for spread-response precoding at high SNR, also applies to linear antenna precoding in the same regime.

Since $\mathcal{P}$ in (1.104) is uniformly lower than the corresponding QPSK bit-error rate with a single element transmitter antenna—i.e., (1.12)—we are able to conclude that with an appropriately chosen linear equalizer, the use of transmitter diversity in the form of linear antenna precoding not only asymptotically eliminates the variance in performance among receivers but significantly improves the *average* level of performance among these receivers as well.

1.6.2.3 Dual-Form Linear Antenna Precoding

Aspects of linear antenna precoding as developed in the preceding section can be potentially problematic in implementations. For example, even with a finite number of antennas, the optimum equalizer has, in general, an infinite length, unrealizable, unit-sample response. And, while finite-length approximations can give a close approximation to the optimum performance in practice, they do so at a cost of excessive delay. These problems can be conveniently circumvented through an alternative form of linear antenna precoding, which is the focus of this section. To distinguish the two forms of precoding, it is convenient to refer to the strategy of Section 1.6.2.1 as *direct-form* linear antenna precoding, and to the one we describe next as *dual-form* linear antenna precoding.

Both direct-form and dual-form precoding can be described within a broader framework for linear antenna precoding in which the linear processing that takes place at each antenna element is generally periodically time-varying (i.e., LPTV filtering). Within this framework, direct-form linear antenna precoding corresponds to the special unit-period case of time-invariant processing (i.e., LTI filtering), and the associated unit-sample responses are the antenna signatures.

In dual-form linear antenna precoding systems [29], the LPTV filtering takes the form depicted in Figure 1.25. In particular, the symbol stream $x[n]$ is first processed by a common LPTV prefilter that is time-varying with some period[22] $K \geq 2$ and has length L, and whose kernel we denote by $g[n; k]$. The result,

$$y[n] = \sum_k g[n; k]x[n - k],\tag{1.105}$$

is then subsequently processed at each of the antennas. Specifically, this prefiltered stream is modulated at each antenna by a different L-periodic sequence, i.e.,

$$y_l[n] = \tilde{h}_l[n]y[n],\tag{1.106}$$

where $\tilde{h}_l[n]$ is the generally complex-valued periodic sequence associated with the lth antenna. The signature $h_l[n]$ of the associated antenna element is defined to be a single period of this modulating sequence, i.e.,

$$h_l[n] = \begin{cases} \tilde{h}_l[n] & 0 \leq n \leq L - 1 \\ 0 & \text{otherwise.} \end{cases}$$

---

[22]Again, the period K in fact plays a relatively minor role in the development.

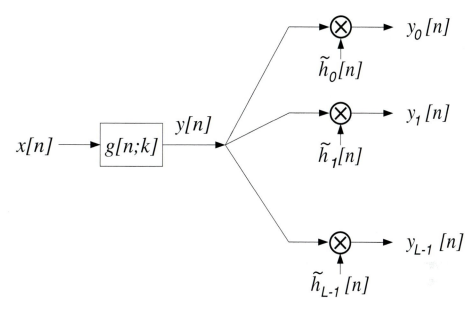

Figure 1.25   Dual-form linear antenna precoding.

We first describe the design of the signatures $h_l[n]$ and then turn our attention to the design of the prefilter kernel $g[n; k]$.

To ensure that the total transmitted power is independent of the number of antenna elements L, the following normalization is imposed. First, we constrain the prefilter to be an orthogonal (energy-preserving) transformation, so that

$$\operatorname{var} y[n] = \operatorname{var} x[n]. \tag{1.107}$$

In turn, using, in order, (1.106) and (1.107), we obtain that the total transmitted power is

$$\sum_{m=0}^{L-1} \operatorname{var} y_m[n] = \operatorname{var} x[n] \sum_{m=0}^{L-1} \left| h_m[n] \right|^2. \tag{1.108}$$

Since this total transmitted power must equal $\operatorname{var} x[n]$, we obtain the condition

$$\sum_{m=0}^{L-1} \left| h_m[n] \right|^2 = 1. \tag{1.109}$$

This form of linear antenna precoding also has a powerful interpretation as a channel transformation strategy, and one that is the dual for that for the direct-form system. In particular, while direct-form linear antenna precoding transforms nonselective fading channels into frequency-selective ones, the dual-form system effectively transforms them into time-selective ones. To see this, it suffices to note that the response of the channel to the prefiltered symbol stream $y[n]$ is, using (1.106) in (1.85),

$$r[n] = \tilde{a}[n]y[n] + w[n], \tag{1.110}$$

where

$$\tilde{a}[n] = \sum_{l=0}^{L-1} a_l \tilde{h}_l[n] \tag{1.111}$$

is an L-periodic fading sequence. For future convenience, $a[n]$ denotes one period of this sequence, i.e.,

$$a[n] = \sum_{l=0}^{L-1} a_l h_l[n]. \tag{1.112}$$

This interpretation leads to some important insights into system design. First, it implies that the prefilter kernel $g[n; k]$ should be so designed as to allow the inherent temporal diversity introduced by the modulation process to be efficiently exploited at the receiver. Second, we see that the maximum time-diversity benefit is obtained when the fading is independent among time samples within a period, and thus the signature sequences should be chosen to ensure that this condition is met.

To develop these concepts further, it is convenient to collect the dual-form signatures into a matrix of the form

$$\mathbf{H} = \begin{bmatrix} h_0[0] & h_0[1] & \cdots & h_0[L-1] \\ h_1[0] & h_1[1] & \cdots & h_1[L-1] \\ \vdots & \vdots & \ddots & \vdots \\ h_{L-1}[0] & h_{L-1}[1] & \cdots & h_{L-1}[L-1] \end{bmatrix} \tag{1.113}$$

as we did in the direct-form case. From (1.112) we see that the coefficients $a[0], a[1], \ldots, a[L-1]$ are zero-mean and jointly Gaussian. Moreover, the correlation between an arbitrary pair of these coefficients is proportional to the inner product between the corresponding columns of $\mathbf{H}$ in (1.113), i.e.,

$$\text{cov}(a[n], a[m]) = E[a[n]a^*[m]] = \sigma_a^2 \sum_{l=0}^{L-1} h_l[n]h_l^*[m]. \tag{1.114}$$

Hence, the coefficients $a[0], a[1], \ldots, a[L-1]$ are statistically independent when the columns of $\mathbf{H}$ are orthogonal. In addition, the normalization constraint (1.109) further constrains the columns of $\mathbf{H}$ to have unit norm. Thus, to achieve the maximum diversity benefit, we require that $\mathbf{H}$ in (1.113) be a unitary matrix. In turn, this unitary condition implies that the antenna signatures (i.e., the rows of **H**) should be chosen to be orthogonal to one another.

From this perspective, it is clear that an unlimited number of signature sets allow the independent fading condition to be met. For example, the choice **H** = **I** corresponds to a strategy in which prefiltered symbols are dealt among the antennas and transmitted in order. One potential disadvantage of this particular choice, however, is the high peak-power requirement. An alternative is the choice **H** = **F**, where **F** is again the DFT matrix. In this case, the result can be interpreted as an efficient discrete-time variant of a phase-sweeping transmitter antenna diversity system explored by both Hiroike et. al. [32] and Weerackody [33]. As a final example, the choice **H** = **Ξ**, where **Ξ** is again the Hadamard matrix, has some particularly attractive characteristics. First, like the DFT-based signatures, the Hadamard-based signatures have minimal peak power requirements. Second, since Hadamard-based signatures are binary-valued, they result in very low computational complexity implementations of (1.106), requiring sign changes but no multiplications.

The design of the prefilter follows immediately from earlier developments in the chapter, specifically those in Section 1.5.7. In particular, the role of the prefilter is to enable the time diversity generated by the signature modulation process to be exploited.[23] As such, the maximally spread LPTV spread-response precoders developed in Section 1.5.7 are naturally suited as prefilters for this application, providing effectively optimum linear diversity benefit with very low computational complexity. As discussed, those corresponding to an LPTV system with minimal period ($K = 2$) suffice.

As in the case of direct-form antenna precoding, we focus on the use of a linear equalizer at the front end of our receiver. In the dual-form system, this equalizer takes the form depicted in Figure 1.26. Specifically,

$$\hat{x}[n] = \sum_k g^*[-n; -k]\hat{y}[n - k], \tag{1.115}$$

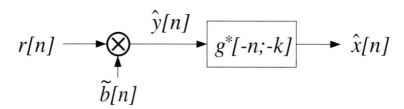

Figure 1.26   Receiver structure for dual-form linear antenna precoding.

---

[23]Note that more generally, this linear prefiltering can be replaced (or augmented) with nonlinear processing in the form of error-correction coding. This can lead to still better performance, though at a computational cost that may be prohibitive, particularly if the number of antennas is very large.

where

$$\hat{y}[n] = \tilde{b}[n]r[n], \tag{1.116}$$

and where $\tilde{b}[n]$ is a suitable equalizer for the time-selective fading. Note that since the prefilter (with kernel $g[n; k]$) is an orthogonal system, $g^*[-n; -k]$ is the kernel of its inverse, which has the implementation depicted earlier in Figure 1.19.

An equivalent-channel characterization for dual-form linear antenna precoding is obtained by exploiting the close connection to spread-response precoding. The key result [29] is that dual-form and direct-form precoding systems share the same input-output characteristics when the number of antenna elements L is large. Specifically, subject to only mild constraints on the equalizer, the postfilter output $\hat{x}[n]$ is related to a white input symbol stream $x[n]$ via (1.97), with μ and $v[n]$ having the same properties and interpretations they did in the direct-form context.

The connection between dual-form linear antenna precoding and spread-response precoding can be further exploited in developing a suitable equalizer for the system. In particular, it follows immediately that the SINR-optimizing equalizer is identical to that which arises in the context of spread-response precoding when the fading is frequency-nonselective, viz.,

$$\tilde{b}[n] \propto \frac{\tilde{a}^*[n]}{1 + |\alpha[n]|^2}, \tag{1.117a}$$

where

$$\alpha[n] = \frac{\mathcal{E}|a[n]|^2}{\mathcal{N}_0 \mathcal{W}_0}. \tag{1.117b}$$

The performance characteristics of these dual-form systems are also readily obtained by adapting results from spread-response precoding. For example, it follows that with the equalizer (1.117), the resulting SINR is identical to that obtained in the direct-form implementation, i.e.,

$$\gamma = \frac{1}{E\left[\dfrac{1}{\alpha[n]+1}\right]} - 1 = \frac{1}{\zeta_0 e^{\zeta_0} E_1(\zeta_0)} - 1, \tag{1.118}$$

with

$$\frac{1}{\zeta_0} = E[\alpha[n]] = \frac{\sigma_a^2 \mathcal{E}}{\mathcal{N}_0 \mathcal{W}_0}. \tag{1.119}$$

As a result, when the number of antenna elements L is large, the QPSK bit-error probability for both direct-form and dual-form systems is well approximated by (1.104) in general, and by (1.44) at high SNR. When smaller numbers of antenna elements are involved, the bit-error behavior can be determined by reinterpreting Figure 1.10. In particular, it suffices to exploit the fact that the average bit-error probability in a dual-form linear antenna precoding system with L antenna elements is identical to that of a spread-response precoding system exploiting L degrees of temporal diversity. Interpreting Figure 1.10 in this context, we see that while additional antennas invariably give better performance, there are clearly diminishing returns beyond a moderate value of L. Moreover, hardware costs and system delay constraints typically limit values of L that can be used in practice.

While dual-form precoding does not provide an improvement in performance over ideal direct-form precoding, it is important to emphasize that dual-form systems have some significant implementational advantages in many contexts. In particular, unlike the direct-form implementations, dual-form implementations have the characteristic that the system delay is finite. Indeed, the fading equalizer (1.117) introduces no delay, whereas the pre- and postfiltering each introduce delay L, so the overall delay is proportional to the number of antennas.

Finally, it is also important to emphasize that transmitter antenna diversity is inherently less effective in improving system performance than receiver antenna diversity with a similar number of antenna elements. In fact, comparing Figures 1.10 and 1.21, we see that even the relative advantage of receiver antenna diversity grows dramatically with the number of antenna elements involved. Ultimately, however, both transmitter and receiver arrays are equally efficient in exploiting the available spatial diversity. The source of the performance gap is a fundamental difference in SNR behavior that exists even in the absence of any fading. Indeed, without fading, transmit arrays cannot improve SNR over single antenna systems, whereas receiver arrays yield a SNR enhancement that is linear in the number of antenna elements.

Nevertheless, in many applications, such as broadcasting, where receiver diversity is inherently less practical, exploiting transmitter diversity with even only a few antennas can provide a substantial reduction in transmit power requirements for a given bit-error rate over a single antenna system, and at a very modest cost in terms of additional system hardware.

1.6.2.4 Incorporating Bandwidth Expansion

Additional performance enhancements can be achieved by the combination of transmitter antenna diversity with bandwidth expansion even on the nonselective channels. In fact, with sufficient bandwidth expansion, it is possible to approach the bit-error rate performance achieved with infinite spectral diversity on the frequency-selective fading channel.[24]

To achieve this result with linear antenna precoding, we follow an approach analogous to that used to enhance performance with spread-response precoding in Section 1.4.2. In particular, prior to direct-form (dual-form) linear antenna precoding, we upsample the symbol stream by an integer bandwidth expansion factor of ρ, replace the SNR $\alpha(\omega)$ with $\alpha(\omega)/\rho$ (the SNR $\alpha[n]$ with $\alpha[n]/\rho$) in the SINR-optimizing equalizer, and finally downsample the output of the receiver by a factor of ρ. The resulting SINR enhancement is identical to that obtained with spread-response precoding as given by (1.45). As such, Figure 1.11 also characterizes the improvement in bit-error probability that is obtained by augmenting a large transmit antenna array with bandwidth expansion. Again, still better performance can be achieved by more efficiently exploiting the excess bandwidth through the use of suitably designed error-correcting codes, albeit generally at the expense of an increase in complexity.

1.7 CONCLUDING REMARKS

This chapter provided an overview of the main forms of diversity that can be used to enhance performance and improve reliability in both single-user and multiuser wireless systems. In particular, we examined spectral, temporal, and spatial diversity and developed—in a unified manner—ways that such diversity can be exploited via linear processing. In the process, our development highlighted a natural and convenient multirate signal processing framework for describing, analyzing, and relating these diversity strategies.

Ultimately, our development serves to underscore the crucial role that sophisticated signal processing algorithms ultimately have to play in realizing and exploiting diversity in wireless communication systems. Clearly, the rapidly escalating growth in demand for wireless services means that future wireless systems will have to be increasingly efficient in their use of all available forms of diversity. The linear techniques described in this chapter can go a long way toward accommodating this growth in demand.

At the same time, still greater capacities can ultimately be achieved by exploitation of diversity through specifically nonlinear algorithms either at the

[24]Again, a simple strategy for achieving such performance on the infinite-bandwidth nonselective channel with transmitter antenna diversity is described in Jakes [28]. In particular, if we let the bandwidth grow in proportion to the number of antennas L and partition this bandwidth into L nonoverlapping subbands of equal size, one to be used by each antenna element, then having each antenna element transmit the symbol stream in its subband will suffice. This strategy also has the feature that it has bounded delay regardless of the number of antenna elements.

receiver or transmitter, or both. Indeed, the growing body of information-theoretic analysis suggests that using nonlinear processing in the form of coding either in conjunction with or as an alternative to the linear methods discussed in this chapter can dramatically enhance system performance or equivalently increase system capacity. More generally, novel and powerful techniques for exploiting diversity through nonlinear signal processing continue to emerge, making this one of several exciting and increasingly active areas of research within this important domain.

REFERENCES

[1] J. G. Proakis, *Digital Communications*. New York: McGraw-Hill, 2nd ed., 1989.

[2] P. A. Bello, "Characterization of randomly time-variant linear channels," *IEEE Trans. Commun. Syst.*, vol. CS-11, pp. 360–393, Dec. 1963.

[3] G. W. Wornell, "Spread-signature CDMA: Efficient multiuser communication in the presence of fading," *IEEE Trans. Inform. Theory*, vol. 41, pp. 1418–1438, Sept. 1995.

[4] E. A. Lee and D. G. Messerschmitt, *Digital Communication*. Boston: Kluwer Academic, 2nd ed., 1994.

[5] R. Price and P. E. Green, Jr., "A communication technique for multipath channels," *Proc. IRE*, vol. 46, pp. 555–570, Mar. 1958.

[6] A. V. Oppenheim and R. W. Schafer, *Discrete-Time Signal Processing*. Englewood Cliffs, NJ: Prentice-Hall, 1989.

[7] D. V. Sarwate and M. B. Pursley, "Crosscorrelation properties of pseudorandom and related sequences," *Proc. IEEE*, vol. 68, pp. 593–619, May 1980.

[8] D. V. Sarwate, "Mean-square correlation of shift-register sequences," *IEE Proceedings, Part F*, vol. 131, pp. 101–106, Apr. 1984.

[9] E. Biglieri, D. Divsalar, P. McLane, and M. Simon, *Introduction to Trellis-Coded Modulation with Applications*. New York: Maxwell-Macmillan, 1991.

[10] G. W. Wornell, "Spread-response precoding for communication over fading channels," *IEEE Trans. Inform. Theory*, vol. 42, pp. 488–501, Mar. 1996.

[11] A. Wittneben, "An energy- and bandwidth-efficient data transmission system for time-selective fading channels," in *Proc. IEEE GLOBECOM*, vol. 3, pp. 1968–1972, 1990.

[12] S. Beheshti and G. W. Wornell, "Iterative interference cancellation and decoding for spread-signature CDMA systems," in *Proc. Vehic. Tech. Soc. Conf.*, vol. 1, pp. 26–30, 1997.

[13] A. Wittneben, "A novel bandwidth efficient analog coding/decoding scheme for data transmission over fading channels," in *Proc. IEEE GLOBECOM*, vol. 1, pp. 34–38, 1994.

[14] S. H. Isabelle and G. W. Wornell, "Recursive multiuser equalization for CDMA systems in fading environments," in *Proc. Allerton Conf. Commun., Contr., Signal Processing*, pp. 613–622, 1996.

[15] A. D. Wyner, "Recent results in the Shannon theory," *IEEE Trans. Inform. Theory*, vol. IT-20, pp. 2–10, Jan. 1974.

[16] R. G. Gallager, "A perspective on multiaccess channels," *IEEE Trans. Inform. Theory*, vol. IT-31, pp. 124–142, Mar. 1985.

[17] G. W. Wornell, "Emerging applications of multirate systems and wavelets in digital communications," *Proc. IEEE*, vol. 84, pp. 586–603, Apr. 1996.

[18] P. P. Vaidyanathan, *Multirate Systems and Filter Banks*. Englewood Cliffs, NJ: Prentice-Hall, 1993.

[19] M. J. E. Golay, "Multislit spectrometry," *J. Opt. Soc. Amer.*, vol. 39, pp. 437–444, 1949.

[20] M. J. E. Golay, "Static multislit spectrometry and its application to the panoramic display of infrared spectra," *J. Opt. Soc. Amer.*, vol. 41, pp. 468–472, 1951.

[21] R. Turyn, "Ambiguity functions of complementary sequences," *IEEE Trans. Inform. Theory*, vol. IT-9, pp. 46–47, Jan. 1963.

[22] Y. Taki, H. Miyakawa, M. Hatori, and S. Namba, "Even-shift orthogonal sequences," *IEEE Trans. Inform. Theory*, vol. IT-15, pp. 295–300, Mar. 1969.

[23] C.-C. Tseng and C. L. Liu, "Complementary sets of sequences," *IEEE Trans. Inform. Theory*, vol. IT-18, pp. 644–652, Sept. 1972.

[24] H. S. Shapiro, "Extremal problems for polynomials and power series," Master's thesis, M. I. T., 1951.

[25] W. Rudin, "Some theorems on Fourier coefficients," *Proc. Amer. Math. Soc.*, vol. 10, pp. 855–859, 1959.

[26] A. Narula, M. D. Trott, and G. W. Wornell, "Information-theoretic analysis of multiple-antenna transmission diversity for fading channels," in *Proc. Int. Symp. Inform. Theory and Appl.*, Sept. 1996.

[27] A. Narula, M. D. Trott, and G. W. Wornell, "Information-theoretic analysis of multiple-antenna transmission diversity," 1996. Submitted to *IEEE Trans. Inform. Theory*.

[28] W. C. Jakes, ed., *Microwave Mobile Communications*. New York: John Wiley and Sons, 1974.

[29] G. W. Wornell and M. D. Trott, "Efficient signal processing techniques for exploiting transmit diversity on fading channels," *IEEE Trans. Signal Processing*, vol. 45, pp. 191–205, Jan. 1997.

[30] A. Wittneben, "Basestation modulation diversity for digital simulcast," in *Proc. Vehic. Tech. Soc. Conf.*, pp. 848–853, 1991.

[31] J. H. Winters, "The diversity gain of transmit diversity in wireless systems with Rayleigh fading," in *Proc. Int. Conf. Commun.*, pp. 1121–1125, 1994.

[32] A. Hiroike, F. Adachi, and N. Nakajima, "Combined effects of phase sweeping transmitter diversity and channel coding," *IEEE Trans. Vehic. Technol.*, vol. 41, pp. 170–176, May 1992.

[33] V. Weerackody, "Diversity for the direct-sequence spread spectrum system using multiple tranmit antennas," in *Proc. Int. Conf. Commun.*, pp. 1775–1779, 1993.

[34] N. Seshadri and J. Winters, "Two signaling schemes for improving the error performance of frequency division duplex (FDD) transmission systems using transmitter antenna diversity," *Int. J. Wireless Inform. Networks*, vol. 1, pp. 49–60, Jan. 1994.

[35] S. Beheshti, S. H. Isabelle, and G. W. Wornell, "Joint intersymbol and multiple-access interference suppression algorithms for CDMA systems," to appear in *Euro. Trans. Telecomm.*, Special Issue on Code-Division Multiple-Access Techniques for Wireless Communication Systems, 1998.

ACKNOWLEDGMENTS

The author thanks Haralabos Papadopoulos for his careful proofreading and feed-back on earlier drafts of this chapter.

This work was supported, in part, by the Defense Advanced Research Projects Agency monitored by ONR under Contract No. N00014-93-1-0686, the Office of Naval Research under Grant No. N00014-96-1-0930, the National Science Foundation under Grant No. MIP-9502885, the Air Force Office of Scientific Research under Grant No. F49620-96-1-0072, and the Army Research Laboratories under Cooperative Agreement No. DAAL01-96-2-0002.

2

Adaptive Interference Suppression

Michael L. Honig

H. Vincent Poor

As discussed in Chapter 1, one of the major features that distinguish modern wireless communication channels from wireline channels is the significant amount of structured interference that must be contended with in wireless channels. This interference is inherent in many wireless systems due to their operation as *multiple-access* systems, in which multiple transmitter/receiver pairs communicate through the same physical channel using nonorthogonal multiplexing. Structured interference also arises because of other nonsystemic features of wireless systems, such as the desire to share bandwidth with other, dissimilar, communication services.

Signal processing plays a central role in the suppression of the structured interference arising in wireless communication systems. In particular, the use of appropriate signal processing methods can make a significant difference in the performance of such systems. Moreover, since many wireless systems operate under highly dynamic conditions because of the mobility of the transceivers and of the random nature of the channel access, *adaptive* signal processing is paramount in this context.

The study of adaptive processing techniques for interference suppression in wireless systems has been a very active area of research in recent years. This chapter introduces the reader both to the basic problems arising in this area and to the key methods that have been developed for dealing with these problems. This presentation focuses primarily on the problem of suppressing multiple-access interference (MAI), which is the limiting source of interference for the wireless systems being proposed for many emerging applications areas such as third-generation mobile telephony [94, 125] and wireless personal communications [19, 70]. However, we also touch briefly on the related and important problems of multipath mitigation and narrowband interference suppression. Multipath mitigation techniques, specifically, are developed in more detail in Chapter 3.

2.1 MULTIPLE-ACCESS SIGNAL MODEL

In treating the problem of MAI suppression, it is useful to consider a general multiple-access signal model that arises in the context of a wireless digital communications network operating with a coherent modulation format. The waveform received by a given terminal in such a network can be modeled as consisting of a set of superimposed modulated data signals observed in additive noise:

$$r(t) = S_t(\mathbf{b}) + n(t), \quad -\infty < t < \infty, \tag{2.1.1}$$

where $S_t(\mathbf{b})$ and $n(t)$ represent the useful signal and the ambient channel noise, respectively.

The useful signal $S_t(\mathbf{b})$ in this model consists of the data signals of K active users in the channel and can be written as

$$S_t(\mathbf{b}) = \sum_{k=1}^{K} A_k \sum_{i=-B}^{B} b_{i,k} s_k(t - iT - \tau_k), \tag{2.1.2}$$

where $2B + 1$ is the number of symbols per user in the data frame of interest, T is the symbol interval, and A_k, τ_k, $\{b_{i,k}\}$ and $\{s_k(t); 0 \le t \le T\}$ denote, respectively, the received amplitude, delay, symbol stream, and normalized modulation waveform (or pulse shape) of the kth user. The matrix $\mathbf{b}$ denotes the $K \times (2B + 1)$ matrix whose (k, i)th element is $b_{i,k}$. The data signals of the individual users may be asynchronous, in which case the relative delays with which the various data signals arrive at the receiver are distinct. However, when considering analytical properties, it is often sufficient to examine the synchronous case (i.e., $\tau_1 = \tau_2 = \cdots = \tau_K$) since asynchronous problems can be viewed as large synchronous problems. It should be noted further that although this model does not explicitly include effects such as fading, multipath, intersymbol interference, or narrowband interference, such effects can be included without loss of tractability. (See also Chapter 3.) A further generalization of this model allows for spatial diversity at the receiver, in which multiple waveforms are observed, each of which contains information about the data sequences. As discussed in Chapter 1 and as further elaborated on below, such a model arises in the consideration of the use of antenna arrays for reception.

The principal feature that distinguishes multiuser formats of the type described in (2.1.1) and (2.1.2) from one another is the choice of the set of signaling waveforms (i.e., the signal constellation, $s_1, s_2, ..., s_K$). We are interested here in problems in which these waveforms are not orthogonal. The principal, although not exclusive, example of such nonorthogonal signaling arises in networks using code-division multiple-access (CDMA) channel-sharing protocols, which have several advantages over time-division and frequency-division multiple-access schemes. However, with asynchronous channel access, it is not possible to maintain orthogonal signaling waveforms with CDMA. One of the most important formats of this type is the direct-sequence, spread-spectrum, multiple-access format, which corresponds to a set of signaling waveforms of the form

$$s_k(t) = \begin{cases} 2^{\frac{1}{2}} a_k(t) \sin(\omega_c t + \varphi_k), & t \in [0, T] \\ 0 & , \quad t \notin [0, T], \end{cases} \tag{2.1.3}$$

where ω_c is a common carrier frequency, φ_k is the phase of the kth user relative to some reference, and the spreading waveforms $a_k(t)$ are of the form:

$$a_k(t) = \sum_{j=0}^{N-1} a_{k,j} \psi(t - jT_c). \tag{2.1.4}$$

Here, $a_{k,0}, a_{k,1}, ..., a_{k,N-1}$ is a signature sequence of +1's and –1's assigned to the kth user, and ψ is a normalized chip waveform of duration T_c (where $NT_c = T$). The signature sequences (or *spreading codes*) and chip waveform are typically chosen to have autocorrelation and cross-correlation properties that reduce multipath, multiple-access interference, and unintended detectability, criteria that generally lead to signaling waveforms with nearly flat spectral characteristics. Note that, in this type of signaling, the bandwidth of the underlying data signal is spread by a factor of N. This particular model, which is sometimes termed direct-sequence code-division, multiple-access (DS-CDMA) signaling, is discussed further below.

It is the nonorthogonality of the signaling waveforms $s_1, s_2, ..., s_K$ that gives rise to the multiple-access interference with which the receiver must contend. In particular, if the receiver wishes to infer the data stream of a given user, say, user 1, then the fact that the other users' signaling waveforms are not orthogonal to s_1 makes it impossible to isolate user 1's signal without diminishing the detectability of user 1's data. However, through proper signal processing, the effects of the interfering signals can be minimized so that little is lost to this source of error. The area of study that deals with such problems is *multiuser detection*, and this chapter is primarily concerned with performing this task efficiently and adaptively.

This treatment is organized as follows. In Section 2.2, we provide a brief review of the elements of multiuser detection, which provides a framework for the development of most of the remainder of the chapter. In Section 2.3, we consider in more detail a specific class of multiuser detectors—linear multiuser detectors—that contains many of the most promising structures for introducing adaptivity into this problem. This treatment generalizes the model (2.1.1)–(2.1.2) to include diversity and other effects. Section 2.4 discusses the particularization of linear multiuser detection to the DS-CDMA format described above. Next, in Section 2.5 we consider the adaptation of linear, multiuser detectors. In particular, we discuss several basic adaptive algorithms in the context of their complexity, convergence and performance characteristics. Section 2.6 considers the implications of some nonideal effects arising in wireless channels and also discusses some systems issues impacting the application of adaptive interference suppression in wireless systems. An extensive, but not exhaustive, bibliography of key sources in this area is also included.

2.2 ELEMENTS OF MULTIUSER DETECTION

Almost by definition, the performance characteristics of multiple-access channels featuring traditional demodulation techniques are limited by multiple-access interference. It can be shown, however, that such limitations are due largely to the use of nonoptimal signal processing in the demodulator and are not due to fundamental characteristics of the channel. Multiuser detection seeks to remove this MAI limitation by the use of appropriate signal processing. Essentially, through the use of multiuser detection (or derivative signal processing techniques), performance in multiple-access channels can be returned to that of corresponding single-access channels, or at least to a situation in which performance is no longer MAI limited. This property is obviously very desirable, even in radio networks using power control or other protocols that seek to limit the effects of MAI.

The basic problem of multiuser detection is that of inferring the data contained in one or more signals embedded in a nonorthogonal multiplex, the entire multiplex of which is received in ambient noise. Equations (2.1.1) and (2.1.2) describe a model of such a received signal. Within this context, multiuser detection refers to the problem of detecting all or part of the symbol matrix $\mathbf{b}$ from the multiplex (2.1.1)–(2.1.2) with nonorthogonal signaling waveforms, such as those arising in DS-CDMA. In the demodulation of any given user in such a multiplex, it is necessary to process the received signal in such a way as to minimize two types of detrimental effects—the multiple-access interference caused by the remaining $K - 1$ users in the channel and the ambient channel noise. In order to focus on the multiple-access interference, the great majority of research on this problem has ascribed the simplest possible model to the ambient channel noise; namely, that the only ambient channel noise is additive white Gaussian noise (AWGN) with fixed spectral height, say, σ^2, and that this noise is independent of the data signals. In the following paragraphs, we also assume this model for the ambient noise.

Within the multiple-access signaling model are two general scenarios of interest: an *uplink* (or *reverse-link*) scenario, in which all signaling waveforms and data timing are known; and a *downlink* (or *forward-link*) scenario, in which sometimes only a single user's waveform and data timing are known. In the uplink scenario, the general characteristics of optimal demodulation schemes for (2.1.1)–(2.1.2) under an AWGN model can be inferred by examining the likelihood function of the observed waveform (2.1.1), conditioned on the knowledge of all data symbols (i.e., conditioned on **b**). On assuming that the received amplitudes are known, this likelihood function can be written via the Cameron-Martin formula [76] as

$$\ell(\{r(t); \ -\infty < t < \infty\} \mid \mathbf{b}) = C \exp\{\Omega(\mathbf{b})/2\sigma^2\}, \tag{2.2.5}$$

where

$$\Omega(\mathbf{b}) = 2\int_{-\infty}^{\infty} S_t(\mathbf{b})r(t)dt - \int_{-\infty}^{\infty} S_t^2(\mathbf{b})dt \tag{2.2.6}$$

and where C is a constant. The part of (2.2.6) that depends on the received waveform can be written as

$$\int_{-\infty}^{\infty} S_t(\mathbf{b})r(t)dt = \sum_{k=1}^{K} A_k \sum_{i=-B}^{B} b_{i,k} y_{i,k}, \tag{2.2.7}$$

where

$$y_{i,k} \equiv \int_{-\infty}^{\infty} s_k(t - iT - \tau_k)r(t)dt \tag{2.2.8}$$

is the output of a filter matched to the kth user's signaling waveform shifted to the ith symbol interval of the kth user.

It follows from (2.2.5)–(2.2.8) that the matrix of matched filter outputs,

$$\mathbf{y} = \{y_{i,k}; \ k = 1,\dots, K; \ i = -B,\dots, B\}, \tag{2.2.9}$$

forms a sufficient statistic for the matrix **b** of data symbols; that is, all information in the received waveform that is relevant to making inferences about **b** is contained in **y**. So, the main job of optimal multiuser detection is to map the matrix **y** of observables to a matrix $\hat{\mathbf{b}}$ of symbol decisions. Thus, the general structure of optimal systems for determining the data symbols from the received waveform consists of an analog front end that extracts the matched filter outputs, followed by a decision algorithm that infers optimal decisions from the collection of these outputs. The nature of the decision algorithm in this process depends on the optimality criterion that one wishes to apply to the decision. If one adopts either a *maximum-likelihood* criterion,

$$\max_{\hat{b}} \ell(\{r(t); -\infty < t < \infty\} \mid \hat{b}), \qquad (2.2.10)$$

or a *minimum-error-probability* criterion,

$$\min_{\hat{b}_{i,k}} P(\hat{b}_{i,k} \neq b_{i,k} \mid \{r(t); -\infty < t < \infty\}), \qquad (2.2.11)$$

then, assuming the signaling waveforms s_k satisfy $s_k(t) = 0$ for $t \notin [0, T]$, this optimal decision algorithm can be implemented as a dynamic program (i.e., a *sequence detector*) having $O(|A|^K)$ time complexity per binary decision (see [120]), where $|A|$ is the size of the symbol alphabet. In the case of synchronous signals, dynamic programming is unnecessary and the optimal detectors essentially involve either exhaustive search over the $|A|^K$ symbol choices in each symbol interval in the case of maximum-likelihood detection, or $O(|A|^K)$ computation of posterior probabilities in the case of minimum-error-probability detection.

Using these techniques, Verdú ([118]) has shown that, for reasonably high symbol-energy-to-ambient-noise ratios, performance very near that of single-user communications is possible with the optimal multiuser detector. This is a considerable performance gain over conventional matched-filter detection (which demodulates $b_{i,k}$ by simple scalar quantization of $y_{i,k}$), which suffers from substantial performance losses in some multiple-access situations. (Such situations include the "near-far" situation, in which interfering users are received with much larger power than are users of interest.) However, the much-improved performance afforded by optimal multiuser algorithms comes at the expense of both *computational* complexity (i.e., the $O(|A|^K)$ computational cost per binary decision); and *informational* complexity due to the need for knowing all delays, amplitudes, and modulation waveforms to extract the matrix sufficient statistic **y**.

During the late 1980s and early 1990s, a significant amount of research addressed the problem of reducing the computational complexity of multiuser detection. A key approach to this problem is to restrict the optimal detector to be of the form of a *linear multiuser detector*, in which the data is demodulated by scalar quantization of a linear mapping on the matrix **y**. In view of the definition of $y_{i,k}$, this type of detector effectively comprises a linear filter applied to the received waveform, followed by a scalar quantizer. (Of course, the filter may depend on both k and i.) Two types of linear detectors of interest are the *decorrelating detector* (or *decorrelator*), which chooses the linear filter to have zero output multiple-access interference [49]; and the *MMSE detector*, which chooses the linear filter to have minimum output energy within the constraint that the response of the filter to $s_k(t - iT - \tau_k)$ is fixed [31, 44, 51, 58, 82, 83, 93, 96, 133]. Such detectors can be shown to also satisfy other optimality criteria. Although such detectors fall short of optimal (maximum-likelihood) detection in terms of error probability, they are still far superior to conventional detection in terms of their error-probability performance in interference-limited environments. Linear detectors form the basis for the results described in the remainder of this chapter, and a detailed description of their properties is found in Sections 2.3 and 2.4.

Several useful nonlinear, lower-complexity, multiuser detectors also have been developed. These are based primarily on various techniques for successive cancellation of interference. Also, methods for combating fading, multipath, etc., have been combined with multiuser detection as well. Some works in these areas include [17, 45, 68, 73, 85, 89, 111, 112, 113, 114, 115, 116, 126, 127, 130, 135, 138, 139, 140, 141, 142], although this list is hardly exhaustive. A survey of basic multiuser detection methods, and a more complete bibliography up to 1993, can be found in [120]. (See also the forthcoming textbook [121].)

The issue of informational complexity in multiuser detection has been addressed through the use of adaptivity. This issue is particularly critical in the context of downlink demodulation, in which the direct implementation of non-adaptive versions of the above-noted detectors is neither practical nor desirable. However, uplink adaptivity is also of interest in practice because of the dynamic nature of practical multiple-access channels. (Some discussion of this issue is found in Section 2.6.) Recent progress on adaptive multiuser detection is the subject of the remainder of this chapter.

2.3 LINEAR INTERFERENCE SUPPRESSION

In this section, we look more carefully at linear multiuser detection, in which linear filtering at the receiver can be used to suppress wideband multiple-access interference. This general approach to interference suppression has its origins in the related problem of linear equalization in the presence of synchronous interfering data signals[1] and seems to have been first studied nearly thirty years ago [23, 43]. There are several reasons why linear interference suppression is attractive for wireless applications:

1. It can suppress both narrowband and wideband multiple-access interference.
2. It has modest complexity relative to other interference suppression and multiuser detection techniques, as noted in the preceding section.
3. The linear filter can be implemented as a digital filter, or tapped-delay line, making it convenient for adaptation by means of conventional adaptive algorithms [25, 32].

[1]The multiuser signaling model of (2.1.1)–(2.1.2) includes as a special case the single-user intersymbol-interference channel, which corresponds to the case $K = 1$ with s_1 non-zero over more than one symbol interval.

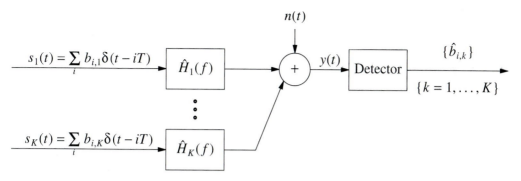

Figure 2.1   Multiple-access channel model.

To discuss this approach, it is convenient to view the multiple-access channel considered in Section 2.1 as a many-input linear system, as shown in Figure 2.1. In this depiction, the (linear) operations of modulation and transmission of the data sequences of the various users are lumped into the linear time-invariant transfer functions $\hat{H}_1(f), \hat{H}_2(f),\ldots,\hat{H}_K(f)$. In particular, (2.1.1)–(2.1.2) is realized by this model if we choose $\hat{H}_k(f)$ to be the system with impulse response

$$H_k(t) = A_k s_k(t - \tau_k). \tag{2.3.1}$$

In this context, we first derive the optimal linear receiver for this type of channel, where the optimality criterion is minimum mean-squared error (MSE). We then discuss some properties of this detector, such as the number of users that can be suppressed completely versus available bandwidth, and the implementation of the detector as a fractionally spaced tapped-delay line. Application to the DS-CDMA model of (2.1.3)–(2.1.4) is discussed in Section 2.4.

2.3.1 Multiple-input/Multiple-output (MIMO) Minimum Mean-Squared Error (MMSE) Linear Detector

To examine the MMSE linear multiuser detector, we generalize the above signaling model slightly. First, in order to treat some nonideal effects later, we will allow the signaling waveforms and received amplitudes to be complex. And, second, we will allow for Mth-order reception diversity, in which we have M observation channels similar to the depiction in Figure 2.1. Usually, these M channels correspond to the outputs of M elements in an antenna array. In this situation, it is convenient to represent the multiple-access channel as a special case of a K-input/M-output linear channel with $M \times K$ transfer function $\hat{\mathbf{H}}(f)$.

Referring to Figure 2.1, the input to the channel $\hat{\mathbf{H}}(f)$ is

$$\mathbf{s}(t) = \sum_i \mathbf{b}_i \delta(t - iT), \tag{2.3.2}$$

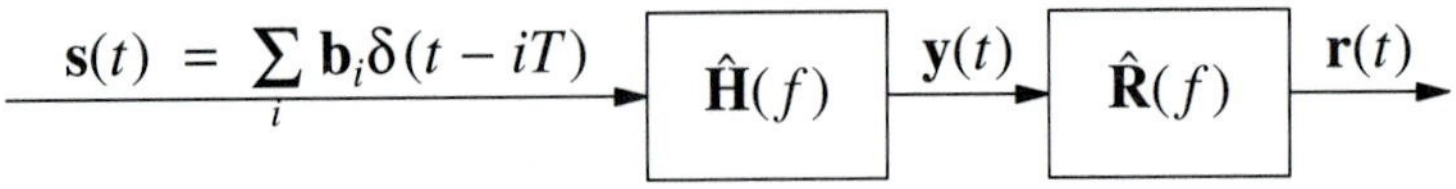

$$s(t) = \sum_i \mathbf{b}_i \delta(t - iT) \longrightarrow \boxed{\hat{\mathbf{H}}(f)} \xrightarrow{\ \mathbf{y}(t)\ } \boxed{\hat{\mathbf{R}}(f)} \xrightarrow{\ \mathbf{r}(t)\ }$$

Figure 2.2   Multi-input/multi-output (MIMO) channel and receiver model.

where $\mathbf{b}_i$ is the K-vector of transmitted symbols at time i (that is, $\mathbf{b}_i$ is the ith column of the matrix $\mathbf{b}$ introduced in Section 2.1) and $1/T$ is the symbol rate, assumed to be the same for all users. The kth component of $\mathbf{b}_i$, denoted as $b_{i,k}$, is the ith transmitted symbol from user k. We assume that the signaling waveforms corresponding to each user, as well as relative amplitudes and delays for each observation channel, are included in the channel transfer function $\hat{\mathbf{H}}(f)$. (This model is general enough to account for both transmitter and receiver diversity of types described in Chapter 1.)

Both channel and receiver filters are illustrated in Figure 2.2. Let $\mathbf{H}(t)$ and $\mathbf{R}(t)$ be the impulse responses associated with the filters $\hat{\mathbf{H}}(f)$ and $\hat{\mathbf{R}}(f)$, respectively. The output of the receiver filter $\hat{\mathbf{R}}$ at time kT is:

$$\mathbf{r}(kT) = \sum_i \{\mathbf{R} * \mathbf{H}[(k-i)T]\}\mathbf{b}_i + \mathbf{R} * \mathbf{n}(kT), \tag{2.3.3}$$

where $*$ denotes convolution, and $\mathbf{n}(t)$ is a noise vector with M components. We wish to find the receiver filter $\hat{\mathbf{R}}$ that minimizes the MSE, $E\{\|\mathbf{r}(0) - \mathbf{b}_0\|^2\}$. This filter has been derived in [30] and [100] and is illustrated in Figure 2.3. It consists of a front-end matched filter with $K \times M$ matrix transfer function $\hat{\mathbf{H}}^*(f)\mathbf{S}_n^{-1}(f)$, where $\mathbf{S}_n(f)$ is the noise spectral density matrix, followed by a MIMO discrete-time filter with $K \times K$ matrix coefficients. The transfer function (i.e., z-transform) of this discrete-time filter is

$$\mathbf{C}(z) = \mathbf{S}_d(z)[\mathbf{S}_H(z)\mathbf{S}_d(z) + \mathbf{I}]^{-1}, \tag{2.3.4}$$

where $\mathbf{S}_H(z)$ is the equivalent discrete-time transfer function that maps the sequence of input symbol vectors $\{\mathbf{b}_i\}$ to the sequence of matched filter outputs $\{\mathbf{r}(iT)\}$, and $\mathbf{S}_d(z)$ is the spectrum of the data sequence. We can therefore write $\mathbf{S}_H(z)$ for z on the unit circle as the aliased version of $\hat{\mathbf{H}}^*\mathbf{S}_n^{-1}\hat{\mathbf{H}}$, i.e., for $z = e^{j2\pi fT}$,

$$\mathbf{S}_H(e^{j2\pi fT}) = \frac{1}{T}\sum_k \hat{\mathbf{H}}^*\left(f - \frac{k}{T}\right)\mathbf{S}_n^{-1}\left(f - \frac{k}{T}\right)\hat{\mathbf{H}}\left(f - \frac{k}{T}\right), \tag{2.3.5}$$

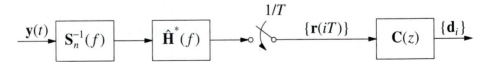

Figure 2.3   MIMO MMSE linear equalizer.

and we have

$$\mathbf{S}_d(z) = \sum_i z^{-i}\rho_i, \tag{2.3.6}$$

where $\rho_i = E\{\mathbf{b}_m \mathbf{b}^*_{m+i}\}$. In Figure 2.3, $\mathbf{d}_i$ denotes the output of the MIMO discrete-time filter $\mathbf{C}$ at symbol time i.

If the noise and data sequences are both uncorrelated in time and have uncorrelated components, then the matched filter becomes $\hat{\mathbf{H}}^*(f)/\sigma_n^2$, where σ_n^2 is the noise variance per channel, and the discrete-time filter becomes

$$\mathbf{C}(z) = \sigma_n^2[\tilde{\mathbf{S}}_H(z) + \xi\mathbf{I}]^{-1} \tag{2.3.7}$$

with

$$\tilde{\mathbf{S}}_H(e^{j2\pi fT}) = \frac{1}{T}\sum_k \hat{\mathbf{H}}^*\left(f - \frac{k}{T}\right)\hat{\mathbf{H}}\left(f - \frac{k}{T}\right); \tag{2.3.8}$$

$\xi = \sigma_n^2/\sigma_b^2$, and σ_b^2 is the data variance per channel. (Note that the only difference between $\tilde{\mathbf{S}}_H$ and $\mathbf{S}_H$ is that $\tilde{\mathbf{S}}_H$ does not include the noise variance σ_n^2. This representation is convenient for what follows.) The minimum value of the MSE associated with the MMSE filter $\hat{\mathbf{R}}$ is then given by

$$\text{MMSE} = \text{trace}\left\{T\int_{-1/(2T)}^{1/(2T)} \mathbf{C}(e^{j2\pi fT})df\right\}. \tag{2.3.9}$$

This form of the MMSE detector for the multiple-access channel is shown in Figure 2.4. Only the detector for user 1 is shown; it consists of a bank of matched filters, symbol-rate samplers, and discrete-time filters. The MMSE transfer function $C_{1,k}(z)$ in Figure 2.4 is the $(1, k)$th component of the matrix $\mathbf{C}(z)$ given by (2.3.7).

In the remainder of this chapter, we assume that the sequences of noise and symbol vectors are both uncorrelated in time and have uncorrelated components, in which case Figure 2.4 gives the canonical, linear MMSE detector structure.

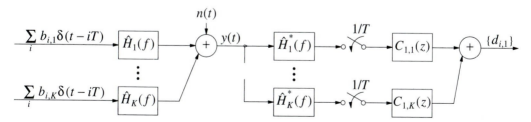

Figure 2.4   MMSE linear detector for the multiple-access channel. Only the detector for user 1 is shown.

2.3.2 Zero-Forcing (Decorrelating) Detector

As the noise variance diminishes to zero, the MMSE linear filter approaches the matched filter $\hat{\mathbf{H}}^*(f)$, followed by a symbol-rate sampler and a discrete-time matrix filter with transfer function

$$\mathbf{C}_{zf}(z) = [\tilde{\mathbf{S}}_H(z)]^{-1}. \tag{2.3.10}$$

The resulting MSE is given by

$$\text{MSE}_{zf} = \sigma_n^2 \, \text{trace}\left\{ T \int_{-1/(2T)}^{1/(2T)} \mathbf{C}_{zf}(e^{j2\pi fT}) df \right\}. \tag{2.3.11}$$

Because the transfer function $\mathbf{C}_{zf}(z)$ inverts the equivalent discrete-time transfer function $\tilde{\mathbf{S}}_H$, which maps the source symbols to the matched filter outputs, it eliminates all intersymbol and multiple-access interference (at the expense of enhancing the background noise). For this reason, the matched filter $\hat{\mathbf{H}}^*$ followed by the transfer function $\mathbf{C}_{zf}(z)$ is known as the *zero-forcing* detector for the MIMO channel $\hat{\mathbf{H}}(f)$. When applied to the multiple-access channel in Figure 2.1, this detector is also known as the *decorrelating detector*, or *decorrelator*, since it removes the correlation among users due to nonorthogonal pulse shapes [49, 50].

It is apparent from (2.3.10) that the zero-forcing solution exists provided that the matrix $\tilde{\mathbf{S}}_H(z)$ is nonsingular for z on the unit circle.[2] This condition has a special interpretation for the multiple-access channel. Specifically, in this case, $\hat{\mathbf{H}}(f)$ is a $1 \times K$ row vector, so that $\hat{\mathbf{H}}^*(f)\hat{\mathbf{H}}(f)$ is an outer product matrix, which has rank one. $\tilde{\mathbf{S}}_H(e^{j2\pi fT})$ in (2.3.8) is therefore the sum of L rank-one matrices, where for each $f \in [-1/(2T), 1/(2T)]$, L is the number of Nyquist zones[3] where $\hat{\mathbf{H}}(f) \neq 0$. Since $\tilde{\mathbf{S}}_H(e^{j2\pi fT})$ is a $K \times K$ matrix, a necessary condition for $\tilde{\mathbf{S}}_H(e^{j2\pi fT})$ to be nonsingular for all $f \in [-1/(2T), 1/(2T)]$ is that $K \leq L$ for each f. This implies that for the zero-forcing solution to exist, there must be at least K Nyquist zones available to the users. This property was first observed by Petersen and Falconer [74] in the context of wire (twisted-pair) channels with crosstalk. (See also [4].) Note that these Nyquist zones can be spread among the users so that (i) the users do not overlap in frequency (namely, Frequency-Division Multiple-Access (FDMA)), (ii) all of the users overlap at all frequencies (CDMA or TDMA), or (iii) some users overlap at some frequencies but not at other frequencies (combined FDMA/TDMA/CDMA).

---

[2]Of course, it is possible that $\mathbf{S}_H(z)$ is singular for some set of z on the unit circle and that the MSE resulting from substituting (2.3.10) into (2.3.11) is finite. However, finite MSE requires that $\mathbf{S}_H(e^{j2\pi fT})$ cannot be singular for f in some interval with positive length.

[3]In this context, a Nyquist zone is a translate of the basic Nyquist interval, $[-1/(2T), 1/(2T)]$ by an integral multiple of $1/T$.

For the multiple-access channel, the availability of K Nyquist zones for K users is necessary but not sufficient to ensure the existence of the zero-forcing solution. That is, it may happen that even with more than K Nyquist zones available, the zero-forcing solution does not exist for f in some positive interval contained in $[-1/(2T), 1/(2T)]$. For sufficiency, there must be at least K vectors $\hat{H}(f - k/T)$ appearing in the sum (2.3.8), that are linearly independent at each f. (This set of K vectors may depend on f.)

Consider the case where the channels for each user shown in Figure 2.1 are the same, i.e., $H_k(f) = H(f)$ for each k. The preceding discussion suggests that each additional Nyquist zone in $H(f)$ can be viewed as an additional "dimension," or "degree of freedom," that can support an additional user without causing interference to existing users. This "dimensionality" interpretation will be useful when the tapped-delay line implementation of the linear MMSE detector is discussed in the next section. Note that for "orthogonal" multiple-access systems such as FDMA and TDMA, this observation is equivalent to stating that each user requires at least one Nyquist zone to ensure the existence of the zero-forcing equalizer [46, Ch. 10].

We now examine the effect of receiver diversity on the preceding results. In this case, each channel $H_k(f)$ in Figure 2.1 becomes an $M \times 1$ column vector, where M is the order of the receiver diversity, so that $\hat{H}(f)$ is an $M \times K$ matrix (K inputs, M outputs). Consequently, the rank of $\hat{H}^*(f)\hat{H}(f)$ is at most M, and we conclude that a necessary condition for $\tilde{S}_H(e^{j2\pi fT})$ to be nonsingular for all $f \in [-1/(2T), 1/(2T)]$ is that the number of users $K \le LM$, where L is again the number of Nyquist zones available to the users. This upper bound can be achieved if the matrices $\hat{H}(f - k/T)$ in the sum (2.3.8) contain LM linearly independent columns at each f. The number of dimensions or degrees of freedom available to suppress users is therefore given by the number of Nyquist zones times the number of antenna elements. (A similar treatment of dimensionality in the frequency and spatial domains is given in [18].)

A final remark about the zero-forcing detector is that even when it does not exist (i.e., $S_H(e^{j2\pi fT})$ is singular), the MMSE detector is still well defined. Namely, it is always possible to select a filter to minimize output MSE. Consequently, the zero-forcing detector can be viewed more generally as the limit of the MMSE detector as the level of background noise tends to zero. This limit always exists even though the zero-forcing solution may not exist. This representation for the zero-forcing detector is useful in situations where the number of interferers exceeds the available dimensions that the detector has to supress multiple-access interference. Although the zero-forcing solution technically does not exist in this situation, the more general representation may still offer a substantial performance improvement relative to a simpler (e.g., matched-filter) detector.

2.3.3 Implementation as a Tapped-Delay Line (TDL)

The preceding formulation of the MMSE and zero-forcing detectors assumes knowledge of the user signaling waveforms along with relative timing and phase, the channel characteristics for each user, and the noise spectral density. In Section 2.5 we show that the MMSE detector can be implemented without this knowledge. This implementation depends on an alternative representation of the MIMO MMSE linear filter as a bank of fractionally spaced tapped-delay lines or discrete-time filters, which we now develop.

A classical result for single-user channels is that the optimal (MMSE) linear equalizer can be implemented as a fractionally spaced tapped-delay line (TDL) [46, Ch. 10], [91]. To see this, let W denote the two-sided bandwidth of the received data signal. Referring to Figure 2.4, the combination of the matched filter $H_k^*(f)$, $1/T$ sampler, and discrete-time filter $C_k(z)$ in each branch of the MMSE detector can be replaced by a lowpass filter $B(f)$ with two-sided bandwidth W, a sampler at rate W, and a discrete-time filter $\tilde{C}_k(z)$ with frequency response that has period W. Stated another way, each front-end continuous-time matched filter in the MMSE linear equalizer can be moved to the associated discrete-time filter, provided that the sampling rate is increased from $1/T$ to at least W.

The preceding discussion implies that each branch of the MMSE detector shown in Figure 2.4 can be replaced by a low-pass filter $B(f)$ followed by a rate W sampler and fractionally spaced TDL, where the tap spacing is $1/W$. Choosing $B(f)$ to be the same for each branch allows the K branches shown in Figure 2.4 to be collapsed into a *single* branch consisting of $B(f)$, a rate W sampler, and discrete-time filter with transfer function given by the sum of the transfer functions for each branch.

To summarize, the MMSE multiuser linear detector for the multiple-access channel can be replaced by the bank of fractionally spaced TDLs as shown in Figure 2.5. The filter $C_k(z)$, $1 \le k \le K$, is selected to minimize MSE for user k. The sampling rate, or tap spacing in the discrete-time filters thus depends on the bandwidth of the received signal. If the zero-forcing solution exists, then the bandwidth W must be at least K/T, where $1/T$ is the symbol rate for the users. To interpret this result another way, observe that to distinguish K symbols transmitted by K (non-cooperative) users, the receiver must sample at least K times per symbol. Furthermore, these K samples must be linearly independent. From Nyquist sampling theory, this implies that the bandwidth of the received signal must be greater than or equal to the sampling rate K/T.

With spatial diversity at the receiver, it is straightforward to show that the MMSE receiver can be implemented by summing the outputs of fractionally spaced TDLs associated with each antenna. This implementation is illustrated in Figure 2.6 for the case of two antennas. Figure 2.6 shows the MMSE detector for user 1, which contains two TDLs. Given M antennas, the MMSE detector has M

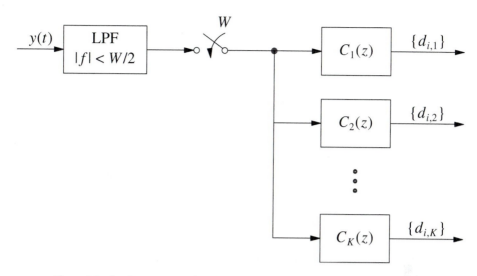

Figure 2.5   Implementation of the linear multiuser detector as a bank of fractionally spaced TDLs.

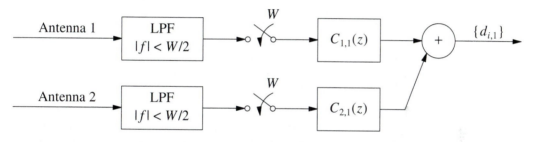

Figure 2.6   MMSE linear filter for user 1 with two-branch diversity.

TDLs for each user (indicating an M-fold increase in computational complexity associated with computing the filter outputs).

In general, the discrete-time impulse response associated with the MMSE filter $C_k(z)$ in Figure 2.5 can be of infinite length. Infinite-length impulse response (IIR) filters are problematic, since they cannot be implemented as TDLs and are difficult to optimize when channel and interference parameters are changing. However, it is always possible to approximate each $C_k(z)$ with a finite-length impulse response (FIR) filter. Of course, there is some performance degradation associated with this truncation, which will depend on how fast the filter impulse response associated with $C_k(z)$ decays to zero.

Finally, we remark that an important benefit of the fractionally spaced TDL implementation is that it eases timing recovery. That is, it is well known that for single-user channels a fractionally spaced adaptive equalizer (with taps spaced at $T/k, k > 1$) is more robust with respect to timing offset than is an adaptive equalizer with T-spaced taps. For DS-CDMA signals, timing recovery can be combined with interference suppression by using the adaptive algorithms discussed in Section 2.5 [52, 102, 130].

2.4 APPLICATION TO DS-CDMA

We now apply the developments of Section 2.3 to the DS-CDMA model of (2.1.3)–(2.1.4). It is convenient to write this signal by using a complex baseband model, in which case the received signal can be written as

$$y(t) = \sum_{k=1}^{K} A_k \left[\sum_i b_{i,k} p_k (t - iT - \tau_k) \right] + n(t), \tag{2.4.1}$$

where the pulse shape p_k for user k is given by

$$p_k(t) = \sum_{n=0}^{N-1} a_{k,n} \psi (t - nT_c); \tag{2.4.2}$$

that is, p_k is s_k from (2.1.3) with $\omega_c = \varphi_k \equiv 0$. Here, as before, T_c is the chip duration and $\psi(t)$ is the normalized chip waveform, which are assumed to be the same for all users, and A_k, τ_k, $\{b_{i,k}\}$, and $\{a_{k,n}\}$ are the received amplitude, delay, bit stream, and spreading sequence of user k. In order to represent the phase differences between signals, we allow the amplitudes A_k to be complex. The noise $n(t)$, representing noise in the complex baseband, is also assumed to be complex. For generality, we also allow ψ to be complex. For the purposes of exposition, in what follows we assume that the desired user to be demodulated is user 1 and that $A_1 = 1$ and $\tau_1 = 0$.

Roughly speaking, the bandwidth spreading factor for DS-CDMA is the processing gain N (assuming nonrectangular, bandwidth-efficient chip waveforms). That is, each user spreads the transmitted bandwidth across N Nyquist bands, so that the receiver has N dimensions available with which to suppress interferers. We therefore conclude that the zero-forcing solution exists provided that $K \leq N$ and that the received pulse shapes are linearly independent.

For the TDL implementation of the MMSE detector, the front-end analog filter must cover the signal bandwidth, which is approximately N/T, where T is the symbol duration. (This assumes a bandwidth-efficient chip waveform. If rectangular chips are used, then the bandwidth is approximately $2N/T$.) The sampling rate is then N/T, and the TDL has taps spaced at T/N. If the TDL is of infinite length, then in principle, it can effectively suppress $N-1$ strong interferers. In what follows we assume that the front-end analog filter is a chip matched filter with impulse response $\psi^*(-t)$, which maximizes the signal-to-noise ratio at the output of this filter in the absence of interference.

2.4.1 Discrete-Time Representation

We first specify the TDL coefficients in terms of the received samples at the output of the chip matched filter. Define the vector of received samples at the output of the chip matched filter during the ith symbol as[4]

$$\mathbf{r}_i^T = [r[iT], r[iT + T_c], \dots, r[iT + (N-1)T_c]], \tag{2.4.3}$$

where

$$r(t) = \int_{-\infty}^{\infty} y(t - s)\psi^*(-s)ds \tag{2.4.4}$$

and $y(t)$ is the channel output given by (2.4.1). If $\psi(t)$ is confined to $[0, T_c]$, then the integral is from t to $t + T_c$. For the time being, we assume that all users are both chip- and symbol-synchronous. That is, referring to (2.4.3), $\tau_k = 0, 1 \le k \le K$. Combining (2.4.1), (2.4.2), and (2.4.4), we can write the vector $\mathbf{r}_i$ as a linear combination of vectors contributed by each of the users plus noise:

$$\mathbf{r}_i = \sum_{k=1}^{L} b_{i,k} A_k \mathbf{p}_k + \mathbf{n}_i, \tag{2.4.5}$$

where the upper index L depends on whether the multiuser data signal is synchronous or asynchronous. For synchronous CDMA, $L = K$ and $\mathbf{p}_k$ is the vector of samples at the output of the chip matched filter in response to the kth user's input waveform. The mth component of $\mathbf{p}_k$ in this case is therefore

$$\mathbf{p}_{k,m} = \int_{-\infty}^{\infty} p_k(mT_c - s)\psi^*(-s)ds. \tag{2.4.6}$$

Assuming zero interchip interference (i.e., $\psi_k(t - iT_c)$ are orthogonal waveforms for different i), then this integral becomes

$$\mathbf{p}_{k,m} = \int_{-\infty}^{\infty} a_{k,m} |\psi(s - mT_c)|^2 ds = a_{k,m} \tag{2.4.7}$$

since $\psi(t)$ is a unit energy pulse, and where $a_{k,m}$ is the mth spreading coefficient for user k. Consequently, for the case of synchronous DS-CDMA, we have

$$\mathbf{r}_i = \sum_{k=1}^{K} b_{i,k} A_k \mathbf{a}_k + \mathbf{n}_i, \tag{2.4.8}$$

---

[4]The superscript ^{T} denotes the transpose operator.

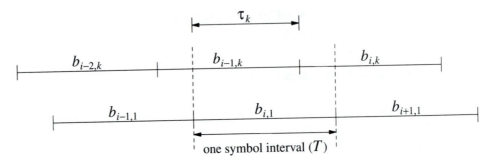

Figure 2.7   Illustration of interference from asynchronous user k. The dashed lines designate the time window spanned by the detector **c**.

where $\mathbf{a}_k$ is the vector of spreading coefficients assigned to user k.

To specify the received samples for asynchronous DS-CDMA, the delay associated with user k is expressed as

$$\tau_k = (\iota_k + \delta_k)T_c \tag{2.4.9}$$

where ι_k is an integer between 0 and $N-1$, and $\delta_k = \tau_k/T_c - \iota_k$ lies in the interval $[0, 1)$. The delay ι_k specifies the number of whole chips by which user k is shifted relative to user 1, and δ_k represents the additional partial chip delay. (In *chip-synchronous* DS-CDMA, $\delta_k = 0$, although $\iota_k \neq 0$ in general.) The computation of $\mathbf{p}_{k,m}$ for asynchronous DS-CDMA is illustrated in Figure 2.7. First note that user k transmits *two* symbols, $b_{i-1,k}$ and $b_{i,k}$, within the time window $(i-1)T$ to iT associated with $b_{i,1}$. This implies that user k contributes *two* vectors to the sum (2.4.5), associated with the left and right parts of $p_k(t - iT - \tau_k)$ within $(i-1)T$ to iT. We therefore rewrite (2.4.8) as

$$\mathbf{r}_i = b_{i,1}A_1\mathbf{p}_1 + \sum_{k=2}^{K} A_k(b_{i-1,k}\mathbf{p}_k^- + b_{i,k}\mathbf{p}_k^+) + \mathbf{n}_i, \tag{2.4.10}$$

where $\mathbf{p}_k^-$ and $\mathbf{p}_k^+$ are the sampled outputs of the matched filter during symbol i in response to $p_k[t + (T - \tau_k)]$ and $p_k(t - \tau_k)$, respectively. The mth components of $\mathbf{p}_k^+$ and $\mathbf{p}_k^-$ are then

$$\mathbf{p}_{k,m}^+ = \int_{-\infty}^{\infty} p_k(x)\psi^*[x - (mT_c - \iota_k) + \delta_k T_c]dx$$

$$= \begin{cases} a_{k,m-\iota_k-1}\varphi_2 + a_{k,m-\iota_k}\varphi_1 & \text{for } m > \iota_k \\ a_{k,0}\varphi_1 & \text{for } m = \iota_k \\ 0 & \text{for } m < \iota_k \end{cases} \tag{2.4.11a}$$

and

$$\mathbf{p}_{k,m}^{-} = \int_{-\infty}^{\infty} p_k(x)\psi^*[x - (m + N - \iota_k)T_c + \delta_k T_c]dx$$

$$= \begin{cases} a_{k,m+N-\iota_k}\varphi_1 + a_{k,m+N-\iota_k-1}\varphi_2 & \text{for } m < \iota_k \\ a_{k,N-1}\varphi_2 & \text{for } m = \iota_k \\ 0 & \text{for } m > \iota_k, \end{cases} \qquad (2.4.11b)$$

where

$$\varphi_1 = \int_{-\infty}^{\infty} \psi(s)\psi^*(s + \delta_k T_c)ds, \qquad \varphi_2 = \int_{-\infty}^{\infty} \psi(s)\psi^*[s - (1 - \delta_k)T_c]ds, \qquad (2.4.12)$$

and where we are accounting only for the contribution of the two chips from $p_k(t - \tau_k)$ (or $p_k[t + (T - \tau_k)]$) centered next to the mth chip of $p_1(t)$.

As an example, suppose that $\psi(t) = 1/\sqrt{T_c}$ for $0 < t < T_c$ and is zero elsewhere (rectangular chips). Then, we have that

$$\mathbf{p}_{k,m}^{+} = \begin{cases} a_{k,m-\iota_k-1}\delta_k + a_{k,m-\iota_k}(1 - \delta_k) & \text{for } m > \iota_k \\ a_{k,0}(1 - \delta_k) & \text{for } m = \iota_k \\ 0 & \text{for } m < \iota_k \end{cases} \qquad (2.4.13a)$$

and

$$\mathbf{p}_{k,m}^{+} = \begin{cases} a_{k,N-(\iota_k-m+1)}\delta_k + a_{k,N-(\iota_k-m)}(1 - \delta_k) & \text{for } m < \iota_k \\ a_{k,N-1}\delta_k & \text{for } m = \iota_k \\ 0 & \text{for } m > \iota_k. \end{cases} \qquad (2.4.13b)$$

Note that except for $m = \iota_k$, if $\mathbf{p}_{k,m}^{+} \neq 0$, then $\mathbf{p}_{k,m}^{-} = 0$, and vice versa.

We therefore conclude that for both synchronous and asynchronous DS-CDMA, the received vector of chip matched-filter outputs during time i can be written as (2.4.5). For synchronous DS-CDMA, $L = K$, and the vectors in the sum (2.4.5) are the spreading sequences assigned to the users. For asynchronous DS-CDMA, $L \leq 2K - 1$, and the vectors in the sum (2.4.5) are given by $\mathbf{p}_k^{+}$ and $\mathbf{p}_k^{-}$.

2.4.2 Computation of MMSE Coefficients

As noted in Section 2.3, the optimal discrete-time filter for MMSE detection is not necessarily an FIR filter. Thus, in order to limit the complexity of the MMSE detector in this setting, it is desirable to truncate the number of taps in the TDL. To consider this issue, let us define an "extended" received vector

$$\overline{\mathbf{r}}_i^{T} = [\mathbf{r}_{i-M}^{T}, \mathbf{r}_{i-M+1}^{T}, \ldots, \mathbf{r}_i^{T}, \ldots, \mathbf{r}_{i+M}^{T}], \qquad (2.4.14)$$

where $2M + 1$ is the width of the truncated processing window to be considered. That is, $\bar{\mathbf{r}}_i$ consists of the vectors $\mathbf{r}_{i-M}, \cdots, \mathbf{r}_{i+M}$ stacked on top of each other and has dimension $N(2M + 1)$. Letting $\mathbf{c}$ denote the vector of TDL coefficients, the output of the TDL at time iT can written as[5]

$$d_i = \mathbf{c}^\dagger \bar{\mathbf{r}}_i. \tag{2.4.15}$$

The estimate of the transmitted symbol $b_{i,1}$ can then be obtained by quantizing this output. In the case of binary transmissions ($b_{i,k} \in \{\pm 1\}$), the detected (uncoded) symbol is $\hat{b}_{i,1} = \text{sgn}(d_i)$.

For the MMSE detector, the TDL coefficient vector is selected to minimize

$$\text{MSE} = E\{|\mathbf{c}^\dagger \bar{\mathbf{r}}_i - b_{i,1}|^2\} = 1 + \mathbf{c}^\dagger \mathbf{R} \mathbf{c} - 2\text{Re}\{\mathbf{c}^\dagger \bar{\mathbf{p}}_1\}, \tag{2.4.16}$$

where

$$\bar{\mathbf{p}}_1^T = [0 \cdots 0 \ \mathbf{p}_1^T \ 0 \cdots 0], \tag{2.4.17}$$

with the number of zeros that precede or succeed $\mathbf{p}_1^T$ equal to NM; the covariance matrix $\mathbf{R}$ is defined as

$$\mathbf{R} = E\{\bar{\mathbf{r}}_i \bar{\mathbf{r}}_i^\dagger\}; \tag{2.4.18}$$

the noise samples at the output of the chip matched filter are white with variance σ_n^2; we normalize $E\{|b_{i,1}|^2\} = 1$; and the transmitted symbols are assumed to be uncorrelated.

Selecting $\mathbf{c}$ to minimize the MSE gives

$$\mathbf{c}_{\text{mmse}} = \mathbf{R}^{-1} \bar{\mathbf{p}}_1 \tag{2.4.19}$$

and

$$\text{MMSE} = 1 - \mathbf{c}_{\text{mmse}}^\dagger \bar{\mathbf{p}}_1 = 1 - \bar{\mathbf{p}}_1^\dagger \mathbf{R}^{-1} \bar{\mathbf{p}}_1. \tag{2.4.20}$$

For synchronous DS-CDMA, we note that $\mathbf{R}$ is a block-diagonal matrix, where each $N \times N$ diagonal block is given by

$$\mathbf{R}_N = \sum_{k=1}^{K} A_k^2 \mathbf{p}_k \mathbf{p}_k^\dagger + \sigma_n^2 \mathbf{I}. \tag{2.4.21}$$

Consequently, the MMSE TDL coefficient vector, specified by (2.4.19), has the form

$$\mathbf{c}_{\text{mmse}} = [0 \cdots 0 \ \tilde{\mathbf{c}}^T \ 0 \cdots 0], \tag{2.4.22}$$

where NM zeros precede and succeed the N-vector $\tilde{\mathbf{c}}^T = \mathbf{R}_N^{-1} \mathbf{p}_1$. We therefore conclude that for synchronous DS-CDMA, the MMSE detector consists of a chip

---

[5]The superscript $\dagger$ denotes the complex conjugate transpose.

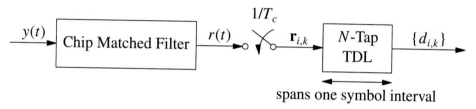

Figure 2.8 N-tap MMSE detector for user k.

matched filter followed by a *finite-length* TDL that spans only one symbol (i.e., we can set $M = 0$ in (2.4.14)).

For asynchronous DS-CDMA, the MMSE TDL is no longer finite in general; however, we can still consider a truncated version that spans one symbol. This detector, assuming T_c-spaced taps and a front-end chip matched filter, is shown in Figure 2.8. The only difference between this "N-tap MMSE detector" and the conventional matched-filter detector is the way in which the TDL coefficients are selected. For the matched-filter detector, $\mathbf{c} = \mathbf{a}_1$ (the spreading coefficients for user 1), whereas for the MMSE detector,

$$\mathbf{c}_{\mathrm{mmse}} = \mathbf{R}^{-1}\mathbf{p}_1, \qquad (2.4.23)$$

where $\mathbf{R} = \mathbf{R}_N$ in (2.4.21). Note that in the absence of background noise, $\mathbf{R}$ is singular if $L < N$. However, it is easily shown that any $\mathbf{c}$ that satisfies $\mathbf{R}\mathbf{c} = \bar{\mathbf{p}}_1$ minimizes MSE even in this singular case.

2.4.3 Geometric Interpretation

Throughout the rest of this section we focus for simplicity on the N-tap detector shown in Figure 2.8. The following discussion is easily generalized to account for a TDL that spans multiple symbol intervals. The vectors $\mathbf{p}_1,\ldots,\mathbf{p}_L$ that appear in the sum (2.4.21) are illustrated in Figure 2.9. The space spanned by these vectors is the *signal subspace*, denoted as $\mathbf{S}$. The *interference subspace*, denoted as $\mathbf{S}_I$, is the space spanned by $\mathbf{p}_2,\ldots,\mathbf{p}_L$. (We continue to assume that user 1 is the user of interest.) If $\mathbf{p}_1,\ldots,\mathbf{p}_L$ are linearly independent, then $\mathbf{S}$ has dimension L, and $\mathbf{S}_I$ has dimension $L - 1$.

We first observe that the MMSE solution $\mathbf{c}$ must lie in $\mathbf{S}$. Otherwise, we can write

$$\mathbf{c} = \mathbf{c}_s + \mathbf{c}_s^{\perp}, \qquad (2.4.24)$$

where $\mathbf{c}_s \in \mathbf{S}$ and $\mathbf{c}_s^{\perp}$ is orthogonal to $\mathbf{S}$. We then have that $\mathbf{p}_k^{\dagger}\mathbf{c}_s^{\perp} = 0$ for each $k = 1,\ldots,K$, and $(\mathbf{c}_s^{\perp})^{\dagger}\mathbf{r}_i = (\mathbf{c}_s^{\perp})^{\dagger}\mathbf{n}_i$. We therefore conclude that the component $\mathbf{c}_s^{\perp}$ in (2.4.24) adds a noise term to the filter output d_i, which increases the MSE. To minimize MSE, we must take $\mathbf{c}_s^{\perp} = \mathbf{0}$.

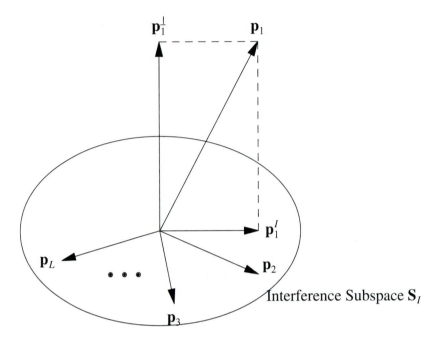

Figure 2.9 Geometric representation of desired signal and interference vectors. $\mathbf{p}_1^I$ is the projection of $\mathbf{p}_1$ onto the interference subspace $\mathbf{S}_I$.

Because the MMSE solution $\mathbf{c}_{mmse} \in \mathbf{S}$, we can express $\mathbf{c}_{mmse}$ as a linear combination of the signal vectors. Let $\mathbf{P}$ denote the $N \times L$ matrix with columns $\mathbf{p}_1, \ldots, \mathbf{p}_L$. Then, from (2.4.5), we can write the received vector as

$$\mathbf{r}_i = \mathbf{PAb}_i + \mathbf{n}_i, \tag{2.4.25}$$

where $\mathbf{A} = \text{diag}[A_1 \ A_2 \ \cdots \ A_K]$ and $\mathbf{b}_i^T = [b_{i,1} \ b_{i,2} \ \cdots \ b_{i,K}]$. Now, define the $N \times K$ matrix $\mathbf{C}$, where the kth column of $\mathbf{C}$ is the vector of TDL coefficients used to demodulate user k. The first column of $\mathbf{C}$ is therefore the vector $\mathbf{c}$ used in (2.4.15) to demodulate user 1. From the previous discussion it follows that each column of $\mathbf{C}_{mmse}$ can be expressed as a linear combination of the signal vectors. We therefore write $\mathbf{C} = \mathbf{P\Gamma}$, where Γ is a $K \times K$ matrix; note that $\mathbf{C}_k = \mathbf{P\Gamma}_k$, where the subscript k denotes the kth column of the matrix. The total MMSE summed over all K users is

$$\begin{aligned}\text{MSE} &= E\{\|\mathbf{Cr}_i - \mathbf{b}_i\|^2\} \\ &= \text{trace}\{(\Gamma^\dagger \mathbf{P}^\dagger \mathbf{PA} - \mathbf{I})(\mathbf{A}^\dagger \mathbf{P}^\dagger \mathbf{P\Gamma} - \mathbf{I}) + \sigma_n^2 \Gamma^\dagger \mathbf{P}^\dagger \mathbf{P\Gamma}\}\end{aligned} \tag{2.4.26}$$

and selecting Γ to minimize this expression gives (cf., [129])

$$\Gamma = \mathbf{A}[\mathbf{AP}^\dagger \mathbf{PA} + \sigma_n^2 \mathbf{I}]^{-1}. \tag{2.4.27}$$

The matrix $\mathbf{P}^\dagger\mathbf{P}$ is the *cross-correlation* matrix for the set of signal vectors $\mathbf{p}_1,\ldots,\mathbf{p}_K$. Since $\mathbf{c} = \mathbf{P}\boldsymbol{\Gamma}_1$, this relation gives an alternative method to (2.4.23) for computing the MMSE solution $\mathbf{c}_{mmse}$. When $K < N$ and the background noise is small, it is better to compute $\mathbf{c}_{mmse}$ via (2.4.27) since the matrix $\mathbf{R}_N$ defined by (2.4.21) is likely to be ill conditioned. Finally, we note that this expression is analogous to the expression (2.3.7) for the matrix (multiuser) discrete-time transfer function with a bank of front-end matched filters.

2.4.4 Zero-Forcing (Decorrelating) Solution

In analogy with the zero-forcing solution for the MIMO MMSE detector discussed in Section 2.3.2, it may be possible to choose the N-vector $\mathbf{c}$ to completely remove multiple-access interference. From Figure 2.9 it is apparent that this zero-forcing, or decorrelating, solution is proportional to the orthogonal projection of the desired user vector $\mathbf{p}_1$ onto the interference subspace $\mathbf{S}_I$. Denoting the zero-forcing solution for $\mathbf{c}$ as $\mathbf{c}_{zf}$, the filter output is given by

$$d_i = \mathbf{c}_{zf}^\dagger \mathbf{r}_i = \mathbf{c}_{zf}^\dagger (b_{i,1}\mathbf{p}_1 + \mathbf{n}_i). \tag{2.4.28}$$

Namely, the output of the zero-forcing filter has only two components, one due to the desired signal and one due to background noise. Let $\mathbf{P}_I$ denote the $N \times (K-1)$ matrix with columns given by $\mathbf{p}_2,\ldots,\mathbf{p}_K$. The orthogonal projection of $\mathbf{p}_1$ onto $\mathbf{S}_I$ is denoted as

$$\mathbf{p}_1^\perp = \mathbf{p}_1 - \mathbf{P}_I(\mathbf{P}_I^\dagger\mathbf{P}_I)^{-1}(\mathbf{P}_I^\dagger\mathbf{p}_1), \tag{2.4.29}$$

and the zero-forcing solution is

$$\mathbf{c}_{zf} = \mathbf{p}_1^\perp/\eta, \tag{2.4.30}$$

where the scale factor $1/\eta$ is selected so that $|\mathbf{c}_{zf}^\dagger\mathbf{p}_1| = |b_1| = 1$. (The quantity η is known as the near-far resistance and has special significance, as explained later.) It is easily shown that $(\mathbf{p}_1^\perp)^\dagger\mathbf{p}_1 = \|\mathbf{p}_1^\perp\|^2$, so that $\eta = \|\mathbf{p}_1^\perp\|^2$.

It is apparent from Figure 2.9 that the zero-forcing solution for $\mathbf{c}$ exists provided that $\mathbf{p}_1$ is not contained in $\mathbf{S}_I$. In that case, the dimension of the signal subspace $\mathbf{S}$ must be no greater than N. If the vectors $\mathbf{p}_1,\ldots,\mathbf{p}_L$ are linearly independent, then we must have $L \le N$. For synchronous DS-CDMA, this implies that the number of users $K \le N$, and for asynchronous DS-CDMA, $2K - 1 \le N$. Of course, even if this latter condition does not hold (as in a heavily loaded cellular system), the MMSE solution is still well defined. (Also, the addition of receiver diversity allows one to increase K beyond this bound, as is discussed below.) As K increases, the performance of the MMSE detector improves relative to the zero-forcing solution.

2.4.5 Asymptotic Behavior of the MMSE Solution

Here, we examine the behavior of the MMSE solution as (i) the noise level diminishes to zero and (ii) the interferers increase in energy. If the noise variance $\sigma_n^2 = 0$, then we observe that the zero-forcing solution, assuming it exists, gives zero MSE. We therefore conclude that the MMSE solution converges to the zero-forcing solution as $\sigma_n^2 \to 0$. It can be shown by matrix manipulations that the zero-forcing solution (2.4.30) is equivalent to the preceding expressions (2.4.23) and (2.4.27), where the noise variance $\sigma_n^2 = 0$.

Now consider what happens as user k's amplitude $A_k \to \infty$. It is easily seen that $\mathbf{c}_{\text{mmse}}^\dagger \mathbf{p}_k \to 0$. Otherwise, we would have $(\mathbf{c}_{\text{mmse}}^\dagger \mathbf{p}_k)^2 > \varepsilon > 0$, which implies $(A_k \mathbf{c}_{\text{mmse}}^\dagger \mathbf{p}_k)^2 > \varepsilon A_k^2$. As $A_k \to \infty$, (2.4.16) implies that MSE $\to \infty$, which contradicts the fact that MMSE ≤ 1 (i.e., $\mathbf{c} = 0$ gives MSE $= 1$). In fact, it can be shown that

$$\lim_{A_k \to \infty} (A_k \mathbf{c}_{\text{mmse}}^\dagger \mathbf{p}_k) = 0, \tag{2.4.31}$$

which implies that as $A_k \to \infty$, the contribution to the MSE from user k diminishes to zero [51, 82].

To generalize (2.4.31), if $A_k \to \infty$ for k in some subset K_s, then (2.4.31) applies for each $k \in K_s$ (assuming $\mathbf{c}$ has enough degrees of freedom to suppress these interferers). If the set K_s contains all $K - 1$ interferers, then clearly $\mathbf{c}_{\text{mmse}} \to \kappa \mathbf{p}_1^\perp$, where κ is a constant. Substituting for $\mathbf{c}_{\text{mmse}}$ in (2.4.28) and selecting κ to minimize MSE gives $\kappa = 1/(\eta + \sigma_n^2)$. We therefore conclude that as the interfering amplitudes $A_k \to \infty$, $k \neq 1$,

$$\mathbf{c}_{\text{mmse}} \to \frac{\mathbf{p}_1^\perp}{\eta + \sigma_n^2}, \qquad \text{MMSE} \to \frac{\sigma_n^2}{\eta + \sigma_n^2}. \tag{2.4.32}$$

Note that the filter $\mathbf{c}_{\text{mmse}}$ gives a biased estimate of $b_{i,1}$.

2.4.6 Performance Measures

In addition to MMSE, two other performance measures of interest are signal-to-interference-plus-noise ratio (SINR) and error probability. The SINR is defined to be the ratio of the desired signal power to the sum of the powers due to noise and multiple-access interference at the output of the filter $\mathbf{c}$. That is,

$$\text{SINR} = \frac{(\mathbf{c}^\dagger \mathbf{p}_1)^2}{\sum_{k=2}^{L} A_k^2 (\mathbf{c}^\dagger \mathbf{p}_k)^2 + \sigma_n^2 \|\mathbf{c}\|^2}. \tag{2.4.33}$$

It can be shown that the MMSE solution $\mathbf{c}$ also maximizes the SINR and that this maximum SINR is

$$\frac{\mathbf{c}_{mmse}^{\dagger}\mathbf{p}_1}{1 - \mathbf{c}_{mmse}^{\dagger}\mathbf{p}_1} = \frac{1}{MMSE} - 1. \tag{2.4.34}$$

To study the error probability, we restrict attention to the case in which all users transmit binary, equally probable symbols. In this case, we have $\Pr\{\hat{b}_1 \neq b_1\}$ $=\Pr\{\hat{b}_1 \neq b_1 \mid b_1 = 1\}$. Conditioning on all users' symbols and assuming white Gaussian noise gives

$$P_{e|\mathbf{b}}(\mathbf{b}) = P(\hat{b}_1 \neq b_1 \mid \mathbf{b}, b_1 = 1) = Q\left(\frac{\mathbf{c}^{\dagger}\mathbf{p}_1 + \sum_{k=2}^{K} b_k A_k (\mathbf{c}^{\dagger}\mathbf{p}_k)}{\sigma_n \|\mathbf{c}\|}\right), \tag{2.4.35}$$

where $Q(x) = (2\pi)^{-\frac{1}{2}}\int_x^\infty e^{-t^2/2}dt$. The *average* error probability is then obtained by averaging (2.4.35) over the distribution for the bit vectors $\mathbf{b}$, i.e., $P_e = E\{P_{e|\mathbf{b}}\}$.

Two additional performance measures related to the asymptotic performance discussed in the preceding section are *asymptotic efficiency* and *near-far resistance*. Let $P_e(\sigma_n)$ denote the average error probability for a specific detector as a function of the noise variance σ_n^2. The asymptotic efficiency of the detector is then defined in [119, 120] as

$$\gamma = \sup\{\kappa : \lim_{\sigma_n \to 0} P_e(\sigma_n)/Q(\sqrt{\kappa}A_1/\sigma_n) > 0\} \tag{2.4.36}$$

and is a limiting measure, as the noise tends to zero, of how well the detector performs in the presence of MAI relative to optimal performance in the absence of MAI. Larger values of γ correspond to more effective MAI suppression. The near-far resistance of the detector is defined in [119] as

$$\eta = \inf_{A_2, \ldots, A_L} \gamma. \tag{2.4.37}$$

That is, the near-far resistance is the asymptotic efficiency evaluated for *worst case* interference energies and is a measure of the robustness of the detector with respect to variations in the received interference energies.

As $\sigma_n \to 0$, the error probability for the MMSE TDL detector satisfies

$$\lim_{\sigma_n \to 0} \frac{P_e(\sigma_n)}{Q(\|\mathbf{p}_1^{\perp}\|/\sigma_n)} = \lim_{\sigma_n \to 0} \frac{\min_{\mathbf{b}} P_{e|\mathbf{b}}(\sigma_n)}{Q(\|\mathbf{p}_1^{\perp}\|/\sigma_n)} = 1. \tag{2.4.38}$$

The asymptotic efficiency of the MMSE detector is therefore $\|\mathbf{p}_1^\perp\|^2$. Since this quantity is independent of the energies of the interference vectors, we also have that

$$\eta = \|\mathbf{p}_1^\perp\|^2. \tag{2.4.39}$$

That is, the near-far resistance of the MMSE detector considered is the squared norm of the component of the desired signal vector that is orthogonal to the space spanned by the interference vectors. From (2.4.39), it is clear that if $\eta > 0$, then the desired vector is not contained in the interference subspace $\mathbf{S}_I$, which in turn implies that the number of vectors contributed by the users $L \leq N$.

From the discussion in Section 2.4.4, we observe that the near-far resistance is closely related to the zero-forcing solution. Specifically, the zero-forcing solution is given by (2.4.30), which includes η as a scale factor. Also, it is easily shown that the MSE corresponding to the zero-forcing solution is

$$\text{MSE}_{zf} = \frac{\sigma_n^2}{\eta}, \tag{2.4.40}$$

so that the noise enhancement associated with the zero-forcing detector is $1/\eta$. Note that $0 \leq \eta \leq 1$ implies that $\sigma_n^2 \leq \text{MSE}_{zf} \leq \infty$. In particular, if $\mathbf{p}_1$ lies in the space spanned by the interferers (i.e., if $\text{rank}(\mathbf{S}_I) > N$), then $\eta = 0$ and $\text{MSE}_{zf} = \infty$.

2.4.7 Space-Time Filtering

It is conceptually straightforward to extend the preceding discussion to combined space-time filtering. Given multiple receiver antennas, the MMSE linear filter for user k is shown in Figure 2.6 (for two antennas) and consists of a chip matched filter and TDL for each antenna. The TDL outputs are simply added together to form the symbol estimate. To compute the TDL coefficients in terms of the received samples on each branch, we define $\mathbf{r}_i^{(m)}$ as the $N \times 1$ received vector of chip matched filter outputs for symbol i on branch m. Then, $\bar{\mathbf{r}}_i$ is the $(MN) \times 1$ vector consisting of $\mathbf{r}_i^{(1)}, \ldots, \mathbf{r}_i^{(M)}$ stacked on top of each other. (This extension assumes $1/T_c$ sampling and that each TDL spans a single symbol. The generalizations to other sampling rates and to multisymbol TDLs are straightforward.)

Let $\mathbf{c}^{(k)}$ be the vector of TDL coefficients associated with the kth branch. As before, the filter output can be expressed as

$$d_i = \bar{\mathbf{c}}^\dagger \bar{\mathbf{r}}_i, \tag{2.4.41}$$

where $\bar{\mathbf{c}}$ is the $(MN) \times 1$ vector of TDL coefficient vectors $\mathbf{c}^{(1)}, \ldots, \mathbf{c}^{(M)}$ stacked on top of each other. The preceding expressions for the MMSE coefficients, zero-forcing solution, and performance measures can therefore be directly applied to this situation. Note, in particular, that with M antennas, a necessary condition for the existence of the zero-forcing solution is that the number of "effective" users $L < MN$. For asynchronous DS-CDMA ($L = 2K - 1$), adding an additional antenna therefore increases the number of strong interferers that can be (completely) suppressed by approximately $N/2$.

Increasing the amount of spatial diversity leads to a substantial increase in system capacity, but at the expense of additional complexity. Specifically, analog front-end filtering, as well as conversion to baseband (if necessary), is needed for each antenna element. The number of TDL coefficients also increases from N to MN, which can adversely affect the performance of the adaptive algorithms discussed in the next section.

2.4.8 Effect of Multipath

Multipath is discussed in more detail in Section 2.6. For now, recall from Chapter 1 that reflections of the transmitted signal off surrounding objects cause the received signal to consist of the sum of weighted and delayed versions of the transmitted signals:

$$r(t) = \sum_{k=1}^{K} \sum_{m=1}^{M_k} \alpha_{k,m} A_k b_k p_k(t - iT - \tau_{k,m}) + n(t),\tag{2.4.42}$$

where M_k is the number of paths associated with user k, and $\tau_{k,m}$ and $\alpha_{k,m}$ are respectively the delay and the (complex) coefficient associated with path m for user k. (Without loss of generality, we assume that $\tau_{k,m} \geq 0$, $m = 1,\ldots, M$; $k = 1, 2,\ldots, K$.) We can write the sampled received vector $\mathbf{r}_i$ defined earlier as

$$\mathbf{r}_i = \sum_{k=1}^{K} A_k \left[\sum_{m=1}^{M_k} \alpha_{k,m} \left(b_{i,k} \mathbf{p}_k^+(m) + b_{i-1,k} \mathbf{p}_k^-(m) \right) \right] + \mathbf{n}_i,\tag{2.4.43}$$

where $\mathbf{p}_{k,m}^+$ and $\mathbf{p}_{k,m}^-$ contain the chip matched-filter output samples within the time window spanned by $\mathbf{r}_i$ in response to the inputs $p_k(t - \tau_{k,m})$ and $p_k(t + T - \tau_{k,m})$, respectively. According to the discussion in Section 2.4.1, the vectors $\mathbf{p}_{k,m}^+$ and $\mathbf{p}_{k,m}^-$ can be computed according to (2.4.11), where $\tau_{k,m}$ replaces τ_k.

Note that we can rewrite (2.4.43) as

$$\mathbf{r}_i = \sum_{k=1}^{K} A_k(b_{i,k} \mathbf{p}_k^+ + b_{i-1,k} \mathbf{p}_k^-) + \mathbf{n}_i,\tag{2.4.44}$$

where

$$\mathbf{p}_k^{\pm} = \sum_{m=1}^{M_k} \alpha_{k,m} \mathbf{p}_k^{\pm}(m).\tag{2.4.45}$$

Consequently, the received vector can once again be expressed as (2.4.5), where the vectors in the sum are computed according to (2.4.11) and (2.4.45). The MMSE and zero-forcing solutions for $\mathbf{c}$ and performance measures previously discussed can then be directly applied. For DS-CDMA applications, it is typically assumed that the path delays for user 1, $\tau_{1,m}$, $m = 1, \ldots, M_1$, span at most a few chips. The inter-symbol interference due to the multipath vectors $\mathbf{p}_1^-(m)$ is then quite small and is typically ignored.[6]

To summarize the preceding discussion, when multipath is present, the geometric interpretation represented by Figure 2.9 applies, where the signal vectors are the *received* vectors, including the effect of multipath. We therefore conclude that the MMSE solution *coherently combines all multipath within the window spanned by the filter* $\mathbf{c}$. Of course, this interpretation applies in practice only when the MMSE solution can be accurately estimated. That is, the estimation algorithm must be able to compensate for the changing multipath amplitudes and phases of all strong users. Techniques for performing combined, linear, multiuser detection and channel tracking are developed in [85], [126], and [127]. These methods are discussed briefly below.

2.5 ADAPTIVE ALGORITHMS

The expressions for the MMSE vector $\mathbf{c}$ given in the preceding section, (2.4.19) and (2.4.27), give the impression that the MMSE receiver requires explicit knowledge of all user and channel parameters (i.e., spreading sequences, relative timing, phase, amplitudes, and multipath parameters). In this section, we show that the MMSE solution for $\mathbf{c}$ can be accurately estimated without this knowledge. In fact, if the user and channel parameters are time-invariant, then the algorithms in this section can estimate $\mathbf{c}_{mmse}$ to arbitrary accuracy (given a sufficient number of received vectors $\mathbf{r}_i$).

The adaptive algorithms in this section require either (i) a training sequence of transmitted symbols, which are known to the receiver for initial adaptation or (ii) accurate knowledge of the received vector corresponding to the *desired* user ($\mathbf{p}_1$) and associated timing. In the absence of multipath, the latter knowledge, which is simply the spreading code of the desired user and associated timing, is also required by the matched-filter receiver. When multipath is present, the received vector can be processed by a RAKE receiver [90, Ch. 7].

---

[6]In high-data-rate systems, such as arise in some indoor wireless applications, ISI can be significant. ISI is also severe in underwater acoustic transmission systems, as developed in Chapter 8. Linear detection techniques for dealing jointly with ISI and MAI in such situations are discussed in detail in Chapter 3 and the references therein; see also [89] and [130].

Three categories of adaptive algorithms are presented herein. The first category consists of the conventional stochastic gradient and least squares algorithms well known in adaptive filtering [25, 32]. These have been applied to obtain MMSE symbol estimates for DS-CDMA in [5, 58, 92]. (Prior to that, MMSE estimation applied to DS-CDMA was considered in [133].) Application of these techniques to narrowband TDMA systems with co- and adjacent-channel interference is reported in [47, 48]. The algorithms in the second category are "blind" in the sense that a training sequence is not required. Instead, knowledge of the received vector $\mathbf{p}_1$ and associated timing is assumed. Finally, the algorithms in the third category are "subspace" algorithms, in which each received vector $\mathbf{r}_i$ is projected onto a lower-dimensional subspace. These techniques are potentially useful when the dimension of the received vectors is much greater than the dimension of the signal subspace. This may be the case when (i) the processing gain is very large relative to the number of users, (ii) an adaptive antenna array is available with TDLs on each branch, or (iii) the filter $\mathbf{c}$ spans multiple symbol intervals.

2.5.1 Stochastic Gradient Algorithm

The stochastic gradient or LMS (least mean squares) algorithm has been successfully applied to many signal processing applications such as noise cancellation, equalization, echo cancellation, and adaptive beamforming [25, 32]. The approach to adaptive interference suppression presented here is, in fact, analogous to adaptive equalization for single-user channels. The main difference between the two applications is that for adaptive equalization, the TDL must span multiple symbols to suppress intersymbol interference (ISI), but can have as few as one tap per symbol. In contrast, for interference suppression, the TDL must have multiple taps per symbol but can span a single symbol interval. Of course, a TDL that spans multiple symbols with multiple taps per symbol can suppress both ISI and multiple-access interference.

Let $\mathbf{c}_i$ denote the TDL vector at symbol time i. The LMS algorithm for updating $\mathbf{c}_i$ is given by

$$\mathbf{c}_i = \mathbf{c}_{i-1} + \mu e_i^* \mathbf{r}_i, \tag{2.5.1}$$

where

$$e_i = b_{i,1} - \mathbf{c}_{i-1}^\dagger \mathbf{r}_i \tag{2.5.2}$$

is the estimation error at time i, and μ is a constant step size, which controls the tradeoff between convergence speed and excess MSE due to random coefficient fluctuations about the mean. Because the LMS algorithm (2.5.1) assumes knowledge of the symbols $b_{i,1}$, it must be implemented as a *supervised* or *decision-directed* algorithm. In practice, $b_{i,1}$ must be generated via a training sequence for initial adaptation (supervision), after which the symbol estimates $\hat{b}_{i,1}$ are used (decision direction).

There have been numerous analyses of the convergence properties of the LMS algorithm (e.g., see [25, 32]). A detailed analysis of this algorithm for the DS-CDMA interference suppression application considered here is given in [59] (see also [31]). A summary of the main results, given a stationary set of interferers and channels, is as follows.

1. Assuming that the received vector $\mathbf{r}_i$ is statistically independent[7] from past vectors $\mathbf{r}_m$, $m < i$, for each i, it can be shown that

$$\delta\mathbf{c}_i = (\mathbf{I} - \mu\mathbf{R})\delta\mathbf{c}_{i-1}, \qquad (2.5.3)$$

where $\delta\mathbf{c}_i = E\{\mathbf{c}_i\} - \mathbf{c}_{mmse}$. The mean coefficient vector therefore converges exponentially to $\mathbf{c}_{mmse}$ according to N normal modes. The time constant associated with the nth mode is $1 - \mu\lambda_n$, where λ_n is the nth eigenvalue of $\mathbf{R}$.

2. An approximate analysis shows that the MSE remains bounded provided that the step-size

$$\mu < \frac{2}{\text{trace}(\mathbf{R})}. \qquad (2.5.4)$$

If $\mathbf{R}$ is given by (2.4.21), then

$$\text{trace}(\mathbf{R}) = \sum_{k=1}^{K} A_k^2 + N\sigma_n^2. \qquad (2.5.5)$$

3. The asymptotic MSE achieved with the LMS algorithm is greater than the MMSE due to random coefficient fluctuations about the mean. Denoting the MMSE as ε_{min}, the *excess* MSE due to these fluctuations can be approximated as

$$\xi_{ex} = \varepsilon_{min} \frac{\frac{\mu}{2} \text{trace}(\mathbf{R})}{1 - \frac{\mu}{2} \text{trace}(\mathbf{R})}, \qquad (2.5.6)$$

where trace $(\mathbf{R})$ is given by (2.5.5).

To interpret the preceding results, suppose that the vectors $\mathbf{p}_1, \ldots, \mathbf{p}_K$ in the sum (2.4.5) are orthonormal. In that case, each of these vectors $\mathbf{p}_k$ is an eigenvector

---

[7]This independence assumption holds for synchronous DS-CDMA but not for asynchronous DS-CDMA. Nevertheless, even for asynchronous DS-CDMA, it gives substantial insight.

of $\mathbf{R}$ with associated eigenvalue $A_k^2 + \sigma_n^2$. The remaining eigenvectors of $\mathbf{R}$ form a basis for the $N - K$ dimensional subspace that is orthogonal to the signal space. Each of these eigenvectors is associated with eigenvalue σ_n^2. If $L < K$, then there are $N - K$ modes of convergence for $E\{\mathbf{c}_i\}$ associated with exponential decay factor $1 - \mu\sigma_n^2$. Typically, σ_n^2 is very small, so that convergence associated with these modes is very slow. If $\mathbf{c}_i$ is in (or close to) the signal space, then this slow convergence is no problem since the dominant modes of convergence lie in the signal space. However, if $\mathbf{c}_i$ lies outside the signal space (such as when the filter has converged to a set of users and a user subsequently departs), then the excess MSE due to the component of $\mathbf{c}_i$ outside the signal space can take a very long time to disappear.

We also note from the preceding discussion that the LMS algorithm is adversely affected by a near-far situation in which A_k is very large. Namely, there will be a slow mode corresponding to an exponential decay factor approximately equal to $1 - \mu/A_k^2$. Also, note that according to (2.5.4) and (2.5.5), the larger the interfering amplitudes, the smaller μ must be for stability. To ensure that μ satisfies the stability condition (2.5.4) in the presence of a changing interference environment, it is useful to *normalize* the step size by an estimate of the input power. Specifically, μ in (2.5.1) can be replaced by $\bar{\mu} = \mu/\hat{\xi}(i)$, where

$$\hat{\xi}(i) = w\hat{\xi}(i-1) + (1-w)\|\mathbf{r}_i\|^2 \tag{2.5.7}$$

is a moving average estimate of the input energy, and w is the averaging constant.

2.5.2 Least Squares (LS) Algorithm

An alternative to the stochastic gradient method is to choose the vector $\mathbf{c}_i$ to minimize the least squares (LS) cost function

$$\varepsilon_{ls}(i) = \sum_{n=0}^{i} w^{i-n} |d_n - \hat{b}_{n,1}|^2, \tag{2.5.8}$$

where $d_n = \mathbf{c}^\dagger \mathbf{r}_n$, and $0 < w < 1$ is an exponential weighting factor that discounts past data. This weighting is important in nonstationary environments where the vector $\mathbf{c}_i$ is computed at each iteration i. The LS solution for $\mathbf{c}_i$ is

$$\mathbf{c}_i = \hat{\mathbf{R}}_i^{-1}\hat{\mathbf{p}}_{i,1}, \tag{2.5.9}$$

where

$$\hat{\mathbf{R}}_i = \sum_{n=0}^{i} w^{i-n} \mathbf{r}_n \mathbf{r}_n^\dagger \tag{2.5.10}$$

and

$$\hat{\mathbf{p}}_{i,1} = \sum_{n=0}^{i} w^{i-n} b_{i,1}^* \mathbf{r}_n. \tag{2.5.11}$$

Note that $\hat{\mathbf{R}}_i$ and $\hat{\mathbf{p}}_{i,1}$ are estimates of $\mathbf{R}$ and $\mathbf{p}_1$, respectively. In the absence of noise, it is possible for $\hat{\mathbf{R}}_i$ to be singular. In that case, any solution to the set of linear equations $\hat{\mathbf{R}}_i \mathbf{c}_i = \hat{\mathbf{p}}_{i,1}$ minimizes the LS cost function $\varepsilon_{ls}(i)$.

The LS criterion is deterministic, as opposed to the stochastic gradient cost criterion (MSE), which is defined in terms of a statistical expectation. In general, LS algorithms converge much faster than do stochastic gradient algorithms but are more complex to implement. Specifically, if the signal vectors $\mathbf{p}_1, \ldots, \mathbf{p}_L$ are linearly independent and $L < N$, then in the absence of noise, the LS solution for the vector $\mathbf{c}_i$ gives $\varepsilon_{ls}(i) = 0$ for $i > L$, provided that the received vectors $\mathbf{r}_0, \ldots, \mathbf{r}_{L-1}$ used to compute $\mathbf{c}$ from (2.5.9) are linearly independent. (This implies that the MSE is zero as well.)

A precise analysis of the convergence properties of the LS algorithm in the presence of noise is quite difficult; however, a useful rule of thumb is that it typically takes approximately $2N$ iterations for the LS algorithm to converge (again assuming stationary noise and interference), where N is the filter order. In contrast, the stochastic gradient algorithm typically takes between 5 to 10 times longer to reach steady-state performance, assuming the spread in eigenvalues of the matrix $\mathbf{R}$ is relatively small. Because the convergence rate of the LS algorithm is insensitive to this eigenvalue spread, a large eigenvalue spread, corresponding to a near-far situation, will lead to a more dramatic difference in performance. Some numerical examples that compare the performance of LS and LMS adaptive algorithms are presented in Section 2.5.5.

A recursive LS (RLS) algorithm computes the LS solution $\mathbf{c}_i$ for each i. In this case, the matrix inverse $\hat{\mathbf{R}}_i^{-1}$ can be propagated in time by using the matrix inversion lemma:

$$\hat{\mathbf{R}}_i^{-1} = \hat{\mathbf{R}}_{i-1}^{-1} - \frac{\mathbf{g}_i \mathbf{g}_i^\dagger}{1 + (\mathbf{r}_i^\dagger \mathbf{g}_i)}, \tag{2.5.12}$$

where $w = 1$ and

$$\mathbf{g}_i = \mathbf{R}_{i-1}^{-1} \mathbf{r}_i. \tag{2.5.13}$$

The vector $\hat{\mathbf{p}}_{i,1}$ can also be updated recursively as

$$\hat{\mathbf{p}}_{i,1} = w \hat{\mathbf{p}}_{i-1,1} + b_{i,1}^* \mathbf{r}_i. \tag{2.5.14}$$

Although the matrix inversion lemma substantially reduces the complexity of a recursive LS (RLS) algorithm, the update (2.5.12) is sensitive to numerical roundoff errors and must therefore be closely monitored or stabilized in some manner. Also, the basic complexity per update is $\mathcal{O}(N^2)$ for RLS, as compared with $\mathcal{O}(N)$ for LMS. (The complexity of RLS can be mitigated through parallelization, as discussed below.) As with the LMS algorithm, the RLS algorithm defined by (2.5.9), (2.5.12)–(2.5.14) requires estimates of the symbols $\{b_{i,1}\}$. This estimate can be accomplished initially through a training sequence and, subsequently, by switching to decision-directed mode.

Rather than compute c_i for each i, it is also possible to update c_i periodically by means of the most recently received data vectors. Specifically, a *block LS* algorithm computes c_i every B iterations, using the data vectors $r_{i-B+1}, \ldots, r_i$. An iterative method for obtaining the estimates $b_{i,1}$ in decision-directed mode is described in [29]. The results in [29] indicate that this type of block decision-directed algorithm can sometimes perform significantly better than the RLS algorithm with exponential weighting.

2.5.3 Orthogonally Anchored (Blind) Algorithms

The decision-directed algorithms presented in the preceding section generally require reliable symbol estimates. Numerical results indicate that the performance of these algorithms begins to degrade when the error rate exceeds 10% [29, 33]. Much higher error rates sustained over many symbols can potentially cause the algorithm to lose track of the desired user. For a mobile wireless channel, this situation can occur when the desired user experiences a deep fade or when a strong interferer suddenly appears. It is therefore desirable to have an adaptive algorithm that does not require symbol estimates. We refer to such an algorithm as a "blind" adaptive algorithm.

An approach to blind adaptation, which was presented in [31], is illustrated in Figure 2.10. The vector c at time i is expressed as

$$c_i = p_1 + w_i, \qquad\qquad (2.5.15)$$

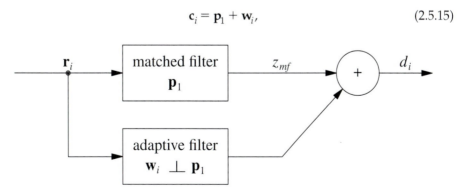

Figure 2.10   Orthogonally anchored adaptive filter.

where $\mathbf{w}_i$ is constrained to be *orthogonal* to $\mathbf{p}_1$ for all i. The filter output is then

$$d_i = \mathbf{c}_i^\dagger \mathbf{r}_i = b_{i,1} + \sum_{k=2}^{L} A_k [\mathbf{p}_1 + \mathbf{w}_i]^\dagger \mathbf{p}_k + (\mathbf{c}^\dagger \mathbf{n}_i), \qquad (2.5.16)$$

where it is assumed that $A_1 \|\mathbf{p}_1\|^2 = 1$. Note that $\mathbf{w}_i$ affects only the interference and noise at the output. Selecting $\mathbf{w}_i$ to minimize the output *variance* $E\{|d_i|^2\}$ therefore minimizes the output interference plus noise energy. In fact, the output MSE is (again, we normalize $E\{|b_{i,1}|^2\} = 1$)

$$E\{|b_{i,1} - d_i|^2\} = 1 + E\{|d_i|^2\} - 2\,\mathrm{Re}\{(\mathbf{p}_1 + \mathbf{w}_i)^\dagger \mathbf{p}_1\}$$
$$= E\{|d_i|^2\} - 1 \qquad (2.5.17)$$

since

$$(\mathbf{p}_1 + \mathbf{w}_i)^\dagger \mathbf{p}_1 = \mathbf{c}^\dagger \mathbf{p}_1 = 1. \qquad (2.5.18)$$

We therefore conclude that selecting $\mathbf{w}_i$ to minimize the output variance $E\{|d_i|^2\}$ also minimizes output MSE. Minimizing the output variance does not require knowledge of the symbol estimates $b_{i,1}$, although it does require knowledge of the desired user's vector $\mathbf{p}_1$ (and associated timing). This minimum variance technique is analogous to the minimum variance technique in adaptive beamforming where the direction of arrival of the desired signal is known [41].

The minimum variance vector $\mathbf{c}_{mv}$ can be derived by defining the Lagrangian

$$L(\mathbf{c}) = E\{|d_i|^2\} - \xi \mathbf{c}^\dagger \mathbf{p}_1, \qquad (2.5.19)$$

where ξ is the Lagrange multiplier, and setting the gradient with respect to $\mathbf{c}$ equal to zero. This gives

$$\mathbf{c}_{mv} = \xi \mathbf{R}^{-1} \mathbf{p}_1 = \xi \mathbf{c}_{mmse}, \qquad (2.5.20)$$

where

$$\xi = E\,|\mathbf{c}_{mv}^\dagger \mathbf{r}_i|^2 = \frac{1}{\mathbf{p}_1^\dagger \mathbf{R}^{-1} \mathbf{p}_1} \qquad (2.5.21)$$

is the constrained minimum output variance. If the signal vectors $\mathbf{p}_1, \ldots, \mathbf{p}_L$ are orthogonal and $L < N$, then the mean output energy is $\xi = 1 + \sigma_n^2$.

Both stochastic gradient and LS adaptive algorithms can be derived based on the preceding minimum variance approach. Before deriving the stochastic gradient algorithm, we note that the constrained, minimum-variance cost function is the intersection of the quadratic form $|\mathbf{c}^\dagger \mathbf{p}_1|^2$ with the hyperplane defined by (2.5.18). This cost function has a unique global minimum, which can be found by gradient search.

Taking the gradient of the output energy with respect to $\mathbf{w}_i$ gives

$$\nabla_{\mathbf{w}_i}(E\{|d_i|^2\}) = 2\,\mathrm{Re}\{E\{d_i^*\mathbf{r}_i\}\}. \tag{2.5.22}$$

To obtain the stochastic gradient algorithm, we drop the expectation and take the orthogonal projection with respect to $\mathbf{p}_1$, which gives

$$d_i^*(\mathbf{r}_i - z_{\mathrm{mf}}^*(i)\mathbf{p}_1), \tag{2.5.23}$$

where

$$z_{\mathrm{mf}}(i) = \mathbf{p}_1^\dagger \mathbf{r}_i \tag{2.5.24}$$

is the matched-filter output. The orthogonally anchored stochastic gradient algorithm is therefore

$$\mathbf{w}_i = \mathbf{w}_{i-1} - \mu d_i^*[\mathbf{r}_i - z_{\mathrm{mf}}^*(i)\mathbf{p}_1]. \tag{2.5.25}$$

The convergence properties of the algorithm (2.5.25) are analyzed in [31]. The main results, which parallel the results for the LMS algorithm in Section 2.3.1, are summarized as follows.

1. Defining

$$\delta \mathbf{c}_I = \mathbf{c}_i - \mathbf{c}_{\mathrm{mv}} \tag{2.5.26}$$

and assuming that the received vector $\mathbf{r}_i$ is statistically independent from past vectors $\mathbf{r}_m$, $m < i$, for each i, it can be shown that

$$\delta \mathbf{c}_i = (\mathbf{I} - \mu \mathbf{R}_{vr})\delta \mathbf{c}_{i-1}, \tag{2.5.27}$$

where

$$\mathbf{v}_i = (\mathbf{I} - \mathbf{p}_1\mathbf{p}_1^\dagger)\mathbf{r}_i \tag{2.5.28}$$

and

$$\mathbf{R}_{vr} = E\{\mathbf{v}_i\mathbf{r}_i^\dagger\} = (\mathbf{I} - \mathbf{p}_1\mathbf{p}_1^\dagger)\mathbf{R}. \tag{2.5.29}$$

The mean coefficient vector therefore converges exponentially to $\mathbf{c}_{\mathrm{mv}}$ according to N normal modes, associated with the eigenvalues of $\mathbf{R}_{vr}$.

2. An approximate analysis shows that the MSE remains bounded provided that the step size satisfies (2.5.4).

3. The excess MSE, defined as the asymptotic MSE minus the MSE associated with $\mathbf{c}_{\mathrm{mv}}$, can be approximated as

$$\xi_{\mathrm{ex}} = \xi_{\mathrm{min}} \frac{\dfrac{\mu}{2}\,\mathrm{trace}\,(\mathbf{R}_{vr})}{1 - \dfrac{\mu}{2}\,\mathrm{trace}\,(\mathbf{R}_{vr})}, \tag{2.5.30}$$

where

$$\text{trace}\,(\mathbf{R}_{vr}) = \sum_{k=1}^{K} A_k^2(1 - |\rho_{1k}|^2) + (N - 1)\sigma_n^2 \qquad (2.5.31)$$

and $\rho_{1k} = \mathbf{p}_1^\dagger \mathbf{p}_k$.

If the vectors $\mathbf{p}_1, \ldots, \mathbf{p}_L$ in the sum (2.4.5) are orthonormal, then each of these vectors is an eigenvector of $\mathbf{R}_{vr}$. The eigenvalue associated with $\mathbf{p}_1$ is zero, whereas the eigenvalue associated with $\mathbf{p}_k, k > 1$ is $A_k^2 + \sigma_n^2$. The remaining eigenvectors of $\mathbf{R}$ form a basis for the $N - K$ dimensional subspace that is orthogonal to the signal space. Each of these eigenvectors is associated with eigenvalue σ_n^2.

Given orthogonal signal vectors, the eigenvalues of $\mathbf{R}_{vr}$ are nearly the same as those for $\mathbf{R}$. The convergence of the mean coefficient vector should therefore be similar for both the minimum variance and standard LMS algorithms, given the same step size μ. However, the excess MSE given by (2.5.30) is substantially larger than the corresponding MSE for the LMS algorithm (2.5.6). Specifically,

$$\frac{\xi_{ex}^{(mv)}}{\xi_{ex}^{(lms)}} = \frac{\xi_{min}}{\varepsilon_{min}} . \qquad (2.5.32)$$

When the signal vectors are approximately orthogonal, it is easily shown that

$$\frac{\xi_{min}}{\varepsilon_{min}} \approx (1 + \sigma_n^2) \cdot \frac{1 + \sigma_n^2}{\sigma_n^2} , \qquad (2.5.33)$$

which can be quite large. Consequently, the blind algorithm (2.5.25) is quite "noisy," and it is best to switch to a decision-directed algorithm once reliable symbol estimates are available.

An LS minimum-variance, adaptive algorithm is obtained by selecting $\mathbf{c}_i$ to minimize the cost function

$$V(i) = \sum_{n=0}^{i} w^{i-n} |\mathbf{c}_i^\dagger \mathbf{r}_i|^2, \qquad (2.5.34)$$

subject to the constraint $\mathbf{c}_i^\dagger \mathbf{p}_1 = 1$. The solution is given by

$$\mathbf{c}_i = \hat{\xi}_i \hat{\mathbf{R}}_i^{-1} \mathbf{p}_1, \qquad (2.5.35)$$

where $\hat{\mathbf{R}}_i$ is given by (2.5.10), and

$$\hat{\xi}_i = (\mathbf{p}_1^\dagger \hat{\mathbf{R}}_i^{-1} \mathbf{p}_1)^{-1}. \qquad (2.5.36)$$

The LS solution for $\mathbf{c}_i$ therefore has the same form as (2.5.20), where expectations are replaced by time averages.

It is interesting to compare the minimum variance LS solution (2.5.35) with the decision-directed LS solution (2.5.9). The only differences are (i) the scale factor $\hat{\xi}_i$ appears in (2.5.35), and (ii) $\hat{\mathbf{p}}_1$ in (2.5.9) is replaced by $\mathbf{p}_1$ in (2.5.35). If the magnitude of $b_{i,1}$ is constant for each i, which corresponds to phase modulation, then the scale factor $\hat{\xi}_i$ in (2.5.35) is irrelevant. That is, omitting $\hat{\xi}_i$ does not affect the error rate. In that case, replacing $\hat{\mathbf{p}}_1$ by $\mathbf{p}_1$ is the only real difference between the minimum variance and decision-directed LS algorithms. As a first-order approximation, the minimum variance LS algorithm performs the same as the decision-directed LS algorithm.

Both block and RLS versions of the minimum variance LS algorithm are possible, depending on how often the matrix $\hat{\mathbf{R}}_i^{-1}$ is updated. The matrix inversion lemma can again be applied to the RLS version to reduce the amount of computation.

A potential problem with the minimum variance approach is that the vector $\mathbf{p}_1$ that appears in the adaptive algorithm may not be exactly equal to the received vector contributed by user 1. This problem may be due to unknown multipath or other types of distortion. This type of receiver *mismatch* can cause a substantial degradation in performance due to suppression of the desired signal.

To illustrate the mismatch problem, suppose that the actual received vector contributed by user 1 is $\mathbf{p}_1$ but that the receiver uses the mismatched estimate $\tilde{\mathbf{p}}_1$. These two vectors are shown in Figure 2.11. According to Figure 2.11, it is possible to choose a vector $\mathbf{w}$, which is orthogonal to $\tilde{\mathbf{p}}_1$ and such that $\tilde{\mathbf{p}}_1 + \mathbf{w}$ is orthogonal to $\mathbf{p}_1$. Consequently, with mismatch, the minimum variance approach attempts to suppress the desired signal down to the level of the interference.

Figure 2.11 indicates that the closer $\tilde{\mathbf{p}}_1$ is to $\mathbf{p}_1$, the longer $\mathbf{w}$ must be to suppress the desired signal. Consequently, one way to mitigate the effect of mismatch is to constrain the length of the vector $\mathbf{w}$. Referring to Figure 2.11, let θ_k denote the angle between $\mathbf{p}_k$ and $\mathbf{p}_1$, and let $\tilde{\theta}$ denote the angle between $\tilde{\mathbf{p}}_1$ and $\mathbf{p}_1$. Typically, we expect that $\theta_k \gg \tilde{\theta}$. As shown in Figure 2.11, $\|\mathbf{w}\|$ corresponding to the $\mathbf{w}$ needed to suppress $\mathbf{p}_k$ is much less than $\|\mathbf{w}\|$ corresponding to the $\mathbf{w}$ needed to suppress the desired signal. We can therefore mitigate the effect of mismatch by incorporating the constraint

$$\|\mathbf{w}_i\|^2 < \chi, \tag{2.5.37}$$

where χ is a constant, into the adaptive algorithm. From the preceding discussion, a reasonable choice for χ is the length of $\mathbf{w}$ needed to suppress the user k corresponding to the smallest value of θ_k. Further discussion and numerical results illustrating how the choice of χ affects the performance of the minimum variance approach are given in [31].

The constraint (2.5.37) is easily incorporated into the minimum variance approach and results in a vector $\mathbf{c}_{\text{mv}}$ that again has the form (2.5.20). The only

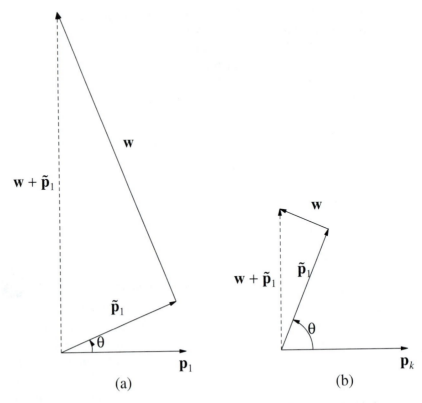

Figure 2.11 Illustration of desired signal suppression with mismatch. (a) shows the vector **w** needed to suppress the desired signal, and (b) shows the vector **w** needed to suppress an interfering signal $\mathbf{p}_k$.

difference is that the noise variance σ_n^2, which appears in the definition of **R** (2.4.21), is replaced by $\sigma_n^2 + v$, where v is a Lagrange multiplier selected to satisfy (2.5.37). The constraint (2.5.37) therefore has the same effect on $\mathbf{c}_{mv}$ as increasing the background noise variance. Similarly, incorporating the constraint (2.5.37) into the LS optimization results in the solution (2.5.35), where

$$\hat{\mathbf{R}}_i = \sum_{n=0}^{i} w^{i-n} \mathbf{r}_i \mathbf{r}_i^{\dagger} + v\mathbf{I}. \tag{2.5.38}$$

Finally, incorporating this constraint into the stochastic gradient algorithm (2.5.25) results in the algorithm

$$\mathbf{w}_i = (1 - \mu v)\mathbf{w}_{i-1} - \mu d_i^* [\mathbf{r}_i - z_{mf}^*(i)\mathbf{p}_1], \tag{2.5.39}$$

which is analogous to the tap-leakage algorithm introduced in [21].

Another possible solution to the mismatch problem, presented in [28], is to combine the orthogonally anchored approach with the decision-directed cost function $E\{|d_i - \text{sgn}(d_i)|^2\}$ (assuming $b_{i,1} \in \{\pm 1\}$). This cost function eliminates the problem of desired signal suppression; however, it can introduce local optima. The stochastic gradient algorithm based on this approach has been observed to perform somewhat better than the analogous minimum variance algorithm in the presence of mismatch.

2.5.4 Projection-Based Approaches

The discussion in preceding sections indicates that the performance (convergence speed) of the adaptive algorithms discussed degrades as the number of filter coefficients increases. Furthermore, increasing the number of filter coefficients generally increases the complexity of the adaptive algorithms. In some situations, it may be desirable to have a TDL c with high dimensionality. For example, c may include TDLs on multiple antennas or may span many symbols. Also, some military applications require a very large processing gain N for covertness. In these situations, it is desirable to reduce the number of *adaptive* coefficients.

One way to reduce the number of adaptive coefficients is to project the received vectors onto a lower-dimensional subspace. Specifically, let $\mathbf{S}_D$ be the $N \times D$ matrix with columns that are the basis vectors for a D-dimensional subspace, where $D < N$. We wish to restrict c to lie in this subspace, so we can write

$$\mathbf{c} = \mathbf{S}_D \boldsymbol{\alpha} \tag{2.5.40}$$

where $\boldsymbol{\alpha}$ is a $D \times 1$ vector of coefficients that must be estimated. Given $\mathbf{S}_D$, it is straightforward to derive stochastic gradient and LS algorithms for estimating $\boldsymbol{\alpha}$. (Note that $\mathbf{c}^\dagger \mathbf{r}_i = \boldsymbol{\alpha}^\dagger \tilde{\mathbf{r}}_i$, where $\tilde{\mathbf{r}}_i = \mathbf{S}_D^\dagger \mathbf{r}_i$ is the projected received vector.)

A few different suggestions for the lower dimensional subspace represented by $\mathbf{S}_D$ have been proposed [35, 51, 101, 106, 129]. For example, in [101] the columns of $\mathbf{S}_D$ are taken to be nonoverlapping segments of the desired spreading sequence, where each segment is of length N/D. (The interpretation is that partial despreading is performed before the adaptive filtering.) Specifically,

$$[\mathbf{S}_D]_m^T = [0 \cdots 0 \; \tilde{\mathbf{p}}_1^T(m) \; 0 \cdots 0], \tag{2.5.41}$$

where $1 \le m \le D$,

$$\tilde{\mathbf{p}}_1^T(m) = [\mathbf{p}_{1,\,(m-1)N/D+1}, \cdots, \mathbf{p}_{1,\,mN/D}], \tag{2.5.42}$$

$(m-1)N/D$ zeros precede $\tilde{\mathbf{p}}_1^T$, $(D-m)N/D$ zeros succeed $\tilde{\mathbf{p}}_1^T$, and N/D is assumed to be an integer. Note that $D = N$ corresponds to the MMSE detector previously discussed (N adaptive coefficients), and $D = 1$ corresponds to the matched-filter detector. Choosing D between 1 and N therefore trades off complexity (D adaptive coefficients) with performance (which is between that of the matched filter and that of MMSE detectors).

If the dimension of the signal space **S** is less than the dimension of **c**, then projecting the received vectors onto the signal space reduces the number of adaptive coefficients without sacrificing optimality (cf. [129]). Generally, this reduction in the number of adaptive components will improve convergence and tracking. Signal subspace methods have received considerable attention in the array processing literature (see [41] and the references within). If the dimension of the signal space is known to be L, then an orthogonal basis for the signal space is given by the L eigenvectors of **R** that correspond to the L largest eigenvalues. In practice, a basis for the signal space can be estimated by forming an eigen-decomposition of the matrix $\hat{\mathbf{R}}_i$ given by (2.5.10). The columns of $\mathbf{S}_D$ in (2.5.40) are then the eigenvectors corresponding to the D largest eigenvalues.

The dimension of the signal space is typically unknown a priori, so that D can either be fixed in advance or be selected as a consequence of the threshold rule

$$\lambda_k(\hat{\mathbf{R}}_i) > \Lambda, \quad \longrightarrow \quad \text{include } \mathbf{v}_k$$
$$\lambda_k(\hat{\mathbf{R}}_i) < \Lambda, \quad \longrightarrow \quad \text{discard } \mathbf{v}_k, \tag{2.5.43}$$

where $\mathbf{v}_k$ is the eigenvector associated with λ_k, and Λ is a constant. More-sophisticated, alternative dimension estimation techniques can also be used [129].

Interference suppression based on a subspace decomposition is discussed in several works, including [22, 29, 35, 109, 129]. Timing estimation for DS-CDMA based on an analogous type of subspace decomposition is presented in [8, 106, 129]. From the viewpoint of adaptivity, the eigen-decomposition needed to estimate the signal space nominally defeats any reduction in complexity achieved by reducing the number of adaptive coefficients. However, very recent work in [129] has shown that low-complexity *subspace-tracking* algorithm of $\mathcal{O}(KN)$ complexity per update can be used to provide subspace-based adaptivity with practical levels of complexity.

2.5.5 Numerical Examples

In this section, we present simulation results that illustrate the performance of some of the adaptive interference-suppression algorithms discussed in the preceding subsections. We first present some convergence results assuming a stationary environment and synchronous users. Figure 2.12 shows averaged SINR as a function of time for each of the following algorithms: (i) decision-directed LMS, (ii) decision-directed

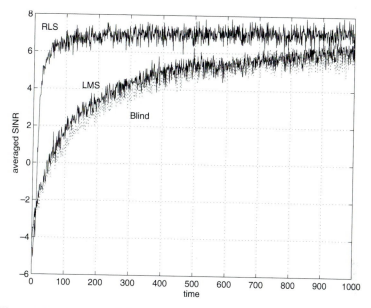

Figure 2.12   Averaged SINR vs. time for the decision-directed LMS, decision-directed RLS, and orthogonally anchored (blind) stochastic gradient algorithms.

RLS, and (iii) orthogonally anchored (blind) stochastic gradient. (The performance of the orthogonally anchored RLS algorithm is nearly the same as the decision-directed algorithm.) The curve for the blind algorithm assumes perfect knowledge of the received pulse shape from the desired user (no mismatch). The convergence curves are obtained by averaging 400 simulation runs, assuming that the spreading codes assigned to all users are fixed. The received amplitudes corresponding to the interferers are twice that of the desired signal. The processing gain is $N = 10$, there are 7 users, and the SNR is 12 dB. In each case, the filter is initialized as the matched filter.

This example shows that the LS algorithm converges much faster than do the stochastic gradient algorithms. Specifically, the LS algorithm converges in approximately 50 iterations, whereas the stochastic gradient algorithms require approximately 700 iterations to converge within 1 dB of the steady-state SINR. The step sizes for both the blind stochastic gradient and LMS algorithms are the same, so that the steady-state SINR for the blind algorithm is somewhat lower than the steady-state SINR for the decision-directed LMS algorithm. This difference in steady-state SINR becomes more significant as the power of the interferers increases [31]. It is therefore desirable to switch to decision-directed mode whenever decisions are reliable.

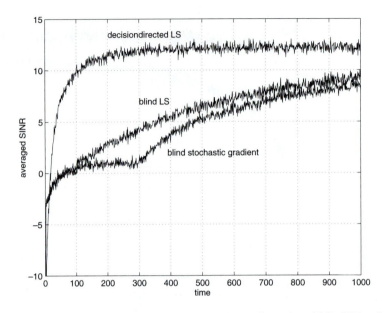

Figure 2.13 Averaged SINR vs. time for the orthogonally anchored (blind) LS and stochastic gradient algorithms with a mismatched anchor. The blind LS and stochastic gradient algorithms switch to decision-directed mode at times 100 and 300, respectively. Also shown is SINR vs. time for the decision-directed RLS algorithm.

Figure 2.13 shows convergence plots for the blind stochastic gradient and LS algorithms in the presence of mismatch. The mismatch was created by adding a single multipath component offset by one chip and attenuated by 3 dB relative to the main component. The blind algorithms are anchored to the strongest path. In each case, the algorithm switches to decision-directed mode after a fixed number of iterations (100 for the LS algorithm and 300 for the stochastic gradient algorithm). The blind algorithms are able to improve the SINR initially but then subsequently suppress the signal. (Figure 2.13 does not show this since the algorithms switch to decision-directed mode before the SINR starts to decrease.) The relatively slow convergence of the LS algorithm is due to the large term added to the diagonal of the matrix $\hat{\mathbf{R}}_i$, given by (2.5.38), which constrains the length of the adaptive filter vector. (Referring to (2.5.38), for this example, $v = 50$.) Note that the additional multipath component in this example improves the asymptotic SINR relative to that shown in Figure 2.12.

Figure 2.14 shows the how the arrival of a new strong interferer affects the performance of the decision-directed RLS algorithm. The scenario used to generate Figure 2.14 is the same as that used to generate Figure 2.12, except that there are only six users initially, and the new (seventh) user appears at iteration 200. The power of the new user is 18 dB above the desired user, representing a severe near-far situation.

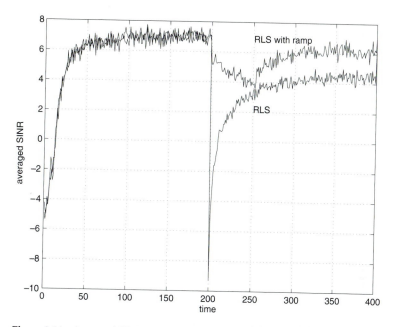

Figure 2.14 Averaged SINR vs. time for the RLS algorithm. A new user appears at time 200. The curve marked "ramp" corresponds to the situation where the new user increases the transmitted power in equal increments (in dB) for a period of 50 iterations.

This example shows that the performance (average SINR) of the LS algorithm is temporarily degraded by the appearance of the new user. This degradation in performance is largely due to the sudden transient created by allowing the new user to begin transmitting with full power. Also shown in Figure 2.14 is the performance curve corresponding to the situation where the new user gradually increases the transmitted power (linearly in dB) to the power limit within 50 iterations. Although this technique mitigates the transient performance degradation, it does add overhead in the form of additional training. (In a packet data system, this additional overhead must be included in each packet.) Adaptive techniques that detect the appearance of a new user are discussed in [34] and [66]. In principle, this additional information can be used to mitigate the degradation in performance caused by the associated transient without additional overhead.

Finally, Figure 2.15 shows a comparison of the subspace tracking algorithm of [129] with RLS. This algorithm makes use of the projection approximation subspace tracking-dilation (PASTd) of [134]. Note that this simulation illustrates the potential gain in SINR that can be obtained by reducing the number of dimensions to be adapted from N to K. (In this example, we have increased the dimension of the received signal to $N = 31$, while keeping the number of users small to illustrate the dimension-reduction advantages of subspace methods.)

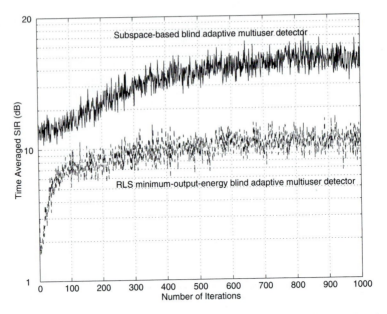

Figure 2.15 Comparison of SINR vs. time for the subspace-tracking and RLS algorithms ($N = 31$ and $K = 6$). Here, there are four MAIs 10 dB above the intended user, and one MAI 20 dB above the intended user. The post-despreading SNR of the intended user is 20 dB.

2.6 FURTHER ISSUES AND REFINEMENTS

In the preceding three sections, we have considered basic elements of adaptive linear multiuser detection. The actual application of these methods requires consideration of a number of further issues, on which we touch in this section. In particular, we discuss briefly some salient features of the mobile wireless communications environment and some additional issues arising with adaptive interference suppression and multiuser detection in this context. Many of these issues apply to multiuser detection in general, although the focus of our discussion is on adaptive linear interference suppression.

2.6.1 The Mobile Wireless Environment

In the preceding sections, we have treated the adaptive multiuser detection problem primarily for the situation in which the parameters of the environment are essentially stable. (An exception is the discussion concerning Figure 2.14, in which the user population changes with time.) One of the primary challenges to multiuser detection offered by the mobile wireless environment is that essentially all user parameters, such as received pulse shapes, amplitudes, relative carrier phase, timing, and whether or not a particular user is active, are time-varying. Consequently, these parameters must be estimated, either directly or indirectly, and the detector must be robust with respect to inaccurate estimates. Also, the ideal model (2.1.1)–(2.1.2) of a multiple-access data signal observed in white Gaussian noise is not necessarily accurate for many situations. Thus, for the practical use of adaptive interference suppression methods, many aspects of channel behavior beyond those described in the preceding sections must be considered. In this section, we briefly describe some of these impairments present in mobile wireless channels. The purpose of this discussion is just to present the models of channel impairments that are commonly used in the literature. More detailed treatments of propagation along with justification for the channel models can be found elsewhere, for example, in [40] or [94].

2.6.1.1 Distance-Related Attenuation and Shadowing

For terrestrial wireless communications, the received signal strength associated with a particular user in general depends on (1) the transmitted power, (2) the distance between the transmitter and receiver, (3) the presence of large objects, such as buildings, foliage, or vehicles, that lie between the transmitter and receiver line-of-sight, and (4) the relative amplitudes and phases of received paths associated with scattering off surrounding objects. The dynamic variation in signal strength due to the motion of the transmitter relative to the receiver (or vice versa) is called *fading*, or, as in Chapter 1, time-selective fading. The fade *rate* is the rate at which the signal experiences fades, depending on the speed of the mobile.

The second and third items listed above are considered *large-scale* effects, whereas item four is a *small-scale* effect [94]. Large-scale effects determine the mean signal strength averaged over a region spanning a few wavelengths in each direction. These cause relatively slow variations in the (mean) signal strength as a mobile moves through space. Small-scale effects cause large swings in signal strength over just a fraction of a wavelength and are superimposed on top of the large-scale effects. As a transmitter moves relative to the receiver, the *mean* received signal strength therefore varies relatively slowly, but the actual signal strength may experience large rapid variations (i.e., 20 to 40 dB) around the mean.

Given an isolated transmitter and receiver in free space separated by distance d, the received signal power is inversely proportional to d^2. Although the presence of buildings and other objects greatly complicates the modes of radio wave propagation, both analysis and measurements indicate that the loss in signal strength, or *path loss*, is proportional to d^n, where n is an integer. The value of n is typically chosen between 2 and 5, depending on the environment considered. In general, the denser the urban environment, the greater the path loss exponent. The value $n = 4$ is typically assumed for modeling urban cellular systems.

In addition to the deterministic path loss due to distance, there is a random component due to the location-dependent, spatial distribution of objects relative to the mobile. That is, the path loss experienced by the mobile depends on both the separation from the transmitter and on the particular placement of surrounding objects that may prevent line-of-sight communications. This latter effect is called *shadowing*. Measurements have shown that the random variations in path loss around the distance-dependent mean can be modeled as a log-normal random variable. That is, the received strength, measured in dB, has a Gaussian distribution with mean specified by the distance-dependent path loss, and standard deviation σ also given in dB. A typical value of σ for urban cellular environments is 8 dB.

Based on the preceding discussion, we can write the received power, taking into account distance-based attenuation and shadow fading, as

$$\overline{P}(d) = \overline{P}_0 \xi \left(\frac{d_0}{d} \right)^n, \tag{2.6.1}$$

where P_0 is the benchmark received power at distance d_0, and ξ is a log-normal Gaussian variable with probability density

$$p_\xi(x) = 10^{-x/10}. \tag{2.6.2}$$

2.6.1.2 Multipath

As discussed previously, multipath is caused by scattering and/or reflections of the transmitted signal off surrounding objects. Given a complex baseband transmitted signal $s(t)$, the effect of multipath is to produce the sum of many delayed and weighted versions of the transmitted signal. Specifically, the received signal (in the absence of noise) is given by

$$y(t) = \sum_{m=1}^{M} a_m s(t - v_m), \tag{2.6.3}$$

where each term in the sum corresponds to a different path, M is the total number of paths, and a_m and v_m are the the path weight and delay associated with path m. Given a complex baseband transmitted signal $s(t)$, the path weights a_m are also complex in general. (Note that the multiuser multipath signal of (2.4.42) consists of the superposition of K such signals, in which for the kth signal we have $M = M_k$, $a_m = \alpha_{k,m} A_k$, and $v_m = \tau_{k,m}$.)

If the delays v_m in (2.6.3) are sufficiently large, then the paths represented by the terms in the sum in (2.6.3) are said to be *resolvable*. That is, the receiver is able to distinguish the different paths and possibly combine them. For two paths to be resolvable, the relative time delay between them, v_m, must be greater than $1/W$, where W is the signal bandwidth. In urban environments where significant scattering occurs, each path in (2.6.3) generally represents the sum of many "micro-paths," which arrive within the resolution time. The relative phases of these scattered paths cause the path weight a_m to fluctuate randomly. If there is no line-of-sight path, as is often the case in an urban environment, then the real and imaginary parts of a_m are typically modeled as Gaussian random variables. In this case, the magnitude of a_m has the Rayleigh probability density

$$
p_R(r) = \begin{cases} \dfrac{r}{\sigma_R^2} e^{-r^2/(2\sigma_R^2)} & r \geq 0 \\ 0 & r < 0 \end{cases} \;,
$$
(2.6.4)

where σ_R^2 determines the mean and variance. The phase of a_m is uniformly distributed. Adding a line-of-sight component to the received path results in an envelope that has a Ricean distribution.

From (2.6.3), the transfer function of a single-user multipath channel can be written as

$$
H(f) = \sum_{m=1}^{M} a_m e^{-2\pi f v_m}.
$$
(2.6.5)

If there is only a single resolvable path, then the magnitude of $H(f)$ is the magnitude of a_1, which is independent of frequency. This type of channel is called a "flat fading" channel because the fading occurs uniformly across the entire signal bandwidth. If there is more than one resolvable path, then the magnitude of $H(f)$ depends on f, so that, as discussed in Chapter 1, this type of channel is called a "frequency selective" fading channel.

The average received power corresponding to each path specifies the *multipath power delay profile* of the multipath channel. This power delay profile can vary considerably, depending on the mobile environment (e.g., rural, urban, hilly). An assumption that is sometimes made for transmission of DS-CDMA signals is that the power delay profile is continuous and decays exponentially. If the received signal is sampled at the chip rate, then the channel is modeled by (2.6.5), where the delays are integer multiples of the chip duration.

It remains to describe how the multipath channel varies with time. The resolvable paths, associated with the delays v_m, tend to change slowly in comparison with the coefficients a_m. Consequently, it is reasonable to assume that the paths are fixed but that the coefficients a_m are time-varying. The time variation in the coefficient a_m is due to Doppler shift, which causes the phases associated with all of the unresolvable paths contributing to path i to vary with time. If the mobile is receiving a carrier with wavelength λ and is traveling with velocity v at an angle θ relative to the transmitter, the change in frequency, or Doppler shift, is $f_d = (v/\lambda) \cos \theta$.

The effect of the Doppler shift is to cause each coefficient a_m to rotate. This model assumes, however, that the mobile receives a single path coming from a specific direction. If the mobile receives many such paths, arriving at different angles, then the time variations of the multipath coefficients becomes more complicated. In that case, $a_m(t)$ is modeled as a random process. For urban mobile cellular systems, uniform scattering is often assumed, which means that the received spatial power density is a constant function of angle. In that case, it can be shown that the power spectral density associated with $a_m(t)$ is given by

$$S_D(f) = \frac{K}{\sqrt{f_d^2 - f^2}} \qquad (2.6.6)$$

for $|f| < f_d$, where K is a constant that determines the power of the random process, and $f_d = v/\lambda$ is the maximum Doppler shift.

For the case of uniform scattering, the coefficients $a_m(t)$ can therefore be modeled as the output of a filter with frequency response $\sqrt{S_D(f)}$ in response to a complex white Gaussian noise input [94]. An alternative, known as Jake's model [40], is to generate $a_m(t)$ by summing complex sinusoids at different frequencies between $-f_d$ and f_d, each weighted by the associated value of $\sqrt{S_D}$. An example of a Rayleigh fading process, which was generated according to the first method, is shown in Figure 2.16.

Consider a wireless multiple access channel in which each user is subject to flat fading. In principle, the MMSE solution for the time-varying linear filter (for a desired user) must take into account the channels associated with all users. However, if the number of users is much less than the processing gain and the background SNR is very high, then the MMSE solution can be approximated by the zero-forcing solution, which does not depend on the channel coefficients. That is, the space spanned by the interferers does not depend on the complex channel coefficients. The adaptive algorithm is therefore relieved from the task of tracking the channels associated with the interferers. Furthermore, the adaptive algorithm does not need to track the flat fading channel associated with the desired user when either differential detection or a pilot signal is used. Consequently, we conclude that for flat fading channels, when the number of users is small relative to the processing gain, the performance of adaptive algorithms should be insensitive to the fade rate (provided that the desired user's channel can be tracked). Simulation results that support this observation are presented in [33].

Now, consider the case where each user experiences frequency selective fading. Recall that the received signal vector contributed by user k after chip matched filtering and sampling is given by (2.4.45) and depends on the time-varying complex coefficients associated with each path. If each path fades independently, then the interference space is *time-varying*, so that the adaptive algorithm attempts to

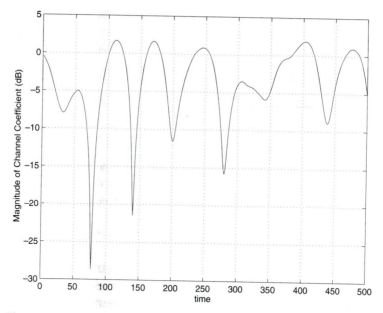

Figure 2.16   Sample path of the magnitude of a Rayleigh fading process.

track the time-varying multipath coefficients associated with *all* users. If the fade rate is sufficiently fast, then the adaptive algorithm is unable to track the combined set of paths for each user and attempts to suppress each path individually. This action, however, degrades the performance of the adaptive filter since it effectively treats each path as a separate interferer. In the worst case, the multipath contributed by the interferers exceeds the number of dimensions (e.g., the processing gain) that the filter has available to suppress interferers, and the performance becomes equivalent to the matched filter. Tracking is therefore a critical issue for frequency-selective fast fading channels.

Results showing the performance of adaptive interference suppression algorithms in the context of DS-CDMA with Rayleigh fading channels are presented in [33, 60, 85, 126, 127]. In [60], a phase predictor is combine with differential coding and detection, whereas in [85], [126], and [127], phase prediction is combined with coherent detection by using a training sequence that must be transmitted periodically. A differential LS algorithm that does not rely on phase prediction is described in [33], which also shows performance results for a cellular type of model with flat Rayleigh fading channels.

2.6.1.3 Delay

Another time variation associated with mobile wireless channels is caused by propagation delay. As the mobiles move, the arrival times of the transmitted signals change, which changes the cross-correlations between received signals. However, this change in delay occurs very slowly relative to the chip duration for chip rates and mobile speeds of interest. (For example, assuming a chip rate of 10^7 chips/s and that the mobile is approaching the base station at a speed of 65 mph, the propagation delay changes by less than one chip/s.)

2.6.1.4 Power Control

Power control is a technique used in currently implemented DS-CDMA mobile telephony systems to alleviate the near-far problem. The basic idea of power control is to provide feedback to mobile transmitters to control their transmitted power levels to yield equal power at the receiver from all mobile transmitters. Since interference suppression techniques can potentially alleviate the near-far problem in DS-CDMA, their use can loosen the requirements on power control. However, as developed in detail in Chapter 5, power control can still benefit performance, with or without interference suppression, and can also reduce the power dissipated by mobile handsets, thereby extending battery life. For the matched-filter detector, the objective of power control for the reverse link is to ensure that all users detected at the base station are received with equal power. For mobile cellular systems, effective power control requires a feedback channel through which the receiver informs the transmitter to raise or lower the transmitted power in small increments (e.g., 1 dB). The effectiveness of the power control depends, of course, on how frequently the power updates are transmitted over the feedback channel and the probability of power control errors (i.e., a "raise" command is received as a "lower" command, or vice versa).

For mobile cellular, it is generally assumed that a practical power control algorithm can respond quickly enough to compensate for shadowing and distance-related attenuation but that it cannot compensate for fast Rayleigh fading due to multipath. Consequently, the received power of a signal that experiences flat Rayleigh fading will experience large, short-term variations in power, but the power averaged over these short-term fades can be set at some target value. It is observed in [123] that for the type of closed-loop adaptive power control previously described, the distribution of the average received power due to variations caused by updates and power control errors is log-normal. The variance (in dB) of the received power reflects how "tight" the power control is. For the matched filter receiver, very tight power control is required for adequate performance, which means that the standard deviation of the received signal power must be approximately 1 to 1.5 dB.

To optimize performance, the power control algorithm should ensure that the error rate for each user is at the maximum acceptable value. For the matched filter receiver, this scheme is the same as equalizing received powers; however, this equivalence is no longer true for receivers with interference suppression. Power control issues and effects are discussed in significantly more detail in Chapter 5.

2.6.1.5 Time-Varying User Population

In addition to time-varying channels, in a mobile cellular environment the set of interferers also varies with time. The interference suppression algorithm must be able to compensate for the appearance of new users, as well as for the disappearance of existing users. "Users" may be associated with calls, in the case of circuit-switched traffic, or with individual packets, in the case of packet-switched traffic. Note that in the latter case, packets may arrive and depart frequently, causing frequent transients in the interference environment that an adaptive algorithm must track. Even in the case of circuit-switched voice traffic, an adaptive algorithm must adapt to the set of users *currently speaking* in order to obtain the potential gains in capacity due to voice inactivity. (In practice, when a user is silent, the power of the transmitted signal is not set to zero but is significantly reduced so that synchronization and channel tracking can be maintained.)

The rate at which users arrive and depart determines the average *traffic load*, measured in *Erlangs per cell* [124]. Specifically, assuming Poisson arrivals at rate λ per cell (assumed to be the same for all cells) and an average service rate (per call or packet) given by μ, the average number of users in the system is $C(\lambda/\mu)$, where C is the number of cells and λ/μ is measured in Erlangs per cell. It is typically the case that in a DS-CDMA system, the average number of users present in the system greatly exceeds the processing gain. This implies that the zero-forcing solutions previously discussed do not exist. However, if the number of *strong* interferers is significantly less than the processing gain, then the linear MMSE detector can effectively suppress these interferers, while treating weak users (e.g., in other cells) as background noise.

We saw from Figure 2.14 that the appearance of a new strong interferer can cause a transient performance degradation in adaptive algorithms. As the traffic load increases, these transients become more frequent. Since, in packet data systems, the appearance of a new user does not necessarily refer to a new call but rather to a new data packet, rapid convergence in response to the appearance of new users is therefore a requirement for adaptive interference suppression in a packet data cellular system. Simulation results for a cellular type of model with stochastic arrivals and departures indicate that an adaptive interference suppression filter using the stochastic gradient algorithm is inadequate for this application, even under moderate traffic loads [29]. Thus, more rapid adaptation techniques are needed for this application.

2.6.1.6 Narrowband Interference

The fact that DS-CDMA systems spread transmitter power over a wide bandwidth allows the possibility that such systems can be overlaid on existing narrowband communication services without undue degradation of either the narrowband or the spread-spectrum service. (The same property allows antijamming capability in military spread-spectrum systems.) Although spread-spectrum communications are inherently resistant to the narrowband interference (NBI) caused by such coexistence with conventional communications, it has been demonstrated that the performance of spread-spectrum systems in the presence of narrowband signals can be enhanced significantly through the use of active NBI suppression prior to despreading. In particular, not only does active suppression improve error-rate performance [9], but it also can lead to increased CDMA cellular system capacity [75] and improved acquisition capability [62].

Over the past two decades, a significant body of research has been concerned with the development of techniques for active NBI suppression in spread-spectrum systems. All of these techniques essentially seek to form a replica of the narrowband signals that can be subtracted from the received signal before data demodulation takes place. The formation of the replica may use predictors or interpolators to explicitly exploit the narrowband nature of the NBI against the wideband nature of the DS-CDMA signal (cf. [61, 98, 122, 131]), or it may use more detailed structural information in the case of digital NBI (which also arises in multirate CDMA systems) [83, 99]. In the latter case, a form of linear multiuser detection is essentially being used. Surveys of advances in this area are found in [61] and [80]. More recently, the adaptive MMSE detection techniques described in Sections 2.3 through 2.5 have been shown to work quite well against combined MAI and NBI of all types [87, 88].

2.6.1.7 Non-Gaussian Ambient Noise

Much of the development and analysis of interference suppression techniques for wireless systems has focused on situations in which the ambient noise is Gaussian. As noted in Section 2.2, this model has allowed the research in this area to focus on the main interference sources, namely structured interference (MAI and NBI). However, for many of the physical channels arising in wireless applications, the ambient noise is known through experimental measurements to be decidedly non-Gaussian. This is particularly true of urban and indoor radio channels [54, 55, 57] and underwater acoustic modem channels [11, 12, 56]. For these channels, the ambient noise is likely to have an impulsive component that gives rise to larger tail probabilities than is predicted by the Gaussian model. When the structured interference dominates, the lack of realism of the ambient noise model is perhaps not crucial. However, with multiuser detection, the MAI-limited nature of multiple-access channels is mitigated and the nature of the ambient noise is more important.

It is widely known in the single-user context that non-Gaussian noise can be quite detrimental to the performance of conventional systems based on the Gaussian assumption. On the other hand, the performance of signaling through non-Gaussian channels can be much better than that for corresponding Gaussian channels if the non-Gaussian nature of the channel is appropriately modeled and ameliorated. (The latter typically involves the use of nonlinear signal processing.) Neither of these properties is surprising. The first is a result of the lack of robustness of linear and quadratic type signal processing procedures to many types of non-Gaussian statistical behavior [42]. The second is a manifestation of the well-known least-favorability of Gaussian channels.

In view of the lack of realism of an AWGN model for ambient noise arising in many practical channels in which multiuser detection techniques can be applied, natural questions arise concerning the applicability, optimization, and performance of multiuser detection in non-Gaussian channels. Although performance indices such as MSE and SINR for linear detectors are not affected by the distribution of the noise (only the spectrum matters), the more crucial bit-error rate can depend heavily on the shape of the noise distribution. The results of an early study of error rates in non-Gaussian DS-CDMA channels are found in [1, 2, 3], in which the performance of conventional and modified conventional detectors is shown to depend significantly on the shape of the ambient noise distribution. In particular, impulsive noise can seriously degrade the error probability for a given level of ambient noise variance in systems designed for Gaussian noise. In the context of multiple-access capability, this implies that fewer users can be supported with conventional detection in an impulsive channel than in a Gaussian channel. However, since non-Gaussian noise can, in fact, be beneficial to system performance relative to Gaussian noise if properly treated, the problem of joint mitigation of structured interference and non-Gaussian ambient noise is of interest [79]. An approach to this problem for NBI in spread-spectrum systems is described in [20]. Some very recent results along these lines for the case of MAI are reported in [81] and [128], the latter of which describes nonlinear adaptive methods that generalize the MMSE approach described in Sections 2.3 through 2.5.

2.6.2 System Issues

In addition to algorithmic issues such as performance and complexity, it is important to determine how adaptive interference suppression will affect other communication system requirements. These system issues are currently not well understood, so the following discussion is necessarily very brief (relative to importance).

2.6.2.1 Coding

As discussed in Chapter 1, coding and interleaving are typically used to achieve reliable communications over fading channels. For example, in the commercial IS-95 standard DS-CDMA air interface, a rate one-third binary convolutional code is used on the uplink and simultaneously serves to spread the bandwidth. Consequently, to achieve the same degree of spreading with coding, as compared to without coding, the length of the pseudorandom (PN) sequence associated with each bit must be reduced by a factor of three. When used with linear interference suppression, this reduction in PN-sequence length reduces the degrees of freedom available to suppress interference. (The trade-off is that the coding may be able to compensate for the residual interference.) In other words, a low-rate code robs "dimensions" from the interference suppression filter. This trade-off does not apply to interference *cancellation* techniques, in which the interference is regenerated and substracted from the received signal [73]. (See also Section 2.2.)

The preceding observation implies that it is best to use high-rate codes with linear interference suppression, rather than low-rate codes. This has also been noted in [97] and [69]. However, existing high-rate codes, such as Ungerboeck codes [110], rely on relatively dense constellations (such as 8-PSK), which may pose additional problems (e.g., with phase tracking) in a fading environment.

Another issue is that if *uncoded* symbols are used for decision-directed adaptation, then a more powerful coding scheme implies a higher decision-error rate in the adaptive algorithm (assuming a fixed target error rate). There is, therefore, a trade-off between the performance gain due to coding and the performance loss due to uncoded decision errors in the adaptive algorithm (assuming that the delay associated with using decoded symbols to direct the adaptation is unacceptable in a fast-fading environment).

2.6.2.2 Power Control

For DS-CDMA systems that use the conventional matched-filter receiver, power control is crucial for mitigating the near-far problem. Also, as noted previously, power control has the advantage of lowering the transmitted power for each user and thereby extends battery life. For the conventional matched-filter receiver, power control is typically used to equalize the received powers [123]. (A further advantage of this use is that it helps avoid saturation of nonlinear receiver elements, such as fixed-point processors.) However, in general, the goal of optimum power control is to adjust the transmitted powers so that the received SINRs corresponding to all users being detected are equal [6]. Although multiuser detection and interference suppression techniques can alleviate power control requirements, power control can enhance the performance of interference suppression techniques. The interaction of adaptive power control with adaptive interference suppression has only begun to receive attention. Chapter 5 develops a promising framework for exploring this interaction.

2.6.2.3 Timing Recovery

The discussion in preceding sections has assumed that the receiver is perfectly synchronized to the desired user. It has been observed that timing offsets can significantly degrade the performance of some multiuser detectors [13, 137]. This observation, however, is implementation dependent. An advantage of the adaptive TDL implementation for linear interference suppression is that it is analogous to an adaptive, fractionally spaced equalizer (for a single-user channel), which is known to be insensitive to timing offset [91]. Furthermore, timing recovery for the multiple-access channel with an adaptive, fractionally spaced TDL can be accomplished analogously to a single-user channel. The combination of timing estimation with adaptive linear interference suppression is studied in [52, 102, 129].

2.6.2.4 Nonuniform Quality of Service

An important difference between linear interference suppression techniques and the matched-filter receiver is that the former relies on the use of short spreading sequences, whereas the latter can use either short or long spreading sequences. A very long spreading sequence, such as used in IS-95, is equivalent to selecting a different random spreading sequence for each bit ("code hopping"). An advantage of code hopping is that each user sees approximately the same performance, assuming perfect power control, since averaging the performance over the sequence of transmitted bits is equivalent to averaging over spreading sequences.

In contrast, the solution for MMSE, given by (2.4.20), implies that the MMSE depends on the particular assignment of spreading codes to users (as well as relative amplitudes and phases). Furthermore, the MMSE will generally be different for different users. Consequently, even though *average* performance (i.e., error probability averaged over all spreading sequences) may be very good, some users may have relatively poor performance because the spreading codes assigned to two users may have relatively high cross-correlation. This observation is made in [117] and is analyzed in [36].

To illustrate the preceding observation, Figure 2.17 shows the distribution (computed via simulation) for SINR assuming that the user signature sequences are selected randomly (i.e., each element of the vectors $\mathbf{p}_k$, $k = 1, \ldots, K$, is determined by a fair coin toss). In this example, the processing gain $N = 30$, there are 10

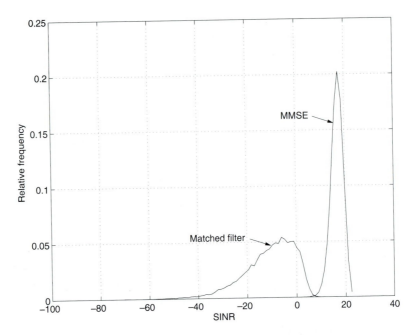

Figure 2.17 Distributions for signal-to-interference ratio assuming random signature sequences. Results for both the linear MMSE detector and the matched filter are shown.

"strong" (i.e., intracell) users, 50 "weak" (i.e., other-cell) users, and the signal-to-background noise ratio is 25 dB. The received powers (for members of each set of strong and weak users) were selected from a log-normal distribution. The SINR distribution for the matched-filter receiver is also shown.

The results in Figure 2.17 indicate that there is a significant spread in performance (> 10 dB) over the user population. Note that if a very long spreading sequence (as in the IS-95 standard) is used, the distribution for the matched-filter becomes a point mass at the average of the distribution shown in Figure 2.17 (-10 dB).

Power control may help to improve the performance of users experiencing poor performance only in some situations. For example, two adjacent users may be assigned "nearby" codes, meaning that their cross-correlation is large. If the adjacent users are transmitting to the same receiver, then power control cannot significantly improve performance for both users. However, if the users are transmitting to different receivers (i.e., if they are in different cells), then it may be possible for one of the users to reduce power, thereby reducing interference to the other user.

Finally, we remark that the significant spread in performance shown in Figure 2.17 assumes a static situation in which the set of users and relative amplitudes and phases are fixed. In a mobile wireless network, these parameters are time-varying, which should alleviate this problem to some degree. In addition, path or space diversity may also reduce the likelihood of having relatively high cross-correlations with neighboring users.

2.6.2.5 Very Long Spreading Sequences

The model that we have proposed in this chapter has addressed primarily the situation in which the received signaling waveform of each user is the same in each symbol interval (aside from fading and other channel effects). As discussed in the previous section, in some current DS-CDMA systems (such as the IS-95 digital cellular standard), this model is not accurate because the period of the spreading waveform spans many bits. Theoretically, this distinction is not overly significant. However, practically, it is quite significant. Since the key parameter determining performance of DS-CDMA systems is the number of chips per symbol, not the number of chips per period of the spreading code, this use of long spreading codes is primarily of value in providing uniform quality of service over the user population, as discussed in the preceding section, and in avoiding the need to assign (or reassign) codes to each new call (or to an existing call that is handed off to an adjacent cell). In the decision whether or not to use short or long codes in future DS-CDMA standards, these advantages should be weighed against the performance advantages offered by practical multiuser detection.

2.6.2.6 Power Consumption

Since mobility is one of the main motivations for using wireless communications, the practicality of many of the techniques described in this chapter depends heavily on the ability to implement them in portable, battery-operated handsets. Thus, the issue of energy consumption is of considerable importance in the development of interference suppression algorithms for wireless systems. In cellular systems, there is an asymmetry with respect to this issue, in that the base station (i.e., the uplink transceiver) is relatively unconstrained by energy consumption, whereas the mobiles (i.e., the downlink transceivers) are severely constrained. So, the use of sophisticated signal processing, such as multiuser detection, at the base station does not pose a serious energy-consumption problem. Since these techniques allow better performance for a given level of received signal energy than do conventional methods, the use of such methods in the base station can reduce required transmitter power at the mobiles, thereby reducing overall battery requirements for portable transceivers. However, the advantages of adaptive linear multiuser detection, multipath mitigation, etc., can significantly enhance the downlink performance as well. (Also, point-to-point systems do not necessarily feature fixed, nonportable transceivers.) Thus, energy-efficient techniques for the linear adaptive algorithms discussed in this chapter are of considerable interest.

One technique for achieving energy efficiency in the MMSE detector is described in [88]. In particular, energy consumption in integrated circuits is an increasing function of gate speed. Since the algorithms of interest here should be implemented at the signaling rate (i.e., at the symbol rate), the algorithms cannot be slowed down to reduce energy consumption. However, as shown in [88], the blind RLS MMSE detection algorithm can be implemented with a slower gate speed by being mapped to a systolic array. This mapping allows the individual gate speed to be reduced significantly without a corresponding reduction in the speed at which the algorithm updates its coefficients.

2.7 CONCLUDING REMARKS

In this chapter, we have discussed the use of adaptive signal processing techniques to suppress structured interference in wireless systems. We have focused on the suppression of multiple-access interference and have primarily considered techniques that are based on the MMSE method of linear multiuser detection. As we have seen, MMSE detection provides many of the performance advantages of optimal multiuser detection, without its attendant complexity. Moreover, the MMSE detector lends itself to a great variety of adaptive methods, and it is relatively robust to other types of interference (such as narrowband interference).

The results presented in this chapter have largely been of a research nature, consistent with the primary mission of the present volume. However, we have also mentioned, in Section 2.6, a variety of other issues that are of concern in bringing these methods to practice. These practical issues present a wealth of other research questions, many of which are currently being addressed by the community.

REFERENCES

[1] B. Aazhang and H. V. Poor,"Performance of DS/SSMA Communications in Impulsive Channels—Part I: Linear Correlation Receivers," *IEEE Trans. Commun.*, Vol. 35, No. 11, pp. 1179–1188, Nov. 1987.

[2] B. Aazhang and H. V. Poor, "Performance of DS/SSMA Communications in Impulsive Channels—Part II: Hard-limiting Correlation Receivers," *IEEE Trans. Commun.*, Vol. 36, No. 1, pp. 88–97, Jan. 1988.

[3] B. Aazhang and H. V. Poor, "An Analysis of Nonlinear Direct-sequence Correlators," *IEEE Trans. Commun.*, Vol. 37, No. 7, pp. 723–731, July 1989.

[4] M. Abdulrahman and D. D. Falconer, "Cyclostationary Crosstalk Suppression by Decision Feedback Equalization on Digital Subscriber Loops," *IEEE J. Select. Areas Commun.*, Vol. 10, No. 3, pp. 640–649, Apr. 1992.

[5] M. Abdulrahman, A. U. H. Sheikh, and D. D. Falconer, "Decision Feedback Equalization for CDMA in Indoor Wireless Communications," *IEEE J. Select. Areas Commun.*, Vol. 12, No. 4, pp. 698–704, May 1994.

[6] S. Ariyavisitakul, "Signal and Interference Statistics of a CDMA System with Feedback Power Control Part II," *IEEE Trans. Commun.* Vol. 42, No. 2/3/4, Feb./Mar./Apr. 1994.

[7] J. Barry, J. Kahn, E. Lee, and D. Messerschmidt, "High-speed Nondirective Optical Communication for Wireless Networks," *IEEE Network Mag.*, Vol. 29, pp. 44–54, Nov. 1991.

[8] S. E. Bensley and B. Aazhang, "Subspace-Based Estimation of Multipath Channel Parameters for CDMA Communication Systems," *Proc. IEEE GLOBECOM/Comm. Theory Mini-Conf.*, pp. 154–157, San Francisco, CA, Dec. 1994.

[9] N. Bershad, "Error Probabilities of DS Spread-spectrum Systems Using an ALE for Narrow-band Interference Rejection," *IEEE Trans. Commun.*, Vol. COM-36, No. 5, pp. 587–595, May 1988.

[10] D. Brady and J. Catipovic, "Adaptive Multiuser Detection for Underwater Acoustical Channels," *IEEE J. Oceanic Engineering*, Vol. 19, No. 2, pp. 158–165, Apr. 1994.

[11] P. L. Brockett, M. Hinich, and G. R. Wilson, "Nonlinear and Non-Gaussian Ocean Noise," *J. Acoust. Soc. Am.*, Vol. 82, pp. 1286–1399, 1987.

[12] P. L. Brockett, M. Hinich, and G. R. Wilson, "Bispectral Characterization of Ocean Acoustic Time Series: Nonlinearity and Non-Gaussianity," in E. J. Wegman, S. C. Schwartz, and J. B. Thomas (eds.), *Topics in Non-Gaussian Signal Processing* (Springer-Verlag: Berlin, 1989).

[13] R. M. Buehrer, N. S. Correal, and B. D. Woerner, "A Comparison of Multiuser Receivers for Cellular CDMA," *Proc. IEEE GLOBECOM*, Vol. 3, pp. 1571–1577, London, UK, Nov. 1996.

[14] J. Catipovic, D. Brady, and A. S. Etchemendy, "Development of Underwater Acoustic Modems and Networks," *Oceanography Mag.*, Vol. 6, pp. 112–119, 1993.

[15] K.-C. Chen, L. J. Cimini, Jr., B. D. Woerner, and S. Yoshida, eds., *IEEE J. Select. Areas Commun.*, Issue on Wireless Local Communications, Vol. 14, No. 3, Apr. 1996.

[16] A. Duel-Hallen, "A Family of Multiuser Decision-Feedback Detectors for Asynchronous Code-Division Multiple-Access Channels," *IEEE Trans. Commun.*, Vol. 43, No. 2/3/4, pp. 421–434, Feb./Mar./Apr. 1995.

[17] A. Duel-Hallen, "Decorrelating Decision-Feedback Multiuser Detector for Synchronous Code-Division Multiple-Access Channel," *IEEE Trans. Commun.*, Vol. 41, No. 2, pp. 285–290, Feb. 1993.

[18] D. D. Falconer, M. Abdulrahman, N. W. K. Lo, B. R. Petersen, and A. U. H. Sheikh, "Advances in Equalization and Diversity for Portable Wireless Systems," *Digital Signal Processing 3*, pp. 148–162, 1993.

[19] V. K. Garg and J. E. Wilkes, *Wireless and Personal Communications Systems.* (Prentice-Hall: Upper Saddle River, NJ, 1996)

[20] L. M. Garth and H. V. Poor, "Narrowband Interference Suppression in Impulsive Channels," *IEEE Trans. Aerosp. Electron. Syst.*, Vol. 28, No. 1, pp. 15–34, Jan. 1992.

[21] R. D. Gitlin, H. C. Meadows, and S. B. Weinstein, "The Tap-Leakage Algorithm: An Algorithm for the Stable Operation of a Digitally Implemented, Fractionally-Spaced, Adaptive Equalizer," *Bell Syst. Tech. J.*, Vol. 61, pp. 1817–1839, Oct. 1982.

[22] A. M. Haimovich and Y. Bar-Ness, "An Eigenanalysis Interference Canceler," *IEEE Trans. Signal Processing*, Vol. 39, No. 1, pp. 76–84, Jan. 1991.

[23] D. J. Harrison, "Adaptive Equalization for Channels with Crosstalk," M. Eng. thesis, Carleton University, Ottawa, Ontario, Canada, Dec. 1969.

[24] Z. J. Hass, R. Alonso, D. Duchamp, and B. Gopinath, eds., *IEEE J. Select. Areas Commun.*, Issue on Mobile and Wireless Computing Networks, Vol. 13, June 1995.

[25] S. Haykin, *Adaptive Filter Theory*, 3rd edition. (Prentice-Hall: Upper Saddle River, NJ, 1996)

[26] T. Helleseth and P. V. Kumar, "Pseudonoise Sequences." In *Mobile Communications Handbook*, J. Gibson, ed. (CRC Press: Boca Raton, Fla., 1996)

[27] J. Holtzman, "A Simple Accurate Method to Calculate Spread-spectrum Multiple Access Error Probabilities," *IEEE Trans. Commun.*, Vol. 40, pp. 461–464, Mar. 1992.

[28] M. L. Honig, "Orthogonally Anchored Blind Interference Suppression Using the Sato Cost Criterion," *Proc. 1995 IEEE Int. Symp. Inf. Theory*, p. 314, Whistler, British Columbia, Canada, Sept. 1995.

[29] M. L. Honig, "Performance of Adaptive Interference Suppression for DS-CDMA With a Time-Varying User Population," *Proc. IEEE Int. Symp. Spread Spectrum Techn. Appl.*, Mainz, Germany, pp. 267–271, Sept. 1996.

[30] M. L. Honig, P. Crespo, and K. Steiglitz, "Suppression of Near- and Far-end Crosstalk by Linear Pre- and Post-filtering," *IEEE J. Select. Areas Commun.*, Vol. 10, No. 3, pp. 614–629, Apr. 1992.

[31] M. L. Honig, U. Madhow, and S. Verdú, "Adaptive Blind Multi-User Detection," *IEEE Trans. Inform. Theory*, Vol. 41, No. 4, pp. 944–960, July 1995.

[32] M. L. Honig and D. G. Messerschmitt, *Adaptive Filters: Structures, Algorithms, and Applications.* (Kluwer Academic Publishers: Boston, 1985)

[33] M. L. Honig, S. L. Miller, L. B. Milstein, and M. J. Shensa, "Performance of Adaptive Linear Interference Suppression for DS-CDMA in the Presence of Flat Rayleigh Fading," *Proc. IEEE Vehic. Techn. Conf.*, Atlanta, GA, May 1996.

[34] M. L. Honig, "Rapid Detection and Suppression of Interference in DS-CDMA," *IEEE Int. Conf. Acoust., Speech, Signal Processing*, Detroit, May 1995.

[35] M. L. Honig, "A Comparison of Subspace Adaptive Filtering Techniques for DS-CDMA Interference Suppression," *Proc. MILCOM*, Monterey, CA, Nov. 1997.

[36] M. Honig and W. Veerakachen, "Performance Variability of Linear Multiuser Detection for DS-CDMA," *Proc. IEEE Vehic. Techn. Conf.*, Vol. 1, pp. 372–376, Atlanta, GA, May 1996.

[37] D. Horwood and R. Gagliardi, "Signal Design for Digital Multiple Access Communications," *IEEE Trans. Commun.*, Vol. COM-23, pp. 378–383, Mar. 1975.

[38] H. C. Huang, *Combined Multipath Processing, Array Processing, and Multiuser Detection for DS-CDMA Channels*, Ph.D. Thesis, Electrical Engineering Dept., Princeton University, Princeton, NJ, 1995.

[39] R. A. Iltis, "An Adaptive Multiuser Detector with Joint Amplitude and Delay Estimation," *IEEE J. Select. Areas Commun.*, Vol. 12, No. 5, pp. 774–785, June 1994.

[40] W. C. Jakes (Ed.), *Microwave Mobile Communications.* (IEEE Press: New York, 1974)

[41] D. H. Johnson and D. E. Dudgeon, *Array Signal Processing—Concepts and Techniques.* (Prentice-Hall: Englewood Cliffs, NJ, 1993)

[42] S. A. Kassam and H. V. Poor, "Robust Techniques for Signal Processing: A Survey," *Proc. IEEE*, Vol. 73, No. 3, pp. 433–481, Mar. 1985.

[43] A. R. Kaye and D. A. George, "Transmission of Multiplexed PAM Signals over Multiple Channel and Diversity Systems," *IEEE Trans. Commun. Technol.*, Vol. COM-18, No. 5, pp. 520–526, Oct. 1970.

[44] A. Klein, G. K. Kaleh, and P. W. Baier, "Zero Forcing and Minimum-mean-square-error Equalization for Multiuser Detection in Code-division Multiple-access Channels," *IEEE Trans. Vehic. Technol.*, Vol. 45, No. 2, pp. 276–287, May 1996.

[45] R. Kohno, "Pseudo-noise Sequences and Interference Cancellation Techniques for Spread-spectrum Systems—Spread Spectrum Theory and Techniques in Japan," *IEICE Trans.*, Vol. E.74, pp. 1083–1092, May 1991.

[46] E. A. Lee and D. G. Messerschmitt, *Digital Communication.* (Kluwer Academic Publishers: Boston, 1994)

[47] N. W. K. Lo, D. D. Falconer, and A. U. H. Sheikh, "Adaptive Equalizer MSE Performance in the Presence of Multipath Fading, Interference, and Noise," *Proc. IEEE Vehic. Techn. Conf.*, pp. 409–413, Chicago, July 1995.

[48] N. W. K. Lo, D. D. Falconer, and A. U. H. Sheikh, "Adaptive Equalization for Co-Channel Interference in a Multipath Fading Environment," *IEEE Trans. Commun.*, Vol. 43, No. 2/3/4, pp. 1441–1453, Feb./Mar./Apr. 1995.

[49] R. Lupas and S. Verdú, "Linear Multi-user Detectors for Synchronous Code-division Multiple-access Channels," *IEEE Trans. Inform. Theory*, Vol. IT-35, No. 1, pp. 123–136, Jan. 1989.

[50] R. Lupas and S. Verdú, "Near-far Resistance of Multi-user Detectors in Asynchronous Channels," *IEEE Trans. Commun.*, Vol. COM-38, No. 4, pp. 496–508, Apr. 1990.

[51] U. Madhow and M. Honig, "MMSE Interference Suppression for Direct-sequence Spread-spectrum CDMA," *IEEE Trans. Commun.*, Vol. 42, pp. 3178–3188, Dec. 1994.

[52] U. Madhow, "MMSE Interference Suppression for Acquisition and Demodulation of Direct-Sequence CDMA Signals," *IEEE Trans. Commun.*, to appear.

[53] N. Madayam and S. Verdú, "Analysis of an Approximate Decorrelating Detector," *Wireless Personal Communications,* Special Issue on Interference in Mobile Wireless Systems, 1997.

[54] D. Middleton, "Man-made Noise in Urban Environments and Transportation Systems: Models and Measurements," *IEEE Trans. Commun.*, Vol. COM-21, pp. 1232–1241, 1973.

[55] D. Middleton, "Statistical-physical Models of Electromagnetic Interference," *IEEE Trans. Electromag. Compat.*, Vol. EMC-19, 106–127, Aug. 1977.

[56] D. Middleton, "Channel Modeling and Threshold Signal Processing in Underwater Acoustics: An Analytical Overview," *IEEE J. Oceanic Eng.* Vol. OE-12, No. 1, pp. 4–28, Jan. 1987.

[57] D. Middleton and A. D. Spaulding, "Elements of Weak-signal Detection in Non-Gaussian Noise," in *Advances in Statistical Signal Processing—Vol. 2: Signal Detection,* H. V. Poor and J. B. Thomas, eds. (JAI Press: Greenwich, CT, 1993)

[58] S. L. Miller, "An Adaptive Direct-Sequence Code-Division Multiple-Access Receiver for Multiuser Interference Rejection," *IEEE Trans. Commun.*, Vol. 43, No. 2/3/4, pp. 1746–1755, Feb./Mar./Apr. 1995.

[59] S. L. Miller, "Training Analysis of Adaptive Interference Suppression for Direct-Sequence Code-Division Multiple-Access Systems," *IEEE Trans. Commun.*, Vol. 44, No. 4, pp. 488–495, Apr. 1996.

[60] S. L. Miller and A. N. Barbosa, "A Modified MMSE Receiver for Detection of DS-CDMA Signals in Fading Channels," *Proc. MILCOM*, Washington, DC, 1996.

[61] L. B. Milstein, "Interference Rejection Techniques in Spread Spectrum Communications," *Proc. IEEE*, Vol. 76, No. 6, pp. 657–671, June 1988.

[62] L. B. Milstein, "Interference Suppression to Aid Acquisition in Direct-sequence Spread-spectrum Communications," *IEEE Trans. Commun.*, Vol. 36, No. 11, pp. 1200–1202, Nov. 1988.

[63] U. Mitra and H. V. Poor, "Neural Network Techniques for Adaptive Multi-user Demodulation," *IEEE J. Select. Areas Commun.*, Vol. 12, No. 9, pp. 1460–1470, Dec. 1994.

[64] U. Mitra and H. V. Poor, "Adaptive Receiver Algorithms for Near-far Resistant CDMA," *IEEE Trans. Commun.*, Vol. 43, No. 2/3/4, pp. 1713–1724, Feb./Mar./Apr. 1995.

[65] U. Mitra and H. V. Poor, "Activity Detection in a Multi-user Environment," *Wireless Personal Communications,* Vol. 3, pp. 149–174, 1996.

[66] U. Mitra and H. V. Poor, "Adaptive Decorrelating Detectors for CDMA Systems," *Wireless Personal Commun.*, Vol. 2, No. 4, pp. 415–550, 1996.

[67] U. Mitra and H. V. Poor, "An Adaptive Decorrelating Detector for Synchronous CDMA Channels," *IEEE Trans. Commun.*, Vol. 44, No. 2, pp. 257–268, Feb. 1996.

[68] L. B. Nelson and H. V. Poor, "Iterative Multiuser Receivers for CDMA Channels: An EM-based Approach," *IEEE Trans. Commun.*, Vol. 44, No. 12, pp. 1700–1710, Dec. 1996.

[69] I. Oppermann, P. Rapajic, and B. S. Vucetic, "Capacity of a Band-Limited CDMA MMSE Receiver Based System When Combined With Trellis or Convolutional Coding," preprint, Aug. 1996.

[70] K. Pahlavan and A. H. Levesque, *Wireless Information Networks.* (Wiley-Interscience: New York, 1995)

[71] K. Pahlavan, T. Probert and M. Chase, "Trends in Local Wireless Networks," *IEEE Commun. Mag.*, Vol. 33, pp. 88–95, Mar. 1995.

[72] D. Parsavand and M. H. Varanasi, "RMS Bandwidth Constrained Signature Waveforms that Maximize the Total Capacity of PAM-synchronous CDMA Channels," *IEEE Trans. Commun.*, Vol. 44, No. 1, pp. 65–75, Jan. 1996.

[73] P. Patel and J. Holtzman, "Analysis of a Simple Successive Interference Cancellation Scheme in a DS/CDMA System," *IEEE J. Select. Areas Commun.*, Vol. 12, No. 5, pp. 796–807, June 1994.

[74] B. R. Petersen and D. D. Falconer, "Minimum Mean Square Equalization in Cyclostationary and Stationary Inteference—Analysis and Subscriber Line Calculations," *IEEE J. Select. Areas Commun.*, Vol. 9, No. 6, pp. 931–940, Aug. 1991.

[75] R. L. Pickholtz, L. B. Milstein, and D. L. Schilling, "Spread Spectrum for Mobile Communications," *IEEE Trans. Vehic. Technol.*, Vol. 40, No. 2, pp. 313–322, May 1991.

[76] H. V. Poor, *An Introduction to Signal Detection and Estimation,* 2nd ed. (Springer-Verlag: New York, 1994)

[77] H. V. Poor, "Adaptivity in Multiple-access Communications," *Proc. IEEE Conf. Dec. Contr.*, New Orleans, 1995.

[78] H. V. Poor, "Adaptive Interference Suppression in CDMA Systems," *Proc. Symp. Interference Rejection, Signal Separation in Wireless Communications,* New Jersey Institute of Technology, Newark, NJ, 1996.

[79] H. V. Poor, "Non-Gaussian Signal Processing Problems in Multiple-access Communications," in *Proc. USC/CRASP Workshop on Non-Gaussian Signal Processing*, Ft. George Meade, MD, May 1996.

[80] H. V. Poor and L. A. Rusch, "Narrowband Interference Suppression in Spread-spectrum Systems," *IEEE Personal Communications*, Vol. 1, No. 3, pp. 14–27, Aug. 1994.

[81] H. V. Poor and M. Tanda, "An Analysis of Some Multiuser Detectors in Impulsive Noise," *Proc. GRETSI Symp. Signal & Image Processing*, Grenoble, France, Sept. 1997.

[82] H. V. Poor and S. Verdú, "Probability of Error in MMSE Multiuser Detection," *IEEE Trans. Inform. Theory*, Vol. 43, No. 3, pp. 858–871, May 1997.

[83] H. V. Poor and X. Wang, "Adaptive Suppression of Narrowband Digital Interferers from Spread Spectrum Signals," *Proc. Int. Conf. Acoustics, Speech Signal Processing,* Vol. 2, pp. 1061–1064, Atlanta, GA, 1996.

[84] H. V. Poor and X. Wang, "Signal Processing for Adaptive Interference Suppression in CDMA Systems," *Proc. Europ. Workshop DSP Applied to Space Communications,* pp. 11.1–11.10, Barcelona, Spain, Sept. 1996.

[85] H. V. Poor and X. Wang, "Adaptive Multiuser Detection in Fading Channels," *Proc. Allerton Conf. Commun., Contr., Computing*, pp. 603–612, University of Illinois, Urbana, IL, Oct. 1996.

[86] H. V. Poor and X. Wang, "Blind Adaptive Suppression of Multiuser Narrowband Digital Interferers from Spread-spectrum Signals," *Wireless Personal Communications,* Special Issue on Interference in Mobile Wireless Systems, Vol. 6, pp. 69–96, 1998.

[87] H. V. Poor and X. Wang, "Code-aided Interference Suppression in DS/CDMA Spread Spectrum Communications—Part I: Interference Suppression Capability," *IEEE Trans. Commun.*, Vol. 45, No. 9, pp. 1101–1111, Sept. 1997.

[88] H. V. Poor and X. Wang, "Code-aided Interference Suppression in DS/CDMA Spread Spectrum Communications—Part II: Parallel Blind Adaptive Implementation," *IEEE Trans. Commun.*, Vol. 45, No. 9, pp. 1112–1122, Sept. 1997.

[89] H. V. Poor and X. Wang, "Blind Adaptive Joint Suppression of MAI and ISI in CDMA Channels," *Proc. Asilomar Conf. Signals, Systems, Computers,* Pacific Grove, CA, Nov. 1997.

[90] J. G. Proakis, *Digital Communications,* 3rd ed. (McGraw-Hill: Boston, 1995)

[91] S. Qureshi, "Adaptive Equalization," *Proc. IEEE*, Vol. 73, No. 9, pp. 1349–1387, Sept. 1985.

[92] P. B. Rapajic and B. S. Vucetic, "Adaptive Receiver Structures for Asynchronous CDMA Systems," *IEEE J. Select. Areas Commun.*, Vol. 12, No. 4, pp. 685–697, May 1994.

[93] P. Rapajic and B. S. Vucetic, "Linear Adaptive Transmitter-receiver Structures for Asynchronous CDMA Systems," *Europ. Trans. Commun.*, pp. 21–27, Jan./Feb. 1995.

[94] T. S. Rappaport, *Wireless Communications: Principles and Practice.* (Prentice-Hall: Upper Saddle River, NJ, 1996)

[95] M. Rupf, F. Tarkoy, and J. L. Massey, "User-separating Demodulation for Code-division Multiple-access Systems," *IEEE J. Select. Areas Commun.,* Vol. 12, No. 5, pp. 786–795, June 1994.

[96] M. Rupf and J. L. Massey, "Optimum Sequence Multisets for Synchronous Code-division Multiple-access Channels," *IEEE Trans. Inform. Theory,* Vol. IT-40, pp. 1261–1266, July 1994.

[97] M. Rupf, F. Tarköy, and J. L. Massey, "User-Separating Demodulation for Code-Division Multiple-Access Systems," *IEEE J. Select. Areas Commun.,* Vol. 12, No. 5, pp. 786–795, June 1994.

[98] L. A. Rusch and H. V. Poor, "Narrowband Interference Suppression in CDMA Spread-spectrum Communications," *IEEE Trans. Commun.,* Vol. 42, No. 4, pp. 1969–1979, Apr. 1994.

[99] L. A. Rusch and H. V. Poor, "Multiuser Detection Techniques for Narrowband Interference Suppression in Spread-spectrum Communications," *IEEE Trans. Commun.,* Vol. 43, No. 2/3/4, pp. 1725–1737, Feb./Mar./Apr. 1995.

[100] J. Salz, "Digital Transmission over Cross-Coupled Linear Channels," *Bell Syst. Tech. J.,* Vol. 64, No. 6, pp. 1147–59, July/Aug. 1985.

[101] R. Singh and L. B. Milstein, "Adaptive Interference Suppression in Direct-Sequence CDMA," submitted to *IEEE Trans. Commun.,* 1997.

[102] R. F. Smith and S. L. Miller, "Code Timing Estimation in a Near-Far Environment for Direct-Sequence Code-Division Multiple-Access," *Proc. IEEE MILCOM,* 1994.

[103] J. M. Straus, ed., *Proc. MILCOM,* San Diego, CA, Nov. 1995.

[104] Y. Steinberg and H. V. Poor, "Multiuser Delay Estimation," *Proc. Conf. Inform. Sci. Syst.,* Baltimore, MD, Mar. 1993.

[105] Y. Steinberg and H. V. Poor, "Sequential Amplitude Estimation in Multiuser Communications," *IEEE Trans. Inform. Theory,* Vol. 40, pp. 11–20, Jan. 1994.

[106] E. G. Strom and S. L. Miller, "A Reduced Complexity Adaptive Near-Far Resistant Receiver for DS-CDMA," *Proc. IEEE GLOBECOM,* pp. 1734–1738, 1993.

[107] E. G. Strom, S. Parkvall, S. L. Miller, and B. E. Ottersten, "Propagation Delay Estimation of DS-CDMA Signals in a Fading Environment," *Proc. IEEE GLOBECOM/Comm. Theory Mini-Conf.,* pp. 85–89, San Francisco, CA, Dec. 1994.

[108] J. Travis, "Dialing up Undersea Data—Long Distance," *Science,* Vol. 263, pp. 1223–1224, Mar. 4, 1994.

[109] D. W. Tufts and A. A. Shah, "Rapid Interference Suppression and Channel Identification for Digital, Multipath Wireless Channels," *Proc. IEEE Vehic. Techn. Conf.* Vol. 2, pp. 1241–1245, Stockholm, Sweden, 1994.

[110] G. Ungerboeck, "Channel Coding with Multilevel/Phase Signals," *IEEE Trans. Inform. Theory,* Vol. IT-28, No. 1, pp. 55–67, Jan. 1982.

[111] M. K. Varanasi, "Noncoherent Detection in Asynchronous Multiuser Channels, *IEEE Trans. Inform. Theory,* Vol. 39, No. 1, pp. 157–176, Jan. 1993.

[112] M. K. Varanasi, "Group Detection for Synchronous Gaussian Code-division Multiple-access Channels, *IEEE Trans. Inform. Theory,* Vol. 41, No. 4, pp. 1083–1096, July 1995.

[113] M. K. Varanasi, "Parallel Group Detection for Synchronous CDMA Communication over Frequency Selective Raleigh Fading Channels," *IEEE Trans. Inform. Theory,* Vol. 42, No. 1, pp. 116–123, Jan. 1996.

[114] M. K. Varanasi and B. Aazhang, "Multistage Detection for Asynchronous Code-Division Multiple-Access Communications," *IEEE Trans. Commun.*, Vol. COM-38, pp. 509–519, Apr. 1990.

[115] M. K. Varanasi and B. Aazhang, "Near-optimum Detection in Synchronous Code-division Multiple-access Systems," *IEEE Trans. Commun.*, Vol. 39, No. 5, pp. 725–736, May 1991.

[116] S. Vasudevan and M. K. Varanasi, "A Near-optimum Receiver for CDMA Communication over the Time-varying Rayleigh Fading Channel," *IEEE Trans. Commun.*, Vol. 44, pp. 1130–1143, 1996.

[117] S. Vembu and A. J. Viterbi, "Two Different Philosophies in CDMA—A Comparison," *Proc. IEEE Vehic. Techn. Conf.*, pp. 869–873, Atlanta, GA, May 1996.

[118] S. Verdú, "Minimum Probability of Error for Asynchronous Gaussian Multiple-access Channels," *IEEE Trans. Inform. Theory*, Vol. IT-32, pp. 85–96, Jan. 1986.

[119] S. Verdú, "Optimum Multi-user Asymptotic Efficiency," *IEEE Trans. Commun.*, Vol. COM-38, No. 4, pp. 496–508, Apr. 1990.

[120] S. Verdú, "Multiuser Detection," in *Advances in Statistical Signal Processing—Vol. 2: Signal Detection*, H. V. Poor and J. B. Thomas, eds. (JAI Press: Greenwich, Conn., 1993).

[121] S. Verdú, *Multiuser Detection.* (Cambridge University Press: Cambridge, UK, 1998, in press)

[122] R. Vijayan and H. V. Poor, "Nonlinear Techniques for Interference Suppression in Spread-Spectrum Systems," *IEEE Trans. Commun.*, Vol. 38, No. 7, pp. 1060–1065, July 1990.

[123] A. J. Viterbi, A. M. Viterbi, and E. Zehavi, "Performance of Power-Controlled Wideband Terrestrial Digital Communication," *IEEE Trans. Commun.*, Vol. 41, No. 4, Apr. 1993.

[124] A. M. Viterbi and A. J. Viterbi, "Erlang Capacity of a Power Controlled CDMA System," *IEEE J. Select. Areas Commun.*, Vol. 11, No. 6, pp. 892–900, Aug. 1993.

[125] A. J. Viterbi, *CDMA: Principles of Spread Spectrum Communications.* (Addison-Wesley: Reading, MA, 1995)

[126] X. Wang and H. V. Poor, "Adaptive Multiuser Diversity Receivers for Frequency-Selective Rayleigh Fading CDMA Channels," *Proc. IEEE Vehic. Techn. Conf.*, Vol. 1, pp. 198–202, Phoenix, AZ, May 1997.

[127] X. Wang and H. V. Poor, "Subspace-based Blind Adaptive Joint Interference Suppression and Channel Estimation in Multipath CDMA Channels," *Proc. Int. Conf. Universal Personal Communications*, San Diego, CA, Oct. 1997. [Also to appear in *IEEE Trans. Signal Processing.*]

[128] X. Wang and H. V. Poor, "Adaptive Multiuser Detection in Non-Gaussian Channels," *Proc. Allerton Conf. Commun., Control, and Computing*, University of Illinois, Urbana, IL, Oct. 1997.

[129] X. Wang and H. V. Poor, "Blind Multiuser Detection: A Subspace Approach," *IEEE Trans. Inform. Theory*, Vol. 44, No. 2, Mar. 1998.

[130] X. Wang and H. V. Poor, "Blind Equalization and Multiuser Detection for CDMA Communications in Dispersive Channels," *IEEE Trans. Commun.*, Vol. 46, No. 1, pp. 91–103, Jan. 1998.

[131] P. Wei, J. R. Zeidler, and W. H. Ku, "Adaptive Interference Suppression for CDMA Overlay Systems," *IEEE J. Select. Areas Commun.*, Vol. 12, No. 9, pp. 1510–1523, Dec. 1994.

[132] Z. Xie, C. K. Rushforth, R. T. Short and T. Moon, "Joint Signal Detection and Parameter Estimation in Multi-user Communications," *IEEE Trans. Commun.*, Vol. 41, pp. 1208–1216, Aug. 1993.

[133] Z. Xie, R. T. Short, and C. K. Rushforth, "A Family of Suboptimum Detectors for Coherent Multi-user Communications," *IEEE J. Select. Areas Commun.*, Vol. 8, No. 4, pp. 683–690, May 1990.

[134] B. Yang, "Projection Approximation Subspace Tracking," *IEEE Trans. Signal Processing*, Vol. 44, No. 1, pp. 95–107, Jan. 1995.

[135] X. Zhang and D. Brady, "Soft-decision Multistage Detection for Asynchronous AWGN Channels," *Proc. Allerton Conf. Commun., Contr., Computing*, Monticello, IL, Sept. 1993.

[136] X. Zhang and D. Brady, "Narrow Bandwidth Waveform Design for Near-far Resistant Multiuser Detection," *Proc. MILCOM*, Ft. Monmouth, NJ, Oct. 1994.

[137] F-C Zheng and S. K. Barton, "On the Performance of Near-Far Resistant CDMA Detectors in the Presence of Synchronization Errors," *IEEE Trans. Commun.*, Vol. 43, No. 12, pp. 3037–3045, Dec. 1995.

[138] L. J. Zhu and U. Madhow, "Adaptive interference suppression for direct sequence CDMA over severely time-varying channels," *Proc. IEEE GLOBECOM*, San Diego, CA, 1997.

[139] L. J. Zhu and U. Madhow, "Performance of adaptive interference suppression for DS-CDMA with Rayleigh fading channel," *Proc. Conf. Inform. Sci. Syst.*, The John Hopkins Univ., Baltimore, MD, Mar. 1987.

[140] Z. Zvonar, D. Brady, and J. Catipovic, "Adaptive Receivers for Cochannel Interference Suppression in Shallow-water Acoustic Telemetry Channels," *Proc. Int. Conf. Acoust., Speech Signal Processing*, Adelaide, Australia, 1994.

[141] Z. Zvonar and D. Brady, "Multiuser Detection in Single-path Fading Channels," *IEEE Trans. Commun.*, Vol. 42, No. 2/3/4, pp. 1729–1739, Feb./Mar./Apr. 1994.

[142] Z. Zvonar and D. Brady, "Differentially Coherent Multiuser Detection in Asynchronous CDMA Flat Rayleigh Fading Channels," *IEEE Trans. Commun.*, Vol. 43, pp. 1252–1255, Feb./Mar./Apr. 1995.

ACKNOWLEDGMENTS

This work was prepared in part under the support of the U.S. Army Research Office under Grant DAAH04-96-1-0378, and in part under the support of the U.S. Office of Naval Research under Grant N00014-94-1-0115.

3

Equalization of Multiuser Channels

Haralabos C. Papadopoulos

In this chapter, we consider the scenario where multiple users are communicating over a wireless channel of fixed bandwidth and focus our attention on the design of receivers that compensate for distortion introduced by the propagation medium. As discussed in Chapter 1, propagation of any of the transmitted signals through the wireless channel results in a collection of several delayed and scaled copies of the transmitted signal at the receiver (often referred to as multipath propagation) giving rise to frequency-selective fading. In addition, due to the changing environmental medium between the transmitters and the receivers, especially in a mobile setting, the channel response varies with time, giving rise to time-selective fading. From the point of view of the receiver design, these effects can be described in terms of a random linear time-varying channel model and give rise to the need for equalization in order to reliably demodulate the information-bearing signals of interest.

A form of interference particular to multiuser communication systems arises from the presence of the signals of additional users; the channel often exacerbates interference between users, and so the receiver must also take into account and compensate for this additional form of interference. As discussed in detail in the preceding chapters, this type of interference, usually referred to as cochannel or multiple-access interference (MAI), is a dominant one in multiuser systems and has often dictated practical system design. For instance, in conventional time-division multiple-access (TDMA) systems where users transmit and receive information in nonoverlapping time slots, guard times are often used to compensate for imperfect synchronization and spreading of transmitted symbols over time due to multipath effects, resulting in limiting the system efficiency. Similarly, MAI is also present in code-division multiple-access (CDMA) systems in which each user's transmitted signal occupies all the available bandwidth at all times. Although in CDMA systems that are synchronous (e.g., the forward link of cellular systems) and that have a limited number of users, MAI can be made zero or negligible by maintaining orthogonality among users, it is hard to maintain synchronization and thus orthogonality in asynchronous CDMA systems (e.g., the reverse link of a cellular system) [1]. Moreover, user nonorthogonality is often allowed in CDMA systems as a method for accommodating higher numbers of users at the expense of MAI. A number of equalization methods that we discuss in this chapter deal effectively with MAI and can potentially result in increased system efficiency in terms of the number of users that can be accommodated in a given bandwidth, as well as system reliability and robustness.

This chapter is structured as follows. In Section 3.1 we briefly describe some convenient and accurate models for fading channels that will serve as a basis for our subsequent development of equalizer structures. In the process, we summarize some of the key parameters that are used to characterize fading channels, suggesting how they impact equalizer design in wireless systems.

Section 3.2 focuses on channel equalization and symbol detection in the scenario that the channel is known at the receiver. This model inherently assumes that reliable channel estimates can be formed at the receiver via training sequences, which remain accurate during equalizer operation. We first summarize the characterization of multiuser detectors over fading channels that minimize the probability of making a symbol sequence error. Although the resulting maximum-likelihood (ML) sequence detectors are, in general, impractical in terms of their computational complexity, they provide insight into the structure of the detection problem. Furthermore, an information-lossless front end arises in their construction, which is employed in a variety of suboptimal but practical detectors. We examine a number of such practical equalizers, including properly constructed multiuser generalizations of common single-user equalizers. In particular, we discuss the single-user matched-filter (MF) detector for time-varying channels, multiuser linear equalizers, and a number of nonlinear detectors including decision-feedback equalizers.

In the design of all equalizers in Section 3.2 it is assumed that the channel is known at the receiver, which, as discussed above, usually implies that some form of training is used, either during initialization or, in the presence of time-selective fading, as part of every transmitted frame. Unavoidably, the use of training results in symbol rate reduction.

In Section 3.3 we discuss equalization and symbol estimation in the absence of training, in which case the channel is effectively treated as unknown. In order to simplify equalization and make it feasible in wireless applications, one class of actively pursued approaches assumes that the receiver uses multiple antennas or, in the context of excess bandwidth transmission, that it oversamples the received signal and subsequently uses this additional information to recover both the symbol stream and the channel responses from short observation intervals. A large number of these approaches currently attracting extensive attention in the community focus on equalization of linear time-invariant (LTI) channels that have finite impulse response (FIR). We present a few representative examples of such approaches in the general single-user case and discuss some challenging issues that can often arise in practice. These techniques are applicable to the forward link of conventional multiuser mobile communication systems where the transmission of the signals of all users is synchronous. Finally, we consider blind equalization on the reverse link and, in particular, in the context of asynchronous CDMA systems, where user signature information is exploited for separating the individual symbol streams.

3.1 CHARACTERIZATION OF WIRELESS CHANNELS

Before we discuss equalizer design for multiple-access communication over wireless channels, it is important to briefly describe the models that are generally used to characterize these channels, as well as some of the model parameters that may play a role in the selection of the equalization strategy. See e.g., [2] for a more extensive introduction in the characterization of fading multipath channels.

Throughout this chapter, we address the problem of equalization in terms of the equivalent low-pass communication system. A real bandpass transmitted signal of bandwidth W is represented by an equivalent complex-valued circularly symmetric baseband signal that is bandlimited in $[-W/2, W/2]$. If the (equivalent) low-pass continuous-time signal $x(t)$ is transmitted over the (equivalent) multipath fading channel, it is received as

$$r(t) = \int_{-\infty}^{\infty} h(t; \tau) x(t - \tau)\, d\tau + w(t), \tag{3.1}$$

where $h(t; \tau)$ denotes the (equivalent) complex-valued response of the medium at time t to an impulse applied at $t - \tau$, and $w(t)$ corresponds to complex-valued measurement noise. The impulse response $h(t; \tau)$ is generally well modeled as a zero-mean complex-valued Gaussian random process, which is stationary as a function of t for any fixed τ. Its envelope $|h(t; \tau)|$ for any t and τ is Rayleigh distributed, and for this reason this channel is often referred to as a Rayleigh fading channel. Examination of the autocorrelation function of $h(t; \tau)$

$$\varphi_{h,h}(\tau_1, \tau_2; \Delta t) = E\left[h(t; \tau_1)h^*(t + \Delta t; \tau_2)\right]$$

reveals a number of key parameters in channel characterization that usually influence system design.[1] In most wireless media, the response with path delay τ_1 is uncorrelated with the response with path delay τ_2, where $\tau_2 \neq \tau_1$, a phenomenon referred to as uncorrelated scattering. Due to uncorrelated scattering, we need only consider $\varphi_{h,h}(\tau_1, \tau_2; \Delta t)$ for $\tau_1 = \tau_2$, since $\varphi_{h,h}(\tau_1, \tau_2; \Delta t) = \varphi_h(\tau_1; \Delta t)\delta(\tau_1 - \tau_2)$, where

$$\varphi_h(\tau; \Delta t) \triangleq E\left[h(t; \tau)h^*(t + \Delta t; \tau)\right].$$

The function $\varphi_h(\tau; 0)$ is generally referred to as the multipath intensity profile, or the delay power spectrum, and corresponds to the average power output of the channel as a function of the delay τ; if an impulse is transmitted at time $t = t_o$, we have

$$E\left[|r(t_o + \tau)|^2\right] = \varphi_h(\tau; 0), \tag{3.2}$$

where $r(t)$ is given by (3.1) in the absence of additive noise. The range of values of τ over which $\varphi_h(\tau; 0)$ is non-zero corresponds to the multipath (or delay) spread of the channel and is henceforth denoted by T_m^h. Equation (3.2) is often used as a basis for estimating the multipath intensity profile. For instance, an estimate of $\varphi_h(\tau; 0)$ for $0 \leq \tau \leq T_m^h$ can be formed by averaging the power of the τ-delayed channel response of each pulse in a sequence of T-spaced pulses of very short duration, where T is chosen so that $T \gg T_m^h$.

The Fourier transform of $\varphi_h(\tau; \Delta t)$ is denoted by $\Phi_h(\Delta f; \Delta t)$ and is typically referred to as the spaced-frequency, spaced-time correlation function. The range of Δf values over which $\Phi_h(\Delta f; 0)$ is non-zero is a measure of the coherence bandwidth of the channel and is denoted by $(\Delta f)_c^h$; due to the relationship between $\Phi_h(\Delta f; 0)$ and $\varphi_h(\tau; 0)$, $(\Delta f)_c^h$ equals the reciprocal of the multipath spread. If the bandwidth of the information-bearing signal is small as compared to the coherence bandwidth $(\Delta f)_c^h$, the channel is said to be frequency-nonselective; the channel frequency response is effectively flat over the transmission bandwidth. If, on the other hand, the bandwidth of the information-bearing signal is larger than $(\Delta f)_c^h$,

---

[1]Throughout this chapter, we use the superscript * to denote complex conjugation.

the channel is termed frequency-selective; the different frequency components of the transmitted signal are affected differently by the channel.

The Doppler spectrum $S_h(\lambda)$ is used in characterizing the time variations in the channel; it corresponds to the Fourier transform of $\Phi_h(\Delta f; \Delta t)$ with respect to the Δt variable evaluated at $\Delta f = 0$. If a pure tone $\exp(j\,2\pi\lambda_o t)$ is transmitted over a channel with Doppler spectrum $S_h(\lambda)$, then in the absence of noise,

$$S_h(\lambda_o) = \int E\left[r(t)r^*(t+\Delta t)\right] d\Delta t. \tag{3.3}$$

In practice, (3.3) may form the basis of a strategy for estimating the Doppler spectrum. Specifically, we can obtain an estimate of $S_h(\lambda_o)$ at a frequency λ_0 by transmitting a pure tone $\exp(j\,2\pi\lambda_o t)$, cross-correlating the received signal with Δt-delayed versions of itself so as to form estimates of $E[r(t)r^*(t+\Delta t)]$, and then using these estimates in (3.3). For channels $h(t; \tau)$ that are time-invariant, $S_h(\lambda)$ reduces to a Kronecker delta function. Therefore, when there are no time variations in the channel, no spectral broadening of the transmitted signal would be observed at the receiver. The Doppler spread B_d^h corresponds to the range of λ values over which $S_h(\lambda)$ is essentially non-zero, and its reciprocal is generally referred to as the coherence time of the channel and is denoted by $(\Delta t)_c^h$. In general, we refer to a channel as slowly time-varying if the intersymbol period T_o of the information-bearing signal is much smaller than the associated coherence time $(\Delta t)_c^h$ or, equivalently, if the Doppler spread is much smaller than the symbol rate. In a number of wireless communication schemes where the fading channel is assumed to be known at the receiver, the channel is slowly time-varying so that its measurements via training sequences remain accurate over reasonably long sequences of symbols.

The product $T_m^h B_d^h$ is usually referred to as the spread factor of the channel and is often taken into account in communication system design. A channel $h(t; \tau)$ is said to be underspread if $T_m^h B_d^h < 1$, and overspread otherwise. Consider, for instance, a channel for which $T_m^h B_d^h \ll 1$. Assume also that given an available bandwidth $\mathcal{W}$, the associated intersymbol time T_o satisfies $T_o \approx 1/\mathcal{W}$. Then, we can choose the bandwidth so that $\mathcal{W} \ll 1/T_m^h$ and $T_o \ll 1/B_d^h$, i.e., so that the channel is frequency-nonselective and slowly time-varying. On the other hand, if the channel is overspread, it is not possible to select T_o and $\mathcal{W}$ so that the resulting channel is frequency-nonselective and slowly time-varying.

3.2 EQUALIZATION OF KNOWN MULTIPATH FADING CHANNELS

In this section, we focus on multiuser equalization techniques for multipath fading channels in the case that the channels are known at the receiver. As discussed earlier, this assumption implies that channel measurements can be made via training that result in reasonably accurate channel estimates, which can then be used in symbol detection. In general, training can be introduced via a variety of methods. For instance, if the channel is slowly time-varying, i.e., the symbol rate T_o^{-1} is larger than the Doppler spread of the channel B_d^h, training can be temporally interspersed with the data; the transmission is temporally partitioned into frames during which the channel can be first identified via a training stage and then assumed known for the next several symbols. Alternatively, training can be interspersed spectrally with the data; a (known) pilot tone occupying a fraction of the available bandwidth may be included as part of the transmitted signal at all times. As discussed in Chapter 2, yet another approach is the one used in a number of practical equalization schemes which are known as decision-directed; there is an initial identification period during which training data are used to identify the channel, followed by an adaptation stage during which current estimates of the channel response are used to obtain symbol estimates and subsequently these symbol decisions are used to update or adapt the current channel estimates.

A number of distinct channel models may arise depending on the bandwidth and the symbol rate of the signaling scheme and the channel parameters. As discussed in Section 3.1, the overall channel model we use is linear and time-varying (LTV). The channel is often very slowly time-varying (i.e., $(\Delta t)_c^h \gg T_o$), so that during the transmission of a long stream of data symbols the channel response remains effectively constant; in this case, the LTI model is often a naturally suited description of the channel. When in addition the multipath spread of the channel is much smaller than that the intersymbol period, i.e., $T_m^h \ll T_o$, an additive white Gaussian noise channel model may adequately capture the key channel characteristics. We discuss equalization in the general case where the channel is LTV, often specializing the resulting processors to LTI and additive white Gaussian noise channels.

We consider several equalizer design criteria that result in distinct systems with diverse performance-complexity characteristics. In particular, we discuss a number of equalizers for use in a multiuser setting that arise as natural extensions or generalizations of known single-user equalizers.

In Section 3.2.1 we describe the general system model and also develop a discrete-time system arising in chip-rate sampling equalizers, which are equalization systems whose front ends consist of low-pass filtering followed by continuous- to discrete-time conversion.

In Section 3.2.2 we describe the maximum-likelihood (ML) sequence detector for the general LTV channel. This detector obtains an intermediate sequence of sufficient statistics from which the ML sequence estimate can be determined. Although their complexity makes ML sequence detectors impractical in general, computation of the sufficient statistics sequence has been used as a front end for a number of suboptimal but practical equalization algorithms.

Section 3.2.3 discusses the matched-filter detector generalization in the LTV case, which is the optimal single-user, single-symbol detector and is typically referred to as the RAKE receiver. The performance of this receiver in a single-user, single-symbol transmission provides an upper bound on the detection performance for any multiuser detector. In a multiuser setting, this low-complexity single-user receiver suffers in general from intersymbol interference (ISI) and MAI.

In Section 3.2.4 we discuss multiuser extensions of the single-user, linear zero-forcing (ZF) equalizer. These linear detectors operate on the sequence of sufficient statistics and completely remove ISI and MAI, but generally suffer from noise enhancement.

Section 3.2.5 discusses the multiuser extension of the single-user, linear minimum mean-square error (MMSE) equalizer, which is the best linear receiver in terms of minimizing the expected mean-square estimation error between the symbol estimates before the decision and the actual symbols. We present a framework based on Kalman filtering for obtaining recursive algorithms for linear MMSE and ZF equalizer design.

Section 3.2.6 discusses recursive implementations of decision-feedback equalizers as well as more general successive cancellation schemes. All these detectors present practical approaches to equalization that are significantly more resistant to MAI than are single-user based receivers.

Finally, in Section 3.2.7 we briefly describe a class of recursive equalizers that correspond to Kalman filtering solutions for properly formulated state-space models arising from the chip-rate sampling front end introduced in Section 3.2.1.

3.2.1 System Model

We consider a multiuser communication scenario where P users are communicating over a common bandwidth $\mathcal{W}$. We assume that the ith user is communicating a symbol stream $s_i[n]$ at symbol rate $\mathcal{W}_o = 1/T_o$ and that the associated transmitted waveform $x_i(t)$ is given as the following sum of T_o-delayed signals

$$x_i(t) = \sum_k c_i(t - kT_o; s_i[k]),$$

where $c_i(t; s)$ is the transmitted waveform of the ith user associated with symbol s. We will mainly focus on linear modulation schemes, in which case the ith symbol stream is linearly modulated on the signature of the ith user, i.e.,

$$x_i(t) = \sum_k s_i[k]c_i(t - kT_o), \tag{3.4}$$

where $c_i(t)$ is the signature of the ith user. The received signal $r(t)$ is a noise-corrupted sum of the received versions of the P transmitted signals $x_i(t)$ after these have propagated through the wireless medium, i.e.,

$$r(t) = \sum_{i=1}^{P} \int_{-\infty}^{\infty} h_i(t; \tau) x_i(t - \tau)\, d\tau + w(t), \tag{3.5}$$

where $w(t)$ is white circularly symmetric, complex-valued stationary Gaussian noise of intensity $\mathcal{N}_o$. The task at the receiver is to estimate the P symbol sequences $s_i[n]$, based on observation of $r(t)$ and assuming knowledge of the channel responses $h_i(t; \tau)$ and the codewords $c_i(t)$.

3.2.1.1 Discrete-Time Signal Model

A particular class of equalization schemes we consider employs sampling of the received waveform $r(t)$ in (3.5) at rate $T = 1/\mathcal{W}$ followed by discrete-time processing. In this case, the sampling rate and the symbol rate are related as $T = T_o/L$, where L is usually a large integer. In general, since P users are sharing a total fixed bandwidth $\mathcal{W} = L\mathcal{W}_o$, the effective bandwidth per user is $L\mathcal{W}_o/P$. In CDMA systems, L equals the number of chips per symbol; for this reason, we refer to these schemes as chip-rate sampling schemes. By using the representation of $x_i(t)$ in terms of its Nyquist T-spaced samples $x_i[n] = x_i(nT)$ (since $x_i(t)$ is bandlimited in $[-\mathcal{W}/2, \mathcal{W}/2]$), (3.5) can be transformed to [2]

$$r(t) = \sum_{i=1}^{P} \sum_{k} h_i(t; k) x_i[k] + w(t), \tag{3.6}$$

where

$$h_i(t; k) \triangleq \int h_i(t; t - \tau) \frac{\sin\left(\pi(\tau - kT)/T\right)}{\pi(\tau - kT)/T}\, d\tau. \tag{3.7}$$

Let $y[n]$ denote the sequence arising from ideal low-pass filtering of $r(t)$ with cutoff frequency $\mathcal{W}/2$, followed by continuous-to-discrete time conversion. Since low-pass filtering is a linear operation on $r(t)$ and using (3.6), we have

$$y[n] = \sum_{i=1}^{P} \sum_{k} h_i[n; k] x_i[n - k] + w[n], \tag{3.8}$$

where $y[n] = r(nT)$, the noise sequence $w[n]$ is zero-mean complex-valued circularly symmetric white Gaussian sequence with variance $\sigma_w^2 = \mathcal{N}_o \mathcal{W}$, and the kernels $h_i[n; k]$ provide a description of the linear transformations of the $x_i[n]$'s that produce $y[n]$ in the absence of noise. The discrete-time model (3.8) arising from chip-rate sampling of $r(t)$ is depicted in Figure 3.1. According to the model, the impulse response of the LTV channel associated with the ith user's sequence $x_i[n]$

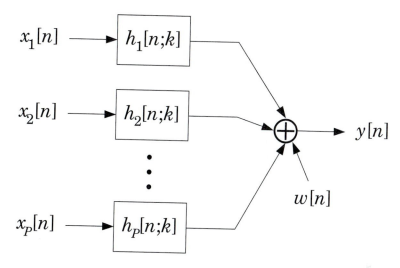

Figure 3.1   A discrete-time multiple-access fading channel model.

is given by the kernel $h_i[n; k]$, which corresponds to the response of the channel at time n to a unit-impulse transmitted by the ith user at time $n - k$.

Clearly, (3.7) and (3.8) are both conceptually convenient expressions involving nonrealizable ideal low-pass filtering. Furthermore, significant information loss can arise from low-pass filtering and sampling $h(t; k)$ in (3.7) to produce $y[n]$ in (3.8), unless the channel Doppler spread is much smaller than the available bandwidth, i.e., $B_d^h \ll W$. In practice, sampling rates higher than $1/W$ can be combined with better-behaved low-pass filters (i.e., with gradually decaying transition bands) with passband including $[-W/2, W/2]$ to provide robust systems that can be reasonably modeled via a discrete-time representation of the form (3.8).

Given the multipath spread of a channel T_m^h and the signal bandwidth of the signaling scheme, there are $\lceil T_m^h W \rceil + 1$ resolvable paths in (3.5) [2]. For a given sampling period T, this model corresponds to an equivalent discrete-time FIR filter kernel $h_i[n; k]$ whose effective spread is $L_c + 1 \approx \lceil T_m^h/T \rceil + 1$. Note, however, that in the presence of additional filtering prior to sampling, the length of the equivalent discrete-time filter may in general be significantly longer than $\lceil T_m^h/T \rceil + 1$.

The model (3.8) for chip-rate sampling also suggests the following discrete-time representation of the linear modulation scheme (3.4):

$$x_i[n] = \sum_k s_i[k] c_i[n - kL], \tag{3.9}$$

which arises from (3.4) and (3.8) by letting $c_i[n] = c_i(nT)$. This equivalent discrete-time baseband model describing the modulation of each symbol stream is depicted in Figure 3.2. The sequence $s_i[n]$ corresponding to the (coded) sequence of symbols

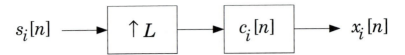

$$s_i[n] \longrightarrow \boxed{\uparrow L} \longrightarrow \boxed{c_i[n]} \longrightarrow x_i[n]$$

Figure 3.2 Modulation of the (coded) symbol sequence of the ith user onto a sequence $c_i[n]$.

transmitted by the ith user is linearly modulated onto a unique signature sequence $c_i[n]$ with support K to produce $x_i[n]$, which is transmitted within the total available bandwidth W. A convenient interpretation of the common multiple-access schemes can be obtained from this discrete-time model. For example, TDMA systems use $c_i[n] = \delta[n - n_i]$ for some integer $0 \le n_i \le L - 1$, where the n_i's are distinct. On the other hand, conventional CDMA systems use binary-valued pseudorandom sequences of length $K = L$, whereas the spread-signature CDMA systems that are developed in Chapter 1 use binary-valued pseudorandom sequences of length $K \gg L$.

In the next section, we discuss processors that perform joint detection of all P symbol streams from observation of $r(t)$ in (3.5) and have the property that they minimize the probability of error in a certain sense.

3.2.2 Limits on Equalizer Performance over Fading Channels—Maximum-Likelihood Sequence Detection

Before discussing practical multiuser equalization strategies over fading channels, it is important to develop some intuition regarding the limits of equalizer performance by examining equalizers that are optimal in a certain useful sense. Specifically, consider the scenario of P users sharing a common bandwidth W, each communicating a sequence of $2N + 1$ symbols over a fading channel. In the limit $N \to \infty$, this scenario reduces to the one considered in Section 3.2.1. Let

$$\mathbf{s}[n] = [s_1[n] \ \ s_2[n] \ \ \cdots \ \ s_P[n]]^T, \tag{3.10}$$

where the superscript ^{T} denotes transposition, and let the vector supersymbol $\mathbf{S}$ denote the collection of all $P(2N + 1)$ transmitted symbols:

$$\mathbf{S} = [\mathbf{s}^T[-N] \ \ \cdots \ \ \mathbf{s}^T[N]]^T. \tag{3.11}$$

We will examine the detector that selects the value of $\hat{\mathbf{S}}$ based on observation of $r(t)$, that maximizes the associated log-likelihood function. This ML sequence detector is optimal in the sense that it minimizes the probability of making a supersymbol error, provided that all possible supersymbols $\mathbf{S}$ are equally likely.

For simplicity, we assume that the coded symbol stream of the ith user $s_i[-N], s_i[-N + 1], \cdots, s_i[N]$ is linearly modulated on the signature waveform (codeword) $c_i(t)$ according to (3.4) and transmitted over the fading channel $h_i(t; \tau)$. Specifically, the ith transmitted signal is given by

$$x_i(t) = \sum_{n=-N}^{N} s_i[n]c_i(t - nT_o).$$

(3.12)

The received signal $r(t)$ consists of the sum of the responses of the P transmitted signals and measurement noise and is given by (3.5). By letting

$$\tilde{h}_i(t; \tau) = \int_{-\infty}^{\infty} h_i(t; \tilde{\tau})c_i(t - \tilde{\tau} - \tau)\, d\tilde{\tau},$$

(3.13)

the response of the ith user's transmitted signal to the ith channel can be recast as follows:

$$u_i(t) = \int h_i(t; \tau)x_i(t - \tau)\, d\tau$$

(3.14)

$$= \sum_{n=-N}^{N} \tilde{h}_i(t; nT_o)s_i[n].$$

(3.15)

Consequently, the sum response $u_S(t) = \sum_i u_i(t)$ satisfies

$$u_S(t) = \sum_{n=-N}^{N} \tilde{\mathbf{h}}_n^T(t)\mathbf{s}[n],$$

(3.16)

where

$$\tilde{\mathbf{h}}_n(t) = [\tilde{h}_1(t; nT_o)\ \tilde{h}_2(t; nT_o)\ \cdots\ \tilde{h}_P(t; nT_o)]^T.$$

(3.17)

Using (3.16), we can then rewrite (3.5) as

$$r(t) = \sum_{n=-N}^{N} \tilde{\mathbf{h}}_n^T(t)\mathbf{s}[n] + w(t).$$

(3.18)

The maximum-likelihood sequence detector for the set of $P(2N + 1)$ symbols summarized by $\mathbf{S}$ based on observation of (3.18) selects the supersymbol $\hat{\mathbf{S}}$ that maximizes the associated log-likelihood function, or equivalently [3],

$$\hat{\mathbf{S}} = \arg\min_{\mathbf{S}} \int_{-\infty}^{\infty} |r(t) - u_S(t)|^2\, dt$$

$$= \arg\max_{\mathbf{S}} 2\,\text{Re}\left\{\sum_{n=-N}^{N} \mathbf{y}^\dagger[n]\mathbf{s}[n]\right\} + \mathcal{E}_{\mathbf{S}},$$

(3.19)

where the superscript † denotes the conjugate transpose operator, $\text{Re}\{\cdot\}$ denotes the real part of its argument, $\mathcal{E}_{\mathbf{S}} = \int_{-\infty}^{\infty} |u_S(t)|^2\, dt$, and

$$\mathbf{y}[n] = \int_{-\infty}^{\infty} r(t)\tilde{\mathbf{h}}_n^*(t)\, dt.$$

(3.20)

The vector sequence $\mathbf{y}[n]$ is a set of sufficient statistics for sequence detection via observation of the waveform $r(t)$. The ML sequence detector first precomputes the vector waveform $\tilde{\mathbf{h}}_n(t)$ for any n, using knowledge of $h_i(t; \tau)$ and $c_i(t)$, then obtains $\mathbf{y}[n]$ at time n as the response of $r(t)$ to a matched filter $\tilde{\mathbf{h}}_n^*(-t)$ at time $t = 0$, and finally solves (3.19). The sufficient statistic sequence $\mathbf{y}[n]$ in (3.20) also satisfies

$$\mathbf{y}[n] = \sum_k R_{\tilde{\mathbf{h}}}[n; k]\mathbf{s}[n-k] + \mathbf{v}[n], \tag{3.21}$$

where the autocorrelation function of $\mathbf{v}[n]$ is given by

$$R_{\mathbf{v}}[n, n-k] = E[\mathbf{v}[n]\mathbf{v}[n-k]] = R_{\tilde{\mathbf{h}}}[n; k],$$

and where the $P \times P$ time-varying kernel $R_{\tilde{\mathbf{h}}}[n; k]$ is given by

$$R_{\tilde{\mathbf{h}}}[n; k] = \int \tilde{\mathbf{h}}_n(t)\tilde{\mathbf{h}}_{n-k}^{\dagger}(t)dt.$$

Given the set of sufficient statistics $\mathbf{y}[n]$, one can, in principle, solve for $\hat{\mathbf{S}}$ in (3.19) by exhaustive search. Since the number of all possible symbol sequences grows exponentially with the number of users P and the number of symbols $2N + 1$, such an exhaustive search method is, in general, computationally impractical. Furthermore, it requires observation of the whole sequence before any decision in made. Consequently, optimal detection is usually impractical. In general, we seek to develop practical equalizers by reducing the complexity as much as possible without significantly sacrificing performance.

As we shall see, many practical detectors operate on $\mathbf{y}[n]$, i.e., employ the (possibly time-varying) matched filter as their front end. In practice, the continuous-time matched filter (3.20) can be implemented in a more robust fashion in discrete time via a method that is a direct generalization of the discrete-time implementation of the continuous-time matched filter for LTI channels [4]. Specifically, band-limited sampling of the continuous-time waveform is first performed at K times the baud rate T_o^{-1}, where K is an integer satisfying $K > 2$, so that the equivalent time-varying channel response can be reconstructed from these samples. The resulting sequence is then passed through a discrete-time filter followed by decimation by a factor K.

3.2.2.1 Sufficient Statistics for LTI Channels

In certain cases, the channel $h_i(t; \tau)$ may be varying slowly enough compared to the symbol rate (i.e., $1/T_o \gg B_d^h$) that it may be convenient to view it as time-invariant over a long sequence of symbols. In the absence of time-selective fading, the ith channel has the following form:

$$h_i(t; \tau) = h_i(\tau). \tag{3.22}$$

The corresponding effective channel $\tilde{\mathbf{h}}_n(t)$ in (3.17) is given by

$$\tilde{\mathbf{h}}(t - nT) \triangleq \tilde{\mathbf{h}}_n(t) = [h_1 * c_1(t - nT_o) \quad h_2 * c_2(t - nT_o) \quad \cdots \quad h_P * c_P(t - nT_o)]^T, \quad (3.23)$$

where $*$ denotes convolution. The continuous-time matched filter (3.20) resulting in the sufficient statistic $\mathbf{y}[n]$ is often used as a front end in a number of practical detectors. In the LTI case, the sequence $\mathbf{y}[n]$ is the response of the filter $\mathbf{h}^*(-t)$ driven by $r(t)$ and sampled at time instant nT_o [5]. In particular, the sequence $\mathbf{y}[n]$ satisfies

$$\mathbf{y}[n] = \sum_k R_{\tilde{\mathbf{h}}}[n - k]\mathbf{s}[k] + \mathbf{v}[n] = R_{\tilde{\mathbf{h}}}[n] * \mathbf{s}[n] + \mathbf{v}[n]. \quad (3.24)$$

The Fourier transform of the sequence $R_{\tilde{\mathbf{h}}}[n]$ is given by the folded spectrum of $\tilde{\mathbf{h}}(t)$ [2, 6]

$$S_{\tilde{\mathbf{h}}}(e^{j\omega}) = \frac{1}{T_o} \sum_{k=-\infty}^{\infty} \tilde{\mathbf{H}}^*(\omega + 2\pi k/T_o)\tilde{\mathbf{H}}^T(\omega + 2\pi k/T_o),$$

and the spectrum of the noise sequence $\mathbf{v}[n]$ satisfies $S_{\mathbf{v}}(z) = \mathcal{N}_o S_{\tilde{\mathbf{h}}}(z)$. As in the single-user case, the frequency domain form of (3.24) will often prove convenient; the z-transform of any sample path $\mathbf{y}[n]$ satisfies

$$\mathbf{y}(z) = S_{\tilde{\mathbf{h}}}(z)\mathbf{s}(z) + \mathbf{v}(z), \quad (3.25)$$

where $S_{\tilde{\mathbf{h}}}(z)$ is the z-transform of the autocorrelation function $R_{\tilde{\mathbf{h}}}[n]$ in (3.24).

As we mentioned in the previous section, the continuous-time matched filter (3.20) can be implemented in a more robust fashion in discrete time. In the special case that the channels are LTI, this implementation is obtained by bandlimited sampling of the continuous-time waveform at K times the baud rate T_o^{-1}, followed by a discrete-time matched filter and decimation by a factor of K, where $K \geq 2$ [4].

3.2.2.2 Sufficient Statistics for Additive White Gaussian Noise Channels

In certain cases where the multipath spread is much smaller than the intersymbol period T_o (or even the chip duration T) and channel is very slowly time-varying compared to the symbol rate, the channel may be well approximated as LTI with impulse response equal to a delayed impulse. We refer to such channels as additive white Gaussian noise (AWGN) channels. The ith channel in the AWGN scenario is a special case of (3.22) and is given by

$$h_i(t; \tau) = \alpha_i \delta(\tau - \tau_i), \quad (3.26)$$

where α_i is the (complex) gain of the ith channels, and τ_i is the delay of the ith user with respect to the receiver. The delay τ_i represents the net effect of propagation delay and transmission delay with respect to the (reference) receiver. Each delay is assumed to be known at the receiver since it is part of the known channel $h_i(t; \tau)$.

We will sometimes consider the synchronous communication case which corresponds to $\tau_1 = \tau_2 = \cdots = \tau_P = \bar{\tau}$.

The associated effective channel $\tilde{\mathbf{h}}_n(t)$ in (3.17) is given by

$$\tilde{\mathbf{h}}_n(t) = [\alpha_1 c_1(t - nT_o - \tau_1) \quad \alpha_2 c_2(t - nT_o - \tau_2) \quad \cdots \quad \alpha_P c_P(t - nT_o - \tau_P)]^T, \tag{3.27}$$

and the ith entry of the sufficient statistic $\mathbf{y}[n]$ satisfies [3]

$$y_i[n] = \alpha_i^* \int_{-\infty}^{\infty} r(t)c_i^*(t - nT_o - \tau_i) \, dt. \tag{3.28}$$

Substituting for $r(t)$ in (3.28), the expression in (3.18) and in conjunction with (3.26) reveals that the contribution of the kth symbol of the jth user to the nth symbol of the ith correlation is proportional to $\alpha_i^* \alpha_j \rho_{i,j}[n, k]$, where

$$\rho_{i,j}[n, k] = \int_{-\infty}^{\infty} c_j(t - kT_o - \tau_j)c_i^*(t - nT_o - \tau_i) \, dt, \tag{3.29}$$

which in the general case will be non-zero for some $n - k$, even if $i \neq j$. Note that $y_i[n]$ is generally not a sufficient statistic for the detection of $s_i[n]$, although the set of all $y_i[n]$ for all i and n forms a set of sufficient statistics for $\mathbf{S}$. In [3], a Viterbi algorithm, which has $O(2^{P-1})$ complexity per binary decision, is presented for the implementation of the ML sequence estimator. In effect, the system is equivalent to a single-user system, where the user codes the nth vector symbol $\mathbf{s}[n] = \boldsymbol{\theta}$ on the sequence

$$p_{\boldsymbol{\theta}}(t) \triangleq \sum_{i=1}^{P} \theta_i \alpha_i c_i(t - \tau_i).$$

In this form, it is readily apparent that ML sequence estimation generally corresponds to a Viterbi algorithm with $|s|^P$ states, where $|s|$ denotes the size of the (common) alphabet from which $s_i[n]$ for any i is drawn. Consequently, ML sequence detection is a combinatorial optimization problem in the number of users P. In [3], the specific structure of $p_{\boldsymbol{\theta}}(t)$ is exploited to obtain an ML sequence detection algorithm that is more efficient in terms of computations but nevertheless still of exponential complexity in the number of users.

Finally, consider the special case where transmission is synchronous and each signature is time-limited in $[0, T_o]$. In this case, $\rho_{i,j}[n, k]$ in (3.29) is zero for all i, j if $n \neq k$. This implies that the nth vector sample $\mathbf{y}[n]$ in (3.20) is a sufficient statistic for the nth vector symbol $\mathbf{s}[n]$, so that (3.19) decouples into a set of separate maximizations for each vector symbol. However, unless $\rho_{i,j} \triangleq \rho_{i,j}[n, n]$ is of the form $\rho_{i,j} = \rho_{i,i}\delta[i - j]$, the complexity of the computation of the ML estimate of each supersymbol is still exponential in the number of users [7].

Next, we focus on suboptimal equalization schemes in the presence of fading, which are significantly less computationally intensive than the ML sequence detector (3.19).

3.2.3 The Matched-Filter Receiver for Time-Varying Channels

A simple receiver often used if the channel is time-varying and known is a generalized form of the matched-filter receiver for linear-time varying channels. As mentioned in earlier chapters, this receiver is usually referred to as the (single-user) RAKE receiver. Although this receiver combats time-selective fading, it does not use knowledge of the surrounding symbols to reduce ISI and treats the signals of the other users in the system as noise, so that it typically suffers from considerable cochannel interference.

The single-user RAKE receiver, invented by Price and Green [8], is optimal in terms of minimizing the probability of a symbol error in the scenario that the channel is known, there is only one user, and only one symbol s is transmitted. Specifically, let the transmitted waveform associated with the symbol s be denoted by $c_s(t)$. The signal $c_s(t)$ passes through a linear channel with kernel $h(t; \tau)$. From (3.5), for any modulation scheme the received signal satisfies

$$r(t) = \int_{-\infty}^{\infty} h(t; \tau) c_s(t - \tau) \, d\tau + w(t). \tag{3.30}$$

Since the channel is known, the signal

$$v_s(t) \triangleq \int_{-\infty}^{\infty} h(t; \tau) c_s(t - \tau) \, d\tau$$

can be computed for every possible value of s. Thus, (3.30) reduces to a test of the following hypotheses:

$$r(t) = v_s(t) + w(t) \quad \text{under } H_s.$$

The RAKE receiver comprises a bank of filters each matched to $v_s(t)$ corresponding to a particular value of s, followed by subsequent energy adjustments and a decision. In this additive white Gaussian noise scenario, it is the detector that minimizes the probability of error.

In the special case that the modulation scheme is linear such as in (3.9), $c_s(t)$ is given by (3.12) for $N = 0$ and $P = 1$. In this case, a single sufficient statistic $y_1[0]$ is obtained from (3.20). The statistic $y_1[0]$ corresponds to the output of $r(t)$ through a matched filter $\tilde{h}_1^*(-t; 0)$ in (3.13) sampled at $t = 0$, which can then be substituted in (3.19) to obtain the ML estimate of s. In general, for linear modulation schemes, the MF detector corresponds to first obtaining the set of sufficient statistics (3.20) and then selecting as the estimate of the nth symbol of the ith user the symbol from the ith alphabet that is closest to the ith element of $\mathbf{y}[n]$.

The single-user RAKE receiver provides one of the most elementary detectors used in time-selective fading channels. Its performance for the model (3.30) provides an upper bound on the performance of all the schemes for which $N > 0$ (ISI is present), there is insufficient knowledge of the channel, and there are also multiple users ($P > 1$). However, this receiver does not compensate for ISI and MAI, and in fact its performance is generally MAI- and ISI-limited; even in the absence of noise, the performance of this receiver can be very poor. Since it does not eliminate MAI, this receiver has poor near-far resistance.[2]

We next briefly consider multiuser extensions of the single-user ZF detectors which result in eliminating both MAI and ISI.

3.2.4 Linear Zero-Forcing Equalizers

Like any linear equalizer, the linear ZF equalizer applies to the received data a linear transformation described by a linear kernel $\mathbf{g}_{\mathrm{LZF}}[n; k]$ followed by a slicer. The linear ZF equalizer has the property that the linear kernel output is exactly the original symbol sequence $\mathbf{s}[n]$ in the absence of noise. This receiver ignores the effects of noise and undoes the effects of the channels, thus completely removing the effects of ISI and MAI. However, in doing so, it generally enhances the noise. The linear ZF equalizer filter kernel for LTV channels naturally arises as a special case of the optimal linear MMSE equalizer, so we defer its development to Section 3.2.5. In the special case that the channel is LTI, considering (3.24) in its z-domain form (3.25) reveals that the ZF equalizer reduces to an LTI filter matrix with frequency response

$$G_{\mathrm{LZF}}(z) = (S_{\tilde{\mathbf{h}}}(z))^{-1}. \tag{3.31}$$

As in the single-user case, the linear ZF equalizer does not always exist; for instance, if for some z on the unit circle $S_{\tilde{\mathbf{h}}}(z) = \mathbf{0}$, then a ZF equalizer does not exist.

We next consider in more detail the ZF equalizer in the special case that the channel is AWGN.

3.2.4.1 AWGN Channel—The Decorrelating Receiver

As discussed in Chapter 2, in the special case that the channel is AWGN, the linear ZF equalizer is usually referred to as a decorrelating receiver [7]. We first briefly discuss synchronous communication, which corresponds to having $\tau_i = \tau_j$ for all i, j in (3.26) and where the α_i's are the received symbol amplitudes. In this case, the decorrelating receiver represents an effective performance/complexity trade-off between the optimal ML receiver and the single-user matched filter [7]. Specifi-

---

[2]In general, as discussed in Chapter 2, a receiver is not near-far resistant if as the additive-noise level goes to zero, there is a choice of MAI power levels that result in non-zero symbol-error probability.

cally, consider the detection of $\mathbf{s} = \mathbf{s}[0]$ from (3.10) at time 0. For convenience, we drop the time-dependence from all variables for the remainder of this section. Let $R_{\tilde{\mathbf{h}}}$ correspond to the $P \times P$ matrix whose (i, j)th entry is given by $[R_{\tilde{\mathbf{h}}}]_{i,j} = \alpha_i^* \alpha_j \rho_{i,j}$, where $\rho_{i,j}$ is given by (3.29). Then, the sufficient statistic for $\mathbf{s}$ given by (3.20) reduces to

$$\mathbf{y} = R_{\tilde{\mathbf{h}}} \mathbf{s} + \mathbf{w}, \tag{3.32}$$

where $\mathbf{w}$ is a zero-mean Gaussian random vector with covariance matrix $\mathcal{N}_o R_{\tilde{\mathbf{h}}}$.

The single-user, matched-filter (RAKE) detector for the ith user treats all the other users as noise and uses solely the ith entry of $\mathbf{y}$, namely y_i, to estimate s_i. Specifically, it selects as its estimate the symbol from the ith alphabet that is closest to y_i in terms of its Euclidean distance. This receiver is optimal in terms of minimizing the probability of error in the single-user case, as well as in the multiuser scenario in which the user's signature waveforms are orthogonal, i.e., $\rho_{i,j} = \rho_{i,i} \delta[i - j]$. In case the codewords are not orthogonal, however, the performance of this receiver is MAI-limited. Even in the absence of noise, the single-user detector may result in unacceptably high probability of error for any given user whose codeword is not orthogonal to all the others, for certain choices of received power levels, given by $w_i = |\alpha_i|^2 \rho_{i,i}$. In this sense, even in the AWGN case, the conventional single-user detector has poor near-far resistance characteristics [7].

Recall from Chapter 2 that the decorrelating receiver is a linear ZF receiver [2] in that it is based on linear transformations of the received signal that result in producing the desired symbols in the absence of noise. Let us assume that the codewords $c_i(t)$ are all linearly independent, in which case $R_{\tilde{\mathbf{h}}}$ in (3.32) is invertible. In the absence of noise, $R_{\tilde{\mathbf{h}}}^{-1} \mathbf{y}$ provides the desired set of symbols $\mathbf{s}$, i.e., it completely removes MAI. In the presence of noise, the detector would first obtain

$$\tilde{\mathbf{s}} = R_{\tilde{\mathbf{h}}}^{-1} \mathbf{y} = \mathbf{s} + R_{\tilde{\mathbf{h}}}^{-1} \mathbf{w} \tag{3.33}$$

and then select as the ith symbol estimate $\hat{s}_i$ the symbol from the ith alphabet that is closest in terms of its Euclidean distance to the ith entry of $\tilde{\mathbf{s}}$. Note that, as is usually the case with ZF equalizers, the noise term in (3.33) is colored, so basing the decision on the ith symbol solely on the ith entry of $\tilde{\mathbf{s}}$ is, in general, suboptimal. If the signatures $c_i(t)$ form a linearly dependent set, one can define decorrelating receivers based on any generalized inverse $R_{\tilde{\mathbf{h}}}^I$ of $R_{\tilde{\mathbf{h}}}$, namely, [7]

$$\tilde{\mathbf{s}} = R_{\tilde{\mathbf{h}}}^I \mathbf{y}. \tag{3.34}$$

In the absence of noise, the detector (3.34) would make correct symbol decisions for any user whose codeword $c_i(t)$ is linearly independent from all the other codewords. However, there is still ambiguity regarding the symbols corresponding to users whose codewords are linearly dependent, and this ambiguity cannot be resolved via linear methods. Note that such ambiguities are often resolvable by ML detectors since these exploit the discrete alphabet property of the symbol streams. A nice feature of the decorrelating receiver is that it achieves the near-far resistance of the ML receiver [7] while maintaining low complexity.

In the asynchronous Gaussian channel case, where the user signatures overlap in time and space, MAI interference is inevitable. As we have seen, due to asynchrony, ML detection is a dynamic programming problem where the number of states grows exponentially with the number of users [3]. A large effort has been devoted to developing practical near-far resistant algorithms for this channel, which often arises in the reverse link of a direct line-of-sight wireless CDMA system. Since the decorrelating receiver is a linear filter which, when applied to the data, reproduces the original symbol sequences in the absence of noise, it clearly constitutes a special case of the linear ZF equalizer for multiple-access LTI channels [9]. Other aspects of the decorrelating receiver are considered in Chapter 2.

We next consider linear equalizers that are optimal in the sense that they result in the minimum mean-square symbol estimation error among all linear equalizers.

3.2.5 Linear MMSE Equalization

A linear equalizer applied on an observed waveform is a linear transformation of the data resulting in a sequence of soft symbol estimates $\tilde{\mathbf{s}}[n]$, followed by a set of P quantizers to obtain the (hard) symbol decisions $\hat{\mathbf{s}}[n]$. In this section, we develop linear MMSE and ZF equalizers operating on the sufficient statistic sequence $\mathbf{y}[n]$, in the case that the channel associated with each user is LTV and known to the receiver. In [10], Klein et. al. develop linear and decision-feedback ZF and MMSE equalizers via a method relying on block processing of long sequences of symbols. Here, we present an alternative approach, which results in recursive equalizers arising as Kalman filters based on appropriately formulated state-space models.

It is convenient to first whiten the noise by means of a linear, causal, invertible transformation of $\mathbf{y}[n]$ in (3.21). Let $\mathbf{b}[n; k]$ denote the causal $P \times P$ matrix kernel that whitens the noise sequence $\mathbf{v}[n]$ for $n = 0, 1, \cdots$, i.e.,

$$\tilde{\mathbf{v}}[n] = \sum_{k \geq 0} \mathbf{b}[n; k]\mathbf{v}[n-k], \tag{3.35}$$

where the noise sequence $\tilde{\mathbf{v}}[n]$ has jointly uncorrelated components and is white with power-spectral density $\mathcal{N}_o$, i.e.,

$$R_{\tilde{\mathbf{v}}}[n, k] = \mathcal{N}_o \mathbf{I}\delta[n-k]. \tag{3.36}$$

The coefficients $\mathbf{b}[n; k]$ can be obtained by Gram-Schmidt orthogonalization starting with $\mathbf{b}[n; 0]$. Note that for any n, $\mathbf{b}[n; k] = 0$ for $k > n$ and $k < 0$, due to the method of its construction. Since no information is lost (i.e., we can retrieve $\mathbf{v}[n]$ from $\tilde{\mathbf{v}}[n]$), we can use the whitening filter as a front end for any equalizer without loss of optimality. Consequently, let $\mathbf{h}[n; k]$ denote the cascade of $\mathbf{b}[n; k]$ with $R_{\tilde{\mathbf{h}}}[n; k]$. Then, the response of $\mathbf{y}[n]$ to the whitening filter $\mathbf{b}[n; k]$ satisfies

$$\tilde{\mathbf{y}}[n] = \sum_k \mathbf{h}[n; k]\mathbf{s}[n - k] + \tilde{\mathbf{v}}[n]. \tag{3.37}$$

Since $R_{\tilde{\mathbf{h}}}[n; k]$ and the whitening filter are known, so is $\mathbf{h}[n; k]$. The linear matrix kernel $\mathbf{b}[n; k]$ is a multiuser, time-varying generalization of the well-known, single-user, whitened matched filter (WMF) for LTI channels.

In the approach that follows, we assume that we can find two integers $N_f, N_b > 0$ such that $\mathbf{h}[n; k] \approx 0$ if $k > N_b$ or $k < -N_f$ for all n. Also, we assume that $\mathbf{s}[n]$ is a zero-mean WSS process with $E[\mathbf{s}[n]\mathbf{s}^\dagger[n]] = \mathcal{E}_s\mathbf{I}$. We can then sketch the form of the linear MMSE equalizer. Let

$$\mathbf{x}[n] = [\mathbf{s}^T[n + N + N_f] \ \cdots \ \mathbf{s}^T[n + 1] \ \mathbf{s}^T[n] \ \cdots \ \mathbf{s}^T[n + \overline{N} - N_b]]^T, \tag{3.38}$$

where N denotes the smoothing window size and satisfies $N \geq 0$, and where $\overline{N}$ equals the minimum of N and N_b. A state-space description for this system is given by the following set of equations:

$$\mathbf{x}[n] = \mathbf{F}\mathbf{x}[n - 1] + \mathbf{G}\mathbf{s}[n + N + N_f] \tag{3.39a}$$

$$\check{\mathbf{y}}[n] = \mathbf{H}[n]\mathbf{x}[n] + \check{\mathbf{v}}[n], \tag{3.39b}$$

where $\check{\mathbf{y}}[n] = \tilde{\mathbf{y}}[n + N]$ and $\check{\mathbf{v}}[n] = \tilde{\mathbf{v}}[n + N]$. The matrix $\mathbf{F}$ in (3.39a) is the following $(N'P) \times (N'P)$, $(P \times P)$–block delay matrix:

$$\mathbf{F} = \begin{bmatrix} 0 & 0 & 0 & \cdots & 0 \\ \mathbf{I} & 0 & 0 & \cdots & 0 \\ 0 & \mathbf{I} & 0 & \cdots & 0 \\ \vdots & \ddots & \ddots & \ddots & \vdots \\ 0 & \cdots & 0 & \mathbf{I} & 0 \end{bmatrix}, \tag{3.40a}$$

where $N' = N + N_f + N_b + 1 - \overline{N}$, and the $(N'P) \times P$ matrix $\mathbf{G}$ satisfies

$$\mathbf{G} = [\mathbf{I}_{P \times P} \ \mathbf{0}_{P \times (N'P - P)}]^T. \tag{3.40b}$$

The measurement equation (3.39b) corresponds to rewriting (3.37) in terms of the vector $\mathbf{x}[n]$ instead of $\mathbf{s}[n]$, and where the matrix $\mathbf{H}[n]$ is composed of blocks of the WMF kernel $\mathbf{h}[n; k]$ and is thus known.

The Kalman filter for the state-space model (3.39) provides the soft symbol estimates of the linear MMSE equalizer and has the following form:

$$\tilde{x}[n\,|\,n] = F\tilde{x}[n-1\,|\,n-1] + \mu[n](\check{y}[n] - H[n]F\tilde{x}[n-1\,|\,n-1]) \tag{3.41a}$$

$$\mu[n] = \Lambda[n\,|\,n-1]H^{\dagger}[n](H[n]\Lambda[n\,|\,n-1]H^{\dagger}[n] + \mathcal{N}_{o}I)^{-1} \tag{3.41b}$$

$$\Lambda[n\,|\,n] = (I - \mu[n]H[n])\Lambda[n\,|\,n-1] \tag{3.41c}$$

$$\Lambda[n\,|\,n-1] = F\Lambda[n-1\,|\,n-1]F^{T} + \mathcal{E}_{s}GG^{T} \tag{3.41d}$$

initialized with $\tilde{x}[-1\,|\,-1] = 0$ and $\Lambda[-1\,|\,-1] = \mathcal{E}_{s}I$. The vector $\tilde{x}[n\,|\,n]$ comprises the equalizer's soft estimates of symbols $s[n - N_{b} + \overline{N}], \ldots, s[n], \ldots, s[n + N + N_{f}]$, given observations up to time $n + N$. For convenience, we use the following notation:

$$\tilde{x}[n\,|\,n] = [\tilde{s}^{T}[n+N+N_{f}\,|\,n+N] \cdots \tilde{s}^{T}[n\,|\,n+N] \cdots \tilde{s}^{T}[n+\overline{N}-N_{b}\,|\,n+N]]^{T}. \tag{3.42}$$

The matrix $\Lambda[n\,|\,k]$ in (3.41) is the error covariance matrix associated with $\tilde{x}[n\,|\,k]$ and thus conveniently provides the mean-square error associated with the soft symbol estimates. For instance, since $\tilde{x}[n\,|\,n]$ and $\tilde{s}[n\,|\,n + N]$ are related via (3.42), the $(N_{f} + 1, N_{f} + 1)$st $P \times P$ block of $\Lambda[n\,|\,k]$ is the error covariance associated with the soft estimate $\tilde{s}[n\,|\,n + N]$.

The hard symbol estimates of $s[n]$ associated with the linear MMSE equalizer are given from the soft estimates in (3.42) via

$$\hat{s}[n] = \arg\max_{s} \|\tilde{s}[n\,|\,n + N] - s[n]\|^{2},$$

which decouples to P separate optimizations

$$\hat{s}_{i}[n] = \arg\max_{s_{i}} \|\tilde{s}_{i}[n\,|\,n + N] - s_{i}[n]\|^{2} \qquad i = 1, 2, \cdots, P. \tag{3.43}$$

The linear ZF equalizer can be obtained by letting $\mathcal{N}_{o} \to 0$ in (3.41b), i.e., it is given by (3.41)–(3.43) where (3.41b) is replaced by

$$\mu[n] = \Lambda[n\,|\,n-1]H^{\dagger}[n](H[n]\Lambda[n\,|\,n-1]H^{\dagger}[n])^{-1}.$$

3.2.5.1 Linear Equalization for LTI Channels

In this section, we examine linear equalization schemes for multiuser LTI channels that can be used to combat the three effects of noise, ISI, and MAI. These equalizers arise as natural multiuser extensions of their single-user counterparts commonly used for equalization in known LTI channels.

As in single-user equalization over LTI channels, it is convenient to revert from a state-space to an input/output problem description. We assume that the input sequence to the channel is the $P \times 1$ wide-sense stationary (WSS) vector sequence $\mathbf{s}[n]$ in (3.10). The received signal is efficiently described by the vector $\mathbf{y}[n]$ given by (3.24), which can be conveniently viewed as providing P observations per vector sample of the information-bearing signal $\mathbf{s}[n]$. In this section, we assume that we know $S_{\tilde{\mathbf{h}}}(z)$, the z-transform of the autocorrelation function $R_{\tilde{\mathbf{h}}}[n]$ in (3.24), and that $S_{\tilde{\mathbf{h}}}(z)$ is rational and stable, i.e., all entries of its z-transform are rational and stable transfer functions.

A linear equalizer for the vector model (3.25) applies a matrix filter kernel to the vector of outputs $\mathbf{y}[n]$ to obtain a sequence $\tilde{\mathbf{s}}[n]$, followed by a set of P quantizers to obtain the symbol decisions $\hat{\mathbf{s}}[n]$. We may then derive the linear MMSE equalizer whose objective is to minimize $E[\|\tilde{\mathbf{s}}[n] - \mathbf{s}[n]\|^2]$, where $\|\mathbf{u}\|$ is the Euclidean norm of the vector $\mathbf{u}$, and $\tilde{\mathbf{s}}[n]$ is the soft estimate of $\mathbf{s}[n]$. Since $\mathbf{s}[n]$ is WSS, from (3.24) the processes $\mathbf{s}[n]$ and $\mathbf{y}[n]$ are jointly WSS. In particular, the spectrum of $\mathbf{y}[n]$ satisfies

$$S_{\mathbf{y}}(z) = S_{\tilde{\mathbf{h}}}(z)S_{\mathbf{s}}(z)S_{\tilde{\mathbf{h}}}(z) + \mathcal{N}_o S_{\tilde{\mathbf{h}}}(z), \tag{3.44}$$

while the cross-spectrum between $\mathbf{y}[n]$ and $\mathbf{s}[n]$ is given by

$$S_{\mathbf{sy}}(z) = S_{\mathbf{s}}(z)S_{\tilde{\mathbf{h}}}(z). \tag{3.45}$$

Since $\mathbf{s}[n]$ and $\mathbf{y}[n]$ are jointly WSS, the resulting linear MMSE filter is time-invariant. Let $G(z)$ be the $P \times P$ matrix MMSE filter, i.e., $\tilde{\mathbf{s}}(z) = G(z)\mathbf{y}(z)$. The orthogonality condition

$$E[(\tilde{\mathbf{s}}[n+k] - \mathbf{s}[n+k])\mathbf{y}^\dagger[n]] = \mathbf{0} \qquad \text{for all } k$$

implies that $S_{\tilde{\mathbf{s}}\mathbf{y}}(z) = S_{\mathbf{sy}}(z)$. Using (3.45) and (3.44), we can deduce that [6]

$$G_{\text{LMMSE}}(z) = S_{\mathbf{s}}(z)[S_{\tilde{\mathbf{h}}}(z)S_{\mathbf{s}}(z) + \mathcal{N}_o \mathbf{I}]^{-1}, \tag{3.46}$$

where $\mathbf{I}$ is the identity matrix. As in the single-user case, it can be shown that there exists $\mathbf{g}[n]$ with transfer function $G_{\text{LMMSE}}(z)$, which is stable [6]. Although conceptually convenient, the equalizer (3.46) is in general noncausal. In particular, in the case that it has infinite impulse response (IIR) and is anticausal, delay constraints imply that only approximate realizations of this equalizer can be implemented. From this point of view, the Kalman filtering algorithm (3.41) appears very attractive both in terms of its recursive implementation and in terms of providing the linear MMSE equalizer output for a given delay constraint (which is directly related to the smoothing factor N). In addition, an estimate for the mean-square error is conveniently available via the covariance matrix $\Lambda[n \mid n]$. Finally, note that by setting $\mathcal{N}_o = 0$ in (3.46), we obtain the linear ZF equalizer (3.31).

3.2.6 Successive Cancellation and Decision-Feedback Equalizers for Multiple-Access Channels

Optimal detection in known multiuser fading channels is prohibitively complex, while the low-complexity, single-user detectors can greatly suffer in performance due to MAI (and possibly ISI) and do not possess any near-far resistance. As we have discussed above, the linear multiuser equalizers presented in Sections 3.2.4–3.2.5 often provide a reasonable trade-off between detector complexity and performance. Successive cancellation schemes may be employed as an alternative to linear multiuser techniques, or, more often, they can be used in conjunction with linear approaches to further improve performance at some additional cost in complexity. These often arise as extensions of single-user techniques in a multiuser setting created in such a way that the resulting equalizer complexity is linear or at most polynomial in the number of users.

To illustrate how a successive cancellation scheme can remove MAI, consider a two-user scenario where user 2 experiences a lot of interference from user 1 due to their nonorthogonality and a large difference in terms of their relative received power levels, i.e., $w_1 \gg w_2$. Let us also assume that only one symbol per user is sent, so that there is no ISI. The ML receiver performs the optimization in (3.19), where the adjustment for the immense power-level difference (and the resulting strong MAI that user 2 is experiencing) is reflected in the term $\mathcal{E}_s$. On the other hand, it is easy to show that a single-user RAKE receiver would perform poorly in this two-user scenario. Specifically, the RAKE receiver of the first user receives negligible interference from user 2, so that at high signal-to-noise ratio (SNR) we have $\hat{s}_1 = s_1$ with high probability. However, the second user's RAKE receiver is MAI-limited since its output strongly depends on the actual symbol that user 1 is transmitting.

An obvious successive cancellation scheme for detecting the signal of user 2 in this scenario would first estimate the symbol of user 1 (by means of a RAKE receiver), subtract its contribution from $r(t)$ (assuming $\hat{s}_1 = s_1$), and set up the single-user RAKE receiver for user 2. If the modulation scheme is linear, user 2 may substitute $\hat{s}_1$ in (3.19) and perform the optimization over s_2. Although in general suboptimal, this scheme does not suffer from the very poor near-far performance experienced by the single-user receiver. Furthermore, this scheme can be easily generalized to the P-user case. The users are first ordered in terms of their power level, i.e., $w_1 \geq w_2 \geq \cdots \geq w_P$. Having detected the symbols of users 1, 2, $\cdots$, $i - 1$, a decision for the ith user's symbol can be formed by first regenerating $x_1(t)$ through $x_{i-1}(t)$ based on the decoded symbols $\hat{s}_1, \hat{s}_2, \cdots, \hat{s}_{i-1}$, stripping off their response from the received signal $r(t)$, and consequently feeding the resulting signal to the ith user RAKE detector.

A number of successive cancellation methods employ several stages of inter-ference rejection to improve performance. Such approaches have their roots in the original multistage MAI rejection method for CDMA systems developed in [11], which is very similar to the successive cancellation approach presented above. Specifically, in [11] Varanasi and Aazhang consider a scenario of P users where ISI from past and future symbols is negligible. At the mth stage, a decision for the ith user symbol is formed by first removing all MAI from the received signal by using the $(m - 1)$st stage symbol estimates and consequently feeding the resulting signal to a single-user receiver for the ith user. The performance of this multistage algo-rithm greatly depends on the choice of the initial symbol estimates (1st stage). In the algorithm described in [11], these estimates are formed via a linear receiver, e.g., a decorrelating receiver.

Although its performance can often be inferior to the one obtained by linear methods and even by conventional single-user receivers, the algorithm in [11] gen-erated considerable interest in the area of successive cancellation. Several other decision-based detectors have subsequently been developed; see, e.g., [12] and the references therein. For instance, a number of other successive cancellation algo-rithms have been proposed for CDMA systems [13–16]. Successive cancellation methods that can approach ML optimal performance in terms of symbol error-rate and near-far resistance have also been developed for spread-signature CDMA systems [17]. More generally, developing robust, low-complexity, multistage-successive-cancellation schemes with improved performance and well-behaved dynamical behavior is a rich and active area of research. In addition, obtaining a better assessment of the performance-complexity trade-offs of successive cancella-tion schemes, as well as of the performance limits as compared to the optimal detectors, constitute issues that are worth further investigation.

Use of successive cancellation to remove MAI is similar to the use of decision-feedback equalization in single-user settings to remove ISI. Specifically, the design of a conventional, single-user, decision-feedback equalizer (DFE) is based on the assumption that past symbol decisions are correct, so that ISI from past symbols can be stripped off before detection of the current symbol. In this sense, DFEs con-stitute a particular class of successive cancellation schemes that remove MAI and ISI based on hard estimates of detected symbols.

We next consider a number of natural multiuser extensions of single-user DFEs that attempt to further improve immunity to MAI and ISI.

3.2.6.1 Decision-Feedback Equalization

A DFE in a multiuser setting uses past symbol decisions from all users in order to remove ISI and MAI, and soft symbol estimates for removal of ISI and MAI from future symbols. When designing DFEs, it is assumed that all the past decisions are correct; this assumption implies that the equalizer is operating at a sufficiently high SNR that the decisions it makes are correct with high probability [2]. As in the single-user case, analysis based on the minimum probability of error criterion is usually difficult. As a result, simpler suboptimal criteria such as MMSE or ZF are usually selected.

In this section, we present DFEs operating on the sufficient statistic sequence $\mathbf{y}[n]$ from (3.20) for known multiuser LTV channels. In developing the MMSE DFE, it is convenient to use the observations in the form of the output of the WMF $\tilde{\mathbf{y}}[n]$ in (3.37), which is an equivalent description of $\mathbf{y}[n]$ in (3.20). In particular, this approach allows us to develop a recursive algorithm for obtaining the soft estimate of the vector symbol $\mathbf{s}[n]$ for any given time index n. Specifically, consider a *fixed* index n. As discussed above, it is assumed that the MMSE DFE hard symbol estimates $\hat{\mathbf{s}}[n-k]$ with $k > 0$ have been obtained and satisfy

$$\hat{\mathbf{s}}[n-k] = \mathbf{s}[n-k] \qquad \text{for all } k > 0. \tag{3.47}$$

Assuming that (3.47) is true, and since the noise sequence $\tilde{\mathbf{v}}[k]$ in (3.37) is white and $\mathbf{h}[n; k] \approx 0$ for $k > -N_f$ or $k < N_b$, we need only consider $\tilde{\mathbf{y}}[n+k]$ for $k \geq -N_f$ in determining the MMSE DFE soft estimate of the vector symbol $\mathbf{s}[n]$. We may thus consider a state-space model where the state at time $k \geq -N_f$ includes $\mathbf{s}[n]$ as well as all the symbols that contribute to $\tilde{\mathbf{y}}[n+k]$. Specifically, consider the following vector sequence as the state of the state-space model

$$\mathbf{x}_n[k] \triangleq [\mathbf{s}^T[n+k+N_f] \quad \mathbf{s}^T[n+k+N_f-1] \quad \cdots$$
$$\mathbf{s}^T[n+k-N_b] \quad \mathbf{s}^T[n+\bar{k}-N_b-1]]^T, \tag{3.48}$$

where

$$\bar{k} = \begin{cases} k & \text{if } -N_f - 1 \leq k \leq N_b + 1 \\ N_b + 1 & \text{if } k > N_b + 1 \end{cases} ;$$

as k increases from $-N_f$ to N_b, the vector symbol of interest, namely $\mathbf{s}[n]$, "slides" from the first to the last $P \times 1$ block of $\mathbf{x}_n[k]$, where it remains for $k \geq N_b + 1$. A state-space description that generates (3.48) consists of the following set of equations:

$$\mathbf{x}_n[k] = \mathbf{F}[k]\mathbf{x}_n[k-1] + \mathbf{G}\mathbf{s}[n+k+N_f] \tag{3.49a}$$

$$\check{\mathbf{y}}_n[k] = \check{\mathbf{H}}_n[k]\mathbf{x}_n[k] + \check{\mathbf{v}}_n[k], \tag{3.49b}$$

where $\check{\mathbf{y}}_n[k] = \tilde{\mathbf{y}}[n+k]$ and $\check{\mathbf{v}}_n[k] = \tilde{\mathbf{v}}[n+k]$. In the recursion (3.49a), $\mathbf{G}$ is given by (3.40b) for $N' = N_b + N_f + 2$. Also, $\mathbf{F}[k]$ is given by (3.40a) with $N' = N_b + N_f + 2$ for $k \leq N_b + 1$, and

$$\mathbf{F}[k] = \begin{bmatrix} 0 & 0 & 0 & \cdots & 0 & 0 \\ \mathbf{I} & 0 & 0 & \cdots & 0 & 0 \\ 0 & \mathbf{I} & 0 & \cdots & 0 & 0 \\ \vdots & \ddots & \ddots & \ddots & \vdots & \vdots \\ 0 & \cdots & 0 & \mathbf{I} & 0 & 0 \\ 0 & \cdots & 0 & 0 & 0 & \mathbf{I} \end{bmatrix} \qquad \text{for } k > N_b + 1.$$

As a result, when initialized with $\mathbf{x}_n[-N_f - 1]$ from (3.48), the recursion (3.49a) generates the entire sequence $\mathbf{x}_n[k]$ in (3.48). Similarly to (3.39b), the set (3.49b) denotes (3.37) rewritten in terms of $\mathbf{x}_n[k]$ instead of $\mathbf{s}[n]$. In particular, by letting $\mathbf{H}_N[n]$ denote the matrix $\mathbf{H}[n]$ used in (3.39b) for a particular N, the matrix $\check{\mathbf{H}}_n[k]$ in (3.49b) satisfies

$$\check{\mathbf{H}}_n[k] = [\mathbf{H}_k[n] \quad \mathbf{0}_{P \times P}].$$

The state-space model (3.49) can be readily used for designing Kalman filtering algorithms employing fixed-point smoothing [18] and is thus naturally suited for the design of the MMSE DFE. Specifically, the Kalman filter for the state-space model (3.49) has the following form:

$$\tilde{\mathbf{x}}_n[k \mid k] = \mathbf{F}[k]\tilde{\mathbf{x}}_n[k-1 \mid k-1] + \boldsymbol{\mu}_n[k](\check{\mathbf{y}}_n[k] \tag{3.50a}$$
$$- \check{\mathbf{H}}_n[k]\mathbf{F}[k]\tilde{\mathbf{x}}_n[k-1 \mid k-1])$$

$$\boldsymbol{\mu}_n[k] = \boldsymbol{\Lambda}_n[k \mid k-1]\check{\mathbf{H}}_n^\dagger[k](\check{\mathbf{H}}_n[k]\boldsymbol{\Lambda}_n[k \mid k-1]\check{\mathbf{H}}_n^\dagger[k] + \mathcal{N}_0\mathbf{I})^{-1} \tag{3.50b}$$

$$\boldsymbol{\Lambda}_n[k \mid k] = (\mathbf{I} - \boldsymbol{\mu}_n[k]\check{\mathbf{H}}_n[k])\boldsymbol{\Lambda}_n[k \mid k-1] \tag{3.50c}$$

$$\boldsymbol{\Lambda}_n[k \mid k-1] = \mathbf{F}[k]\boldsymbol{\Lambda}_n[k-1 \mid k-1]\mathbf{F}^T[k] + \mathcal{E}_s\mathbf{G}\mathbf{G}^T. \tag{3.50d}$$

With proper initialization, the algorithm (3.50) provides the MMSE DFE soft estimate of $\mathbf{s}[n]$ for the particular index n. Specifically, the assumption that all past symbol decisions are correct can easily be incorporated in the form of prior information; the Kalman filter (3.50) is initialized with

$$\tilde{\mathbf{x}}_n[-N_f - 1 \mid -N_f - 1] = \hat{\mathbf{x}}_n[-N_f - 1 \mid -N_f - 1], \tag{3.51a}$$

where

$$\hat{\mathbf{x}}_n[-N_f - 1 \mid -N_f - 1] = [\hat{\mathbf{s}}^T[n-1] \quad \cdots \quad \hat{\mathbf{s}}^T[n - N_b - N_f - 1] \; \hat{\mathbf{s}}^T[n - N_b - N_f - 2]]^T$$

and

$$\boldsymbol{\Lambda}_n[-N_f - 1 \mid -N_f - 1] = \mathbf{0}. \tag{3.51b}$$

The initialization step (3.51) is an equivalent description of (3.47). As a result, the MMSE DFE soft estimate of the vector symbol $\mathbf{s}[n]$ given all observations up to and including time $n + N$ is given by

$$\tilde{\mathbf{s}}[n \mid n + N]$$

$$= \begin{cases} [\mathbf{0}_{P \times (N_f + N)P} \ \ \mathbf{I}_{P \times P} \ \ \mathbf{0}_{P \times (N_b - N + 1)P}] \tilde{\mathbf{x}}_n[N \mid N] & \text{if } -N_f \leq N \leq N_b \\ [\mathbf{0}_{P \times (N_f + N_b + 1)P} \ \ \mathbf{I}_{P \times P}] \tilde{\mathbf{x}}_n[N \mid N] & \text{if } N \geq N_b + 1, \end{cases} \qquad (3.52)$$

where $\tilde{\mathbf{x}}_n[N \mid N]$ is given by (3.50)–(3.51). In practice, the smoothing window size N may be dictated by physical constraints, e.g., delay constraints or limitations in computational complexity. The MMSE DFE makes P symbol decisions at time n by solving the P separate optimizations (3.43), where $\tilde{\mathbf{s}}[n \mid n + N]$ is obtained via (3.50)–(3.52). In agreement with the conventional single-user MMSE DFE for LTI channels, the detector described by (3.50)–(3.52) and (3.43) removes (MAI and ISI) interference from past symbols, based on their hard estimates $\hat{\mathbf{s}}[n - m]$ for $m > 0$ (incorporated in (3.50) via $\tilde{\mathbf{x}}_n[-N_f - 1 \mid -N_f - 1]$ from (3.51)), and interference from current symbols (MAI) and future symbols (MAI and ISI), based on the soft estimates $\tilde{\mathbf{s}}[n + m \mid n + N]$ for $m > 0$.

The matrix $\mathbf{\Lambda}_n[N \mid N]$ is precisely the error covariance matrix that would be associated with $\tilde{\mathbf{x}}_n[N \mid N]$ from (3.50) in the case that $\hat{\mathbf{x}}_n[-N_f - 1 \mid -N_f - 1] = \mathbf{x}_n[-N_f - 1 \mid -N_f - 1]$. Consequently, the diagonal $P \times P$ block entry of $\mathbf{\Lambda}_n[N \mid N]$ that is associated with $\mathbf{s}[n]$ often provides an accurate estimate of the error covariance associated with $\tilde{\mathbf{s}}[n \mid n + N]$ at high SNR.

We can readily obtain the corresponding ZF DFE by letting $\mathcal{N}_o \to 0$ in (3.50b). Specifically, the ZF DFE is given by (3.50)–(3.52) and (3.43), where (3.50b) is replaced by

$$\mathbf{\mu}_n[k] = \mathbf{\Lambda}_n[k \mid k - 1] \check{\mathbf{H}}_n^\dagger[k] (\check{\mathbf{H}}_n[k] \mathbf{\Lambda}_n[k \mid k - 1] \check{\mathbf{H}}_n^\dagger[k])^{-1}.$$

Note that we can also construct successive cancellation DFE structures that make decisions on one symbol at a time (rather than a block of P symbols at a time). Specifically, we can rearrange the users in order of decreasing relative received signal power, i.e., $w_1 \geq w_2 \geq \cdots \geq w_P$, and view the sequence of vector symbols as a time sequence of single-user symbols

$$\cdots, s_1[n], s_2[n], \cdots, s_P[n], s_1[n + 1], s_2[n + 1], \cdots, s_P[n + 1], \cdots.$$

As (3.36) reveals, the kernel $\mathbf{b}[n; k]$ also whitens the noise for this equivalent single-user formulation, in which the observed sequence is a sequence of scalar observations

$$\cdots, \tilde{y}_1[n], \tilde{y}_2[n], \cdots, \tilde{y}_P[n], \tilde{y}_1[n + 1], \tilde{y}_2[n + 1], \cdots, \tilde{y}_P[n + 1], \cdots.$$

Consequently, we can determine the corresponding Kalman filter for the appropriately modified "single user" state-space model (3.49). At any step, the resulting DFE makes a single decision on a single symbol of a specific user. This DFE removes MAI and ISI interference from past symbols, based on hard estimates, and MAI and ISI from future symbols, based on soft estimates. Regarding ISI from current symbols: in detecting the symbol of any given user, this DFE uses *hard* estimates to remove MAI from users with higher, relative, received signal power, and soft estimates otherwise.

Although optimal in a certain sense, DFEs of the form (3.50)–(3.52) are in general computationally intensive since they often require a modest number of steps (i.e., $N + N_f$) of the Kalman filtering algorithm to produce a single symbol decision per user. Obtaining lower-complexity DFE structures (or successive cancellation solutions, in general) which also possess high performance characteristics is, as discussed above, an active and rich area of research.

3.2.6.2 Adaptive Equalization Algorithms

The concepts behind single-user adaptive DFE structures can be used to suppress MAI in a multiuser setting in slowly time-varying scenarios. Specifically, stochastic gradient algorithms can be constructed for this purpose [19–21]. Such algorithms may employ a training stage where the equalizer taps are first determined, followed by an adaptation stage where symbol estimates are instead used in the stochastic gradient algorithm for adjusting the equalizer filter taps. Similarly, adaptive MAI-suppressing equalizers can be designed based on least-squares algorithms. Examples of these and other decision-directed approaches are described in Chapter 2.

3.2.6.3 Decision-Feedback Equalizers for LTI channels

As in the single-user case, if the channels are all LTI, the pre- and post-cursor (matrix) filters associated with the MMSE and ZF DFEs are time-invariant. A typical block diagram for the DFE is shown in Figure 3.3. The z-transform of $\tilde{\mathbf{s}}[n]$ (denoting the soft estimate of the vector symbol $\mathbf{s}[n]$ before the decision) satisfies

$$\tilde{\mathbf{s}}(z) = A(z)\mathbf{y}(z) - B(z)\hat{\mathbf{s}}(z),$$

where $A(z)$ is a $P \times P$ feed-forward matrix filter, $B(z)$ is a $P \times P$ strictly causal feedback matrix filter, and $\hat{\mathbf{s}}[n]$ is the nth symbol estimate. As in the LTV case, when designing a DFE, we assume that it is operating at a high enough SNR so that the decisions are correct with high probability. Then, the design of $A(z)$ and $B(z)$ is based on placing a criterion for quality on the symbol estimated before the decision.

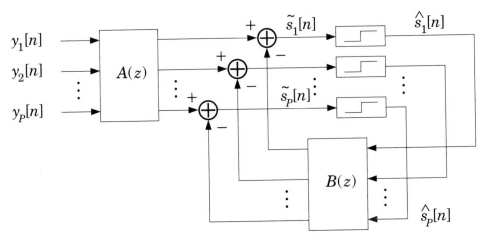

Figure 3.3 DFE block diagram for multiple-access LTI channels.

For illustration, we next consider consider the ZF DFE [22]. Similarly, a multi-user MMSE DFE for LTI channels that is a generalization of its single-user counterpart can be derived [6, 23]. The ZF DFE is the equalizer that eliminates all ISI and MAI at the input of the decision device, assuming that all past decisions were correct. By use of the spectral factorization theorem for matrix spectra that are non-singular on the unit circle, we can write

$$S_{\tilde{\mathbf{h}}}(z)W^{-1} = F(z)F^{\dagger}(z^{-1}),$$

where $F(z)$ is a $P \times P$ causal and stable matrix filter with a stable and causal inverse. The matrix W is a diagonal matrix such that the coefficients multiplying z^0 along the diagonal of $S_{\tilde{\mathbf{h}}}(z)W^{-1}$ are all equal to one. Its significance is that it denotes the received power of the signal of each user, i.e.,

$$[R_{\tilde{\mathbf{h}}}[0]]_{i,i} = [W]_{i,i} = w_i.$$

Based on the assumption that $\hat{\mathbf{s}}[n] = \mathbf{s}[n]$, we must also have

$$\tilde{\mathbf{s}}(z) = \mathbf{s}(z) + \mathbf{n}(z),$$

where $\mathbf{n}(z)$ is a noise process with spectrum $A(z)\mathcal{N}_o S_{\tilde{\mathbf{h}}}(z)A^{\dagger}(z^{-1})$. Since $B(z)$ is a strictly causal filter, the feed-forward filter must remove all the ISI and MAI caused by future samples, i.e.,

$$A(z)\mathbf{y}(z) = M(z)W\mathbf{x}(z) + \mathbf{n}(z),$$

where the $M(z) = \sum_{k=0}^{\infty} M[k]z^{-k}$ is a causal filter. For a given $A(z)$, the corresponding feedback filter $B(z)$ can be written as

$$B(z) = (M(z) - \text{diag } M[0])W.$$

We are interested in the choice of $A(z)$ that would maximize the SNR before the decision or, equivalently, minimize the noise variance for a fixed signal energy. Let $C(z) = M(z)F^{-1}(z)$, and set $\text{diag}(M[0]) = \text{diag}(F[0])$ (without loss of generality). Note that $\text{diag}(C[0]) = \mathbf{I}$ and that $A(z) = C(z)F^\dagger(1/z)$, which implies that the spectrum of $\mathbf{n}(z)$ is given by $\mathcal{N}_o C(z)C^\dagger(1/z)$. The resulting noise variance is minimized by $C(z) = \mathbf{I}$, i.e., the ZF DFE that minimizes the mean-square error is given by $A(z) = (F^\dagger(1/z))^{-1}$, and $B(z) = (F(z) - \text{diag } F[0])W$.

Detection of the current symbol ith user can, in principle, be enhanced if in addition to stripping ISI and MAI from all past symbols, we strip MAI from the current symbol of users whose power level is larger than w_i [6].

3.2.7 Chip-Rate, State-Space Approaches for Time-Varying Channels

The equalizer schemes we have considered in Sections 3.2.2–3.2.6 perform equalization based on discrete-time processing of the set of sufficient statistics $\mathbf{y}[n]$ in (3.20). An alternative approach consists of oversampling the continuous-time signal $r(t)$ many times the symbol rate and obtaining an alternative discrete-time description of the original signal, such as the one in (3.8). Although not a sufficient statistic in general, the resulting sequence $y[n]$ has uncorrelated noise components, which result in significant simplification of equalization. Furthermore, higher oversampling factors will generally result in smaller "information loss" in the A/D stage, while the noise components remain uncorrelated.

In such scenarios, Isabelle and Wornell [24] have developed a collection of state-space methods that recast the equalization problem into one where Kalman filtering methods can be readily applied. We consider the general multiuser channel depicted in Figure 3.1 and described by (3.8).

In the context of this section, we use notation $\mathbf{p}[n]$ and $\mathbf{q}[n; k]$ to denote the Lth order polyphase decompositions of a sequence $p[n]$ and a time-varying channel response $q[n; k]$, respectively, i.e.,

$$\mathbf{p}[n] \triangleq [p[nL] \; p[nL + 1] \; \cdots \; p[nL + L - 1]]^T$$

$$\mathbf{q}[n; k] \triangleq [q[nL; kL] \; q[nL + 1; kL + 1] \; \cdots \; q[nL + L - 1; kL + L - 1]]^T$$

Specifically, we can rewrite (3.8) as

$$\mathbf{y}[n] = \sum_k \tilde{\mathbf{h}}[n; k]\mathbf{s}[n - k] + \mathbf{w}[n], \tag{3.53}$$

where $\mathbf{s}[n]$ is given by (3.10), $\tilde{\mathbf{h}}[n; k]$ is given by

$$\tilde{\mathbf{h}}[n; k] = [\tilde{\mathbf{h}}_1[n; k] \; \tilde{\mathbf{h}}_2[n; k] \; \cdots \; \tilde{\mathbf{h}}_p[n; k]],$$

and where $\mathbf{y}[n]$, $\mathbf{w}[n]$, and $\tilde{\mathbf{h}}_i[n; k]$ denote the Lth order polyphase decompositions of the received signal $y[n]$, the noise sequence $w[n]$, and

$$\tilde{h}_i[n; k] = \sum_\ell h_i[n; \ell]c_i[k - \ell],$$

respectively. From the model (3.8), $w[n]$ is a white sequence with autocorrelation function $R_{\mathbf{w}}[n] = \sigma_w^2 \mathbf{I}\delta[n]$. Note that (3.8) assumes chip-rate sampling (i.e., the vector $\mathbf{y}[n]$ does not denote the sufficient statistic sequence from (3.20)). It is also assumed that all filters are causal and FIR, i.e., for all n we have $h_i[n; k] = 0$ for $k > L_c$ and $k < 0$. Based on (3.53), we can construct a state-space model describing the evolution of the state [24]

$$\mathbf{x}[n] \triangleq [\mathbf{s}^T[n] \quad \mathbf{s}^T[n - 1] \quad \cdots \quad \mathbf{s}^T[n - N + 1]]^T,$$

where N can be viewed as a smoothing window and must be selected at least as large as the effective length of the polyphase components of the time-varying kernel responses $\tilde{h}_i[n; k]$, i.e., $N \geq \lceil (L_c + K)/L \rceil$. Specifically, (3.53) can be written as

$$\mathbf{x}[n + 1] = \mathbf{F}\mathbf{x}[n] + \mathbf{G}\mathbf{s}[n + 1] \tag{3.54a}$$

$$\mathbf{y}[n] = \mathbf{H}[n]\mathbf{x}[n] + \mathbf{w}[n], \tag{3.54b}$$

where $\mathbf{F}$ denotes the $(NP) \times (NP)$, $(P \times P)$–block delay matrix of the form (3.40a), $\mathbf{G}$ is given by (3.40b) for $N' = N$, and $\mathbf{H}[n]$ is the following filtering matrix containing all the channel and codeword information

$$\mathbf{H}[n] = [\tilde{\mathbf{h}}[n; 0] \quad \tilde{\mathbf{h}}[n; 1] \quad \cdots \quad \tilde{\mathbf{h}}[n; N - 1]]^T.$$

The linear MMSE estimator for the state-space model (3.54) can be recursively computed via the Kalman filtering algorithm. The vector $\tilde{\mathbf{x}}[n \,|\, k]$ denotes the estimate of the vector $\mathbf{x}[n]$, using all observations $\mathbf{y}[n]$ up to and including time k, and the matrix $\mathbf{\Lambda}[n \,|\, k]$ to denote the error covariance associated with $\tilde{\mathbf{x}}[n \,|\, k]$. The Kalman filter algorithm corresponding to (3.54) is described by the following set of equations [18]:

$$\tilde{\mathbf{x}}[n \,|\, n] = \mathbf{F}\tilde{\mathbf{x}}[n - 1 \,|\, n - 1] + \boldsymbol{\mu}[n](\mathbf{y}[n] - \mathbf{H}[n]\mathbf{F}\tilde{\mathbf{x}}[n - 1 \,|\, n - 1]) \tag{3.55a}$$

$$\boldsymbol{\mu}[n] = \mathbf{\Lambda}[n \,|\, n - 1]\mathbf{H}^\dagger[n](\mathbf{H}[n]\mathbf{\Lambda}[n \,|\, n - 1]\mathbf{H}^\dagger[n] + \sigma_w^2\mathbf{I})^{-1} \tag{3.55b}$$

$$\mathbf{\Lambda}[n \,|\, n] = (\mathbf{I} - \boldsymbol{\mu}[n]\mathbf{H}[n])\mathbf{\Lambda}[n \,|\, n - 1] \tag{3.55c}$$

$$\mathbf{\Lambda}[n \,|\, n - 1] = \mathbf{F}\mathbf{\Lambda}[n - 1 \,|\, n - 1]\mathbf{F}^T + \mathcal{E}_s\mathbf{G}\mathbf{G}^T \tag{3.55d}$$

initialized with $\tilde{\mathbf{x}}[-1 \,|\, -1] = \mathbf{0}$ and $\mathbf{\Lambda}[-1 \,|\, -1] = \mathcal{E}_s\mathbf{I}$. The soft estimate $\tilde{\mathbf{s}}[n]$ of the nth vector symbol $\mathbf{s}[n]$ is given by

$$\tilde{\mathbf{s}}[n] = [\mathbf{0}_{P \times (N-1)P} \quad \mathbf{I}_{P \times P}]\tilde{\mathbf{x}}[n + N - 1 \,|\, n + N - 1]. \tag{3.56}$$

Finally, the linear MMSE equalizer based on observation of (3.8) selects as the estimate of the nth symbol of the ith user the symbol from the ith alphabet that is closest in Euclidean distance to the ith element of $\tilde{\mathbf{s}}[n]$ in (3.56).

Other related implementation structures can be obtained from (3.55) by appropriate selection of the gain $\boldsymbol{\mu}[n]$. For example, the linear ZF and MF equalizers can be obtained by replacing the gain in (3.55b) with

$$\boldsymbol{\mu}[n] = \boldsymbol{\Lambda}[n\,|\,n-1]\mathbf{H}^{\dagger}[n](\mathbf{H}[n]\boldsymbol{\Lambda}[n\,|\,n-1]\mathbf{H}^{\dagger}[n])^{-1}, \qquad (3.57)$$

and

$$\boldsymbol{\mu}[n] = \boldsymbol{\Lambda}[n\,|\,n-1]\mathbf{H}^{\dagger}[n]/\sigma_w^2,$$

respectively [24]. In these cases, however, the associated matrices $\boldsymbol{\Lambda}[n\,|\,n-1]$ no longer correspond to the associated estimation-error covariance matrices except for limiting cases: $\sigma_w^2 \to 0$ in the ZF case and $\sigma_w^2 \to \infty$ in the MF case. Finally, we should emphasize that the framework presented in this section can be easily extended to perform equalization in the context of systems employing antenna diversity techniques [24].

3.2.7.1 Decision-Feedback Equalization

DFE extensions of the recursive equalizer structures for linear time-varying channels operating on chip-rate sampling can be easily developed. Specifically, observation of the update equation for the state estimate (3.55a) reveals that this estimate consists of two terms. The first term corresponds to prediction of the state based on the state estimate at the previous symbol time. The second term is a correction term based on the difference between the predicted estimate and the most recently received observation. Performance can often be enhanced (especially at high SNR) by replacing the soft estimates $\tilde{\mathbf{x}}[n\,|\,n]$ with the associated decisions obtained by thresholding. By letting $\hat{\mathbf{x}}[n]$ denote the vector of hard decisions resulting from passing each element of the vector $\tilde{\mathbf{x}}[n\,|\,n]$ through the slicer function, we obtain a recursive DFE structure given by (3.55), where (3.55a) is replaced by

$$\tilde{\mathbf{x}}[n\,|\,n] = \mathbf{F}\hat{\mathbf{x}}[n-1\,|\,n-1] + \boldsymbol{\mu}[n](\mathbf{y}[n] - \mathbf{H}[n]\mathbf{F}\hat{\mathbf{x}}[n-1\,|\,n-1]). \qquad (3.58)$$

ZF extensions can also be obtained by replacing (3.55b) with (3.57).

3.3 Blind Equalization in Multipath, Slowly Time-Varying Channels

Blind equalization refers to equalizing the effects of an unknown channel so that the information-bearing signal can be recovered without the use of training. Traditional blind-equalization approaches involving observation of an information-bearing signal through a single unknown channel rely on higher-order statistics

[25–28]. Many of these techniques require a large number of samples to perform equalization reliably and are consequently inadequate for most wireless channels. Recently, Tong et. al. [29] have shown that observations of the same information-bearing signal via multiple unknown channels possess enough structure to render methods based on second-order statistics feasible. These techniques have the potential to allow practical channel equalization with notably fewer observation samples per channel and constitute a very active area of research.

In Section 3.3.1, we present several different formulations of the single-input multiple-channel blind-deconvolution problem. These arise in the forward link of mobile communication channels, in which a single transmitted signal corresponds to the sum of multiple information-bearing signals. We also discuss some of the issues that may potentially attract attention and briefly highlight a few of several different directions that have been taken by the research community to address some of these issues.

In Section 3.3.2, we focus our attention on the reverse link where each user's transmitted waveform is received via a number of distinct channels. In particular, we examine a representative blind-deconvolution method in the context of asynchronous CDMA systems.

3.3.1 The Forward Link: Blind Equalization of Single-Input, Multiple-Output FIR Channels

We first discuss blind equalization in the forward link of a multiuser wireless system. Due to the synchronous transmission in the forward direction, it is conceptually simpler to view symbol detection in two stages. In the first stage, the receiver of any given user attempts to determine a single signal, namely, the sum of the P information-bearing signals; the second stage involves the recovery of the particular symbol stream via despreading. In this section, we focus our attention only on the first stage since the second is straightforward.

In an effort toward providing computationally efficient schemes based on short observation sequences, Tong et. al. [29] suggest identification of a white WSS data sequence via its observation through multiple unknown LTI channels. The general situation is depicted in Figure 3.4. The problem reduces to estimating $h_1[n], \cdots, h_M[n]$ and the information-bearing signal $s[n]$, based on observation of $x_1[n], \cdots, x_M[n]$, without making any overly restrictive assumptions about the input sequence $s[n]$.

In a mobile communication setting, this multichannel approach can result either from using multiple receiver antennas or, in the context of excess bandwidth transmission, from simply oversampling the received signal. In the single-user case, oversampling corresponds to sampling faster than the baud rate of the signal. This case is depicted in Figure 3.5, where $h[n]$ is the oversampled system impulse response, and $y[n]$ the corresponding output. In CDMA and other multiuser systems,

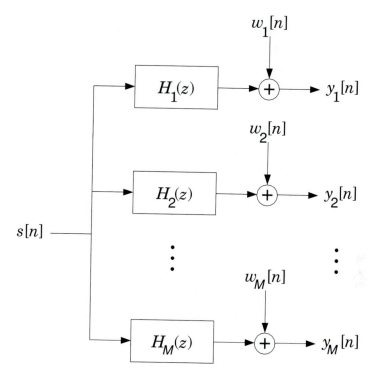

Figure 3.4   Observation of a single information-bearing sequence via multiple channels.

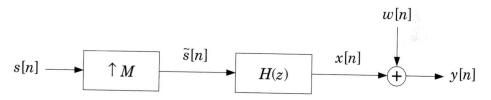

Figure 3.5   Equivalent block diagram for single-sensor oversampling system.

where $s[n]$ corresponds to the aggregate transmitted signal from which all the individual user streams can be decoded, oversampling is equivalent to sampling faster than the chip rate of the signal. The received signal $y[n]$ can be written as

$$y_i[n] \triangleq y[Mn + i] = \sum_k s[k]h_i[n - k] + w_i[n], \qquad i = 1, \cdots, M, \qquad (3.59)$$

where $h_i[n] \triangleq h[Mn + i]$ and $w_i[n] \triangleq w[Mn + i]$. The set of (3.59) reveals that both the oversampled and the multiple sensor representations can be addressed with the same mathematical framework, i.e.,

$$y_i[n] = x_i[n] + w_i[n] = s[n] * h_i[n] + w_i[n] \qquad n = 0, 1, \cdots, N_s - 1. \qquad (3.60)$$

The original approach [29] sparked a lot of interest since it is based on forming second-order statistics from the observed data in (3.60). To appreciate how second-order statistics can be adequate for this deconvolution problem, consider the following example. Suppose, for instance, a white WSS sequence $s[n]$ is observed via two particular FIR channels, $h_1[n]$ and $h_2[n]$, with length 2. Figure 3.6 shows the pole-zero diagrams associated with the spectra of the output sequences $y_1[n]$ and $y_2[n]$ and with the cross-spectrum $S_{y_1 y_2}(z)$. In practice, accurate estimates of these spectra can be obtained from sufficiently large data sets (assuming the FIR filter lengths are known). From these pole-zero diagrams, we can easily determine the zero associated with $H_1(z)$ by finding the zero that $S_{y_1 y_1}(z)$ and $S_{y_1 y_2}(z)$ have in common. Similarly, we can also determine the response of $H_2(z)$ within a scaling constant. Clearly, identification is possible via this method only if the channels have distinct zeros. This approach also holds if noise is present, i.e., in the case that $w_1[n]$ and $w_2[n]$ are white WSS noises uncorrelated with one another and the data $s[n]$, and where the noise variances are known; in this case, we can still obtain estimates of the associated pole-zero diagrams such as those depicted in Figure 3.6 by removing from each spectrum estimate the DC component associated with the particular additive noise term. Techniques of this form can also be extended to identification of $H_i(z)$'s with rational transfer functions that do not share any common zeros and poles and for which the maximum numerator and denominator orders are known [30].

We next examine a representative sample of the approaches that followed and improved on [29], both in terms of exposition of the structure of the blind-deconvolution problem and in terms of performance.

3.3.1.1 The Cross-Relation Method

In many common communication scenarios, the transmitted data samples may be correlated, and, furthermore, the exact statistical description of $s[n]$ may not be available. Moreover, a statistical description of the data may be inadequate such as when very few data samples are available, e.g., in relatively fast time-varying channel scenarios. This scenario can be best dealt with by considering the transmitted signal as a particular *unknown* deterministic signal [31]. In this section, we assume that all filters are FIR with maximum length $L_c + 1$, i.e., $h_i[n] = 0$ for all i, for $n < 0, n > L_c$, and there exists an i such that $h_i[0] \neq 0$ and $h_i[L_c] \neq 0$. Furthermore, the noise vector $\tilde{\mathbf{w}}[n] = [w_1[n] \quad w_2[n] \quad \cdots \quad w_M[n]]^T$ is a zero-mean WSS process with unknown second-order statistics and is independent of the unknown sequence $s[n]$.

It is interesting to note that in the noise-free case (i.e., $w_i[n] = 0$ for all i), the set of (3.60) can in principle be used directly to solve for the coefficients $h_i[n]$ and the sequence $s[n]$. There are MN_s equations and $M(L_c + 1) + N_s + L_c$ unknowns. Thus, if $M \geq 2$ and we pick a large enough N_s, the resulting system of equations

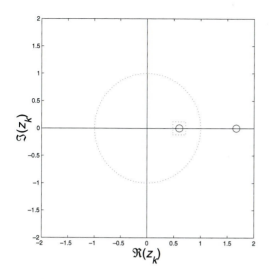

(a) Pole-zero diagram for $S_{y_1 y_1}(z)$.

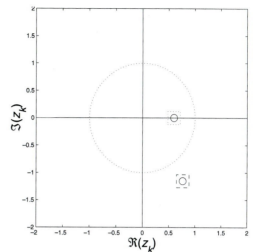

(b) Pole-zero diagram for $S_{y_1 y_2}(z)$.

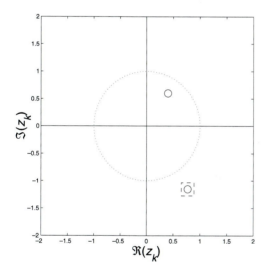

(c) Pole-zero diagram for $S_{y_2 y_2}(z)$.

Figure 3.6   Pole-zero diagram for $S_{y_1 y_1}(z)$, $S_{y_1 y_2}(z)$, and $S_{y_2 y_2}(z)$. In this example, the input stream $s[n]$ is IID, and the two channels $H_1(z)$, $H_2(z)$ are FIR with length 2.

will be overdetermined. This nonlinear set of equations can be transformed to a linear one by convolving both sides of (3.60) with $h_j[n]$ for a given $j \neq i$ (and setting $w_i[n] = 0$), which results in

$$y_i[n] * h_j[n] = h_i[n] * y_j[n]. \qquad (3.61)$$

Equation (3.61) is referred to as the cross-relation (CR) property. Its significance lies in that it is linear in the unknown channel coefficients $h_i[n]$ and $h_j[n]$, and thus standard linear methods can be used. In matrix notation, (3.61) can be written in terms of $\mathbf{h}$, given by

$$\mathbf{h}_i = [h_i[0]\ h_i[1]\ \cdots\ h_i[L_c]]^T, \tag{3.62}$$

and the measurements as

$$\left[\tilde{\mathcal{Y}}_i(L_c + 1)\ \ -\tilde{\mathcal{Y}}_j(L_c + 1)\right]\begin{bmatrix}\mathbf{h}_j \\ \mathbf{h}_i\end{bmatrix} = \mathbf{0},$$

which holds for every $i \neq j$, $1 \leq i, j \leq M$. The matrix $\tilde{\mathcal{Y}}_i(L_c + 1)$ is the $(N - L_c) \times (L_c + 1)$ Hankel matrix associated with the vector

$$\mathbf{y}_i = [y_i[N_s - 1]\ y_i[N_s - 2]\ \cdots\ y_i[0]]^T,$$

that is,

$$\tilde{\mathcal{Y}}_i(L_c + 1) = \begin{bmatrix} y_i[N_s - 1] & y_i[N_s - 2] & \cdots & y_i[N_s - L_c - 1] \\ y_i[N_s - 2] & y_i[N_s - 3] & \cdots & y_i[N_s - L_c - 2] \\ \vdots & \vdots & & \vdots \\ y_i[L_c] & y_i[L_c - 1] & \cdots & y_i[0] \end{bmatrix}$$

$$= \begin{bmatrix} \mathbf{y}_i^T[N_s - 1] \\ \mathbf{y}_i^T[N_s - 2] \\ \vdots \\ \mathbf{y}_i^T[L_c] \end{bmatrix}, \tag{3.63}$$

and where $\mathbf{y}_i[n]$ is the following $(L_c + 1) \times 1$ vector

$$\mathbf{y}_i[n] \triangleq [y_i[n]\ y_i[n - 1]\ \cdots\ y_i[n - L_c]]^T. \tag{3.64}$$

Let $Y_2 = [\tilde{\mathcal{Y}}_2\ -\tilde{\mathcal{Y}}_1]$, and

$$Y_i = \begin{bmatrix} & Y_{i-1} & & 0 \\ \tilde{\mathcal{Y}}_i & & & -\tilde{\mathcal{Y}}_1 \\ & \ddots & & \vdots \\ & & \tilde{\mathcal{Y}}_i & -\tilde{\mathcal{Y}}_{i-1} \end{bmatrix} \tag{3.65}$$

for $3 \leq i \leq M$. The set of equations that arises from the cross-relation property can then be summarized as follows

$$Y\mathbf{h} = \mathbf{0}, \tag{3.66}$$

where we write $Y = Y_M$ for simplicity, and where $\mathbf{h}$ is given by

$$\mathbf{h} = \begin{bmatrix} \mathbf{h}_1^T & \mathbf{h}_2^T & \cdots & \mathbf{h}_M^T \end{bmatrix}^T, \tag{3.67}$$

with $\mathbf{h}_i$ given by (3.62). If the null space of Y has dimension 1, then $\mathbf{h}$ (and thus the $h_i[n]$'s) can be uniquely identified within a complex scaling constant.

It is important to determine under what circumstances the channels are identifiable based on (3.66) in the absence of noise. In principle, there are two fundamental conditions that may render a channel identifiable. The first condition is related to whether the sequence $s[n]$ is "rich" enough to make identification of the channels possible and can typically be satisfied by increasing the length of the observation window [31]. The second and most important condition regarding channel identifiability is related to whether the transfer functions $H_1(z), \cdots, H_M(z)$ share any common zeros. Assuming that the symbol sequence $s[n]$ is rich enough in terms of its linear complexity, channel identifiability reduces to [31]

channel identifiability $\Leftrightarrow$ the $H_i(z)$'s share no common zeros.

We can gain insight regarding this identifiability condition by considering the case where all the $H_i(z)$'s contain common zeros, which is depicted in Figure 3.7. The common zeros are grouped in the filter term $G(z)$, i.e.,

$$h_i[n] = g[n] * f_i[n],$$

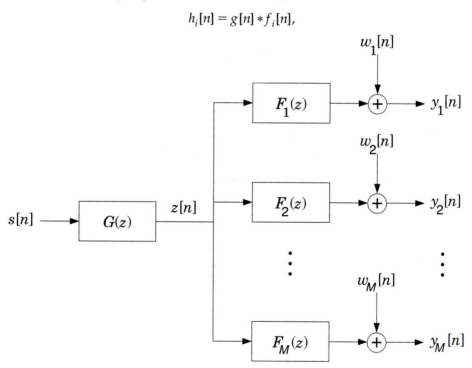

Figure 3.7 Equivalent block diagram for a multichannel system where the $H_i(z)$'s share common zeros.

where $f_1[n], \cdots, f_M[n]$ are all FIR filters sharing no common zeros, and the FIR filter $g[n]$ is at least 2 samples long. If we assume that $z[n]$ in Figure 3.7 can be uniquely determined from the $y_i[n]$'s, the identification problem reduces to determining $s[n]$ from $z[n]$ without knowing $g[n]$. Assuming we have the following measurements

$$z[n] = \sum_{k=0}^{L_g} g[k]s[n-k] \qquad n = 0, 1, \cdots, N_s - 1,$$

we have N_s equations where $g[0], \cdots, g[L_g]$, and $s[-L_g], \cdots, s[N_s]$ are all unknown, i.e., we have an underdetermined systems of equations; the system cannot be solved uniquely without additional knowledge about $s[n]$.

In the presence of additive noise, the least-squares solution to (3.66) provides a fairly computationally efficient, although generally suboptimal, solution. Specifically, the cross-relation algorithm selects $\hat{\mathbf{h}}_{CR}$ so as to minimize the energy in $Y\mathbf{h}$, i.e., [31]

$$\hat{\mathbf{h}}_{CR} = \arg\min_{\|\mathbf{h}\|=1} \mathbf{h}^\dagger Y^\dagger Y \mathbf{h}. \tag{3.68}$$

Consequently, in the presence of noise, $\hat{\mathbf{h}}_{CR}$ is given by the eigenvector associated with the smallest eigenvalue of the matrix $Y^\dagger Y$.

3.3.1.2 Subspace-Based Methods

The subspace-based methods originally introduced by Moulines et. al. [32] provide additional perspective to the single-input multiple-output deconvolution problem of Figure 3.4. Initially, for the exposition of these methods, we may assume that $s[n]$ is a WSS process whose statistics are not known and that $\tilde{\mathbf{w}}[n]$ is a zero-mean WSS process with known second-order statistics, independent of $s[n]$. Then, we consider the scenario where $s[n]$ is deterministic and consider subspace-based methods that are natural extensions of the original algorithm presented in [32].

We introduce an additional variable N, which corresponds to the length of the smoothing window in the matrix formulation of the problem. Specifically, let $\mathbf{y}_i[n]$ be the $N \times 1$ vector resulting from substituting N for $L_c + 1$ in (3.64), and

$$\mathbf{y}[n] \triangleq [\mathbf{y}_1^T[n] \ \ \mathbf{y}_2^T[n] \ \cdots \ \mathbf{y}_M^T[n]]^T. \tag{3.69}$$

For convenience, we denote by $\mathcal{V}_i(R; K)$ the $R \times (R+K)$ single-channel filtering matrix associated with the vector $\mathbf{v}_i = [v_i[0] \ \cdots \ v_i[K]]^T$, i.e.,

$$\mathcal{V}_i(R) = \mathcal{V}_i(R; K) = \begin{bmatrix} v_i[0] & \cdots & v_i[K] & 0 & \cdots & 0 \\ 0 & v_i[0] & \cdots & v_i[K] & \ddots & \vdots \\ \vdots & \ddots & \ddots & & \ddots & 0 \\ 0 & \cdots & 0 & v_i[0] & \cdots & v_i[K] \end{bmatrix}, \tag{3.70}$$

and let $\mathcal{V}(R; K)$ be the M-channel filtering matrix associated with $\mathbf{v} = [\mathbf{v}_1^T \ \cdots \ \mathbf{v}_M^T]$, i.e.,

$$\mathcal{V}(R) = \mathcal{V}(R; K) = \begin{bmatrix} \mathcal{V}_1(R) \\ \vdots \\ \mathcal{V}_M(R) \end{bmatrix}.$$

We can then recast the original problem (3.60) as

$$\mathbf{y}[n] = \mathcal{H}(N)\mathbf{s}[n] + \mathbf{w}[n] \qquad \text{for } n = N, \cdots, N_s, \tag{3.71}$$

where $\mathbf{w}[n]$ is defined in similar fashion to $\mathbf{y}[n]$ in (3.69), the data vector $\mathbf{s}[n]$ is given by

$$\mathbf{s}[n] = [s[n] \ s[n-1] \ \cdots \ s[n-N-L_c+1]]^T,$$

and $\mathcal{H}(N)$ is the $N \times (N + L_c)$ M-channel filtering matrix associated with $\mathbf{h}$ from (3.67).

For simplicity (and without loss of generality), we focus on the white noise case, i.e., let $R_{\mathbf{w}} = \sigma_w^2 \mathbf{I}$. Since $\mathbf{s}[n]$ and $\mathbf{w}[n]$ are WSS (and independent), $\mathbf{y}[n]$ is also WSS and its autocorrelation function is given by

$$R_{\mathbf{y}} = \mathcal{H}(N)R_{\mathbf{s}}\mathcal{H}^\dagger(N) + \sigma_w^2 \mathbf{I}, \tag{3.72}$$

where the $(L_c + N) \times (L_c + N)$ matrix R_s denotes the autocorrelation of $\mathbf{s}[n]$. Equation (3.72) forms the basis of the subspace-based blind equalization algorithms. Assuming for the moment that one knows $R_{\mathbf{y}}$, the question is whether one can obtain from it $\mathcal{H}(N)$ and subsequently identify $\mathbf{h}$. If the filters $h_i[n]$ share no common zeros, then $\mathcal{H}(N)$ has full column rank (i.e., $N + L_c$). If R_s also has full rank, the subspace associated with the signal part in $R_{\mathbf{y}}$ has dimension $N + L_c$ and is spanned by the columns of $\mathcal{H}(N)$. Since $\mathcal{H}(N)R_s\mathcal{H}^\dagger(N)$ is nonnegative definite and the eigendecomposition of $\sigma_w^2 \mathbf{I}$ is invariant to rotation, the eigenvectors associated with the $(N + L_c)$ largest eigenvalues of $R_{\mathbf{y}}$ span the signal subspace, i.e.,

$$R_{\mathbf{y}} = S \,\text{diag}(\lambda_1, \lambda_2, \cdots \lambda_{L_c+N})S^\dagger + \sigma_w^2 U U^\dagger,$$

where $\lambda_i > \sigma_w^2$ for $i = 1, \cdots, L_c + N$. Then, since $\mathcal{H}(N)$ spans S and if $\mathbf{u}_n$ denotes the nth column of U, we have

$$\mathbf{u}_n^\dagger \mathcal{H}(N) = \mathbf{0}. \tag{3.73}$$

Equivalently, due to the commutativity of convolution,

$$\mathbf{h}^\dagger \mathcal{U}_n(L_c + 1) = \mathbf{0}, \tag{3.74}$$

where (with a little abuse of notation) $\mathcal{U}_n(L_c + 1)$ is the $(L_c + 1) \times (N + L_c)$ M-channel filtering matrix associated with the vector $\mathbf{u}_n$. It can be shown [32] that if (3.73) is satisfied by both $\mathcal{H}(N)$ and $\mathcal{H}'(N)$, then $\mathcal{H}(N) = \alpha \mathcal{H}'(N)$, and $\mathbf{h} = \alpha \mathbf{h}'$, where α is a complex scalar. Thus, from an eigen-decomposition of R_y, S and U can be recovered and subsequently $\mathcal{H}(N)$ and $\mathbf{h}$. Moulines et. al. [32] define their estimator as the vector $\hat{\mathbf{h}}_{ss}$ that has the minimum energy contribution to the empirical noise subspace $\hat{U}$ obtained from an estimate of R_y. Specifically, let $\hat{Q} \triangleq \sum_n \hat{U}_n \hat{U}_n^\dagger$, then

$$\hat{\mathbf{h}}_{ss} = \underset{\|\mathbf{h}\|=1}{\arg \min} \mathbf{h}^\dagger \hat{Q} \mathbf{h}, \qquad (3.75)$$

i.e., $\hat{\mathbf{h}}_{ss}$ is the eigenvector associated with the smallest eigenvalue of $\hat{Q}$.

The question that remains to be addressed is how one selects $\hat{R}_y$, the estimate of the correlation matrix of the data, upon which the subspace decomposition will be based. A good choice for $\hat{R}_y$ that would provide the desired decomposition in the absence of noise would have to satisfy an equation of the form

$$\hat{R}_y = \mathcal{H}(N) \hat{R}_s \mathcal{H}^\dagger(N), \qquad (3.76)$$

since in that case its rank can be readily seen to be upper bounded by $(N + L_c)$. This can be accomplished by using a sample covariance matrix estimator, i.e.,

$$\hat{R}_y(N) \triangleq \frac{1}{N_s - N + 1} \sum_{n=N-1}^{N_s-1} \mathbf{y}[n]\mathbf{y}^\dagger[n]. \qquad (3.77)$$

In the absence of noise, (3.77) can be conveniently expressed in the form (3.76), where

$$\hat{R}_s \triangleq \frac{1}{N_s - N + 1} \sum_{n=N-1}^{N_s-1} \mathbf{s}[n]\mathbf{s}^\dagger[n].$$

Inspection of (3.76) suggests that in the absence of noise, $\hat{R}_y$ is singular (its rank is upper bounded by $(N + L_c)$), and if it is used in (3.75), it results in $\hat{\mathbf{h}}_{ss} = \mathbf{h}$ as desired. In fact, we may consider the *deterministic* counterpart of (3.72), namely,

$$\hat{R}_y = \mathcal{H}(N)\hat{R}_s\mathcal{H}^\dagger(N) + \mathcal{H}(N)\hat{R}_{s,w} + \hat{R}_{w,s}\mathcal{H}^\dagger(N) + \hat{R}_w, \qquad (3.78)$$

where $\hat{R}_{s,w}$ and $\hat{R}_w$ are the sample signal-noise cross-covariance and sample noise covariance matrices, respectively. Furthermore, in the presence of zero-mean white noise independent of the data $s[n]$, (3.72) holds for $\hat{R}_s$, since

$$E[\hat{R}_y] = R_y, \quad E[\hat{R}_{w,s}] = 0, \quad E[\hat{R}_{s,w}] = 0.$$

This property can be readily verified by replacing R_s with $\hat{R}_s$ in (3.72), substituting the expression (3.71) for $\mathbf{y}[n]$ in (3.78) and taking the expectation of both sides. The noise-signal subspace decomposition still holds, even if $s[n]$ is *not* WSS, i.e., even if $s[n]$ is *unknown*. The only conditions that we still need in order for this decomposition to provide an estimator structure resulting in perfect estimation as $\sigma_w^2 \to 0$ is that $\mathbf{w}[n]$ is a zero-mean WSS process independent of $s[n]$.

An interesting question is how to make a judicious choice for N. Larger N values correspond to noise subspaces of higher dimension. Presumably, depending on the method used to compute $\hat{Q}$, the more noise eigenvectors, the higher the accuracy of the estimate $\hat{Q}$ in the presence of noise. However, larger N values also correspond to obtaining $\hat{R}_y(N)$ based on fewer measurement vectors, as (3.77) reveals. So, as N becomes comparable to N_s, estimates of $\hat{R}_y(N)$ become more and more noisy, resulting in a deterioration in performance. Furthermore, the case $N = L_c + 1$ is the least expensive in terms of computations, so unless increasing N provides significant performance improvement, it is not recommended. In fact, for $M = 2$, the $\hat{\mathbf{h}}_{CR}$ method is a special case of the $\hat{\mathbf{h}}_{SS}$ method, corresponding to the choice $N = L_c + 1$.

In certain cases such as the oversampling receiver depicted in Figure 3.5, the noise samples are correlated (provided the front-end filter has rejected all out-of-band noise), i.e., $E[w_i[n]w_k^*[n]] \neq 0$. The case of $R_w = \sigma_w^2 \overline{R}_w$ where $\overline{R}_w$ is full rank and known (not an unrealistic assumption in certain cases such as the one depicted in Figure 3.5), can be easily accommodated [32]. We can whiten the noise by pre- and post-multiplying (3.72) by $V^\dagger$ and V, respectively, where $V = \overline{R}_w^{-1/2}$. Consequently, if we simply substitute $\overline{\mathbf{u}}_n = V\hat{\mathbf{u}}_n$ for $\hat{\mathbf{u}}_n$ in (3.74), the rest of the analysis applies.

3.3.1.3 Direct Symbol Estimation

In the two preceding algorithms, estimates of the channel parameters are initially formed and are then used to equalize the channels and obtain $\hat{s}[n]$. However, in multiuser communication, we are interested in estimating the symbols without necessarily obtaining a channel estimate. Liu and Xu [33] have suggested a method based on second-order statistics, where the dependence on $h_i[n]$ is eliminated and thus a relationship involving only the measurements and $s[n]$ is formed. This approach may be attractive in fast time-varying channels, where estimates of the current channel coefficients may not be useful in equalization of many subsequent data samples.

We first rewrite (3.60) in the following matrix form:

$$\tilde{\mathcal{Y}}^T(N) = \tilde{\mathcal{H}}(N)S^T(N),$$

where $S(N)$ is the $N_s \times (L_c + 1)$ Hankel matrix associated with the vector

$$\mathbf{s} = [s[N_s - 1] \quad \cdots \quad s[-L_c]]^T.$$

The Hankel matrix $\tilde{\mathcal{Y}}(N)$ is given by the right-hand side of (3.63) if we substitute

$$\tilde{\mathbf{y}}[n] = [y_1[n] \quad y_2[n] \quad \cdots \quad y_M[n]]^T$$

for $y_i[n]$. The matrix $\tilde{\mathcal{H}}(N)$ corresponds to the vector-channel filtering matrix given by the right-hand side of (3.63) if we substitute

$$\tilde{\mathbf{h}}[n] = [h_1[n] \quad h_2[n] \quad \cdots \quad h_M[n]]^T$$

for $\mathbf{v}_i$ and for $K = L_c$.

If $s[n]$ is rich enough in linear complexity and the channels do not share any common zeros, the vector $\mathbf{s}$ is given as the unique nontrivial solution to [33]

$$V\mathbf{s} = \mathbf{0}. \tag{3.79}$$

The matrix V in (3.79) has block diagonal structure and consists of $N + L_c$ blocks of $V_o(N)$, i.e., $V = \text{diag}\{V_o(N), V_o(N), \cdots, V_o(N)\}$, and where $V_o(N)$ is the null space of the row vectors of $\tilde{\mathcal{Y}}(N)$ [33]. It is straightforward to show that

$$V_o(N)S(N) = \mathbf{0}$$

implies (3.79) due to the Hankel nature of $S(N)$.

In the presence of noise, the least-squares solution to (3.79) can be employed. In that case, $V_o(N)$ is formed by selection of the r least significant eigenvectors of $\tilde{y}^*(N)\tilde{y}^T(N)$. The number r denotes the dimension of the noise subspace and can be readily determined from the noise-free case to be

$$r = N_m + L_c - (2L_c + K) + 1.$$

Finally, the estimate of the information-bearing signal is given by

$$\hat{\mathbf{s}} = \arg\min_{\mathbf{t}} \mathbf{t}^\dagger V^\dagger V \mathbf{t}, \tag{3.80}$$

i.e., $\hat{\mathbf{s}}$ is the least significant eigenvector of $V^\dagger V$. For instance, in the forward link of a CDMA system, the estimate $\hat{s}[n]$ arising from (3.80) can be despread to obtain the particular symbol subsequence of interest.

If the number of samples in $s[n]$ is comparable to the number of channel parameters, this approach may be advantageous since it bypasses the channel estimation step. Furthermore, this method outperforms the cross-relation method in terms of mean-square estimation error in the data [33]. A main disadvantage of this approach, however, is that for any subsequent set of data, either the procedure must be repeated or an estimate of each $h_i[n]$ must be formed.

3.3.1.4 Issues in the Multiple FIR, Channel-Deconvolution Problem

Since the publication of [29], the problem of blind deconvolution of multiple FIR systems has been addressed from several different perspectives. Even though many conceptually appealing formulations have been formed, several challenging issues remain to be addressed before such methods can be practical. As a result, there continues to be considerable activity in this area.

Selecting the right filter order, L_c, is of primary importance in the algorithms we have presented. Let L_c be the assumed filter order and L_{max} denote the actual maximal order among the filters $H_1(z), \cdots, H_M(z)$. Also, assume that if L_c is picked correctly, the system is uniquely identifiable based on the cross-relation-based method by Xu et. al. [31]. It can be readily shown [34] that in the absence of noise, if $L_c > L_{max}$, then dim null(Y) = $L_c - L_{max} + 1$. In other words, the zero-padded filters have common zeros which render them unidentifiable. If $L_c < L_{max}$, then Y has full rank, i.e., precise identification is not possible. In principle, we can identify L_{max} by increasing L_c until dim null(Y) = 1. However, a single, common zero would be indistinguishable from a delay. In general, all parametric FIR methods are sensitive to the choice of L_c [31]. Furthermore, the time-varying nature of the channel can only complicate matters. Assume, for example, that one of the filters is given by $h_i[n] = \delta[n - n_o]$ (direct line of sight scenario), corresponding to $h_i(t) = \delta(t - n_o T)$. If the actual delay of the channel response changes from $n_o T$ to $n_o T + T/2$, the corresponding discrete-time sampled filter can have considerable length.[3] It would be very appealing if we could circumvent the channel identification step and bypass issues such as the maximum filter order ambiguity problem that make such parametric methods sensitive to the modeling parameters. Not surprisingly, if more information is available about $s[n]$, better solutions can be obtained. For example, when $s[n]$ is discrete-valued (such as in digital communication systems), more channels can be identifiable. First, the channel identifiability problem in the case that $H_i(z)$'s share common zeros can be reduced to the single-channel problem with a shorter FIR filter deconvolution problem. Provided that L_g is very small, methods that are based on blind deconvolution through LTI channels driven by discrete-valued alphabet inputs are possible [35]. Furthermore, in [36], Tong presents a blind sequence estimation method, which exploits second-order statistical information of $s[n]$ and is very appealing in time-varying channel contexts. A collection of the recent developments in the problem of the blind multiple-channel equalization is given by Liu et. al. in [37].

Another limitation of many existing algorithms in this area is their nonrecursive structure, which effectively limits their applicability to batch mode operation. For instance, in a time-selective fading environment, any of the discussed algorithms could be used to initialize an equalizer. However, in the case of a long fade, during which the equalizer may lose track of the time-varying channel, such an algorithm may suffer. Structures that recursively compute the new symbol estimates by considering past channel and symbol estimates may be naturally suited for such channels. Giannakis and Halford [38] develop a recursive approach to blind deconvolution by means of an FIR equalizer, where the taps are selected so as to satisfy the ZF criterion, i.e.,

$$h[n] * \hat{g}[n] = C \delta[n - m]$$

[3]The discrete-time filter associated with a half-sample delay in the context of ideal low-pass filtering has infinite length. Furthermore, the filter tails decay to zero very slowly. However, since most practical systems employ excess bandwidth modulation schemes, conversion from continuous to discrete time involves filters whose transition band is fairly wide. In this case, the tails of the associated discrete-time half-sample delay filter decay to zero much faster. Although less pronounced in these systems, the variable filter length problem still occurs.

for some complex scalar C and some integer m, and where $\hat{g}[n]$ denote the equalizer taps. In the absence of noise, this method recursively obtains a ZF equalizer for the model depicted in Figure 3.5, and (3.59). In the presence of noise, [38] also presents FIR solutions that minimize the mean-square error in the estimate subject to a ZF constraint.

3.3.2 Blind Equalization in the Reverse Link via Multiple Observations

The reverse link of a multiuser wireless communication system is depicted in Figure 3.1, where the channels are assumed to be unknown and LTI. Again, the additional channels can arise from the use of additional antennas, oversampling, or, in general, both. The blind equalization problem in the multiuser context where the codes are unknown is more challenging than its single-user counterpart. Specifically, the linear interaction among the multiple sources may be impossible to resolve with algebraic relationships [33]. For instance, a natural extension of the direct symbol estimation method (presented in the preceding section) results in the construction of a matrix V that is orthogonal to all input vectors $\mathbf{s}_1, \cdots, \mathbf{s}_P$, i.e., it has null space dimension equal to P. If no more constraints are provided, the input vectors $\mathbf{s}_1, \cdots, \mathbf{s}_P$, cannot be determined from the null space of V.

Several methods have been proposed that use additional information about the transmitted data in order to resolve the ambiguity with reasonably low complexity. In an extension to their blind symbol estimation algorithm, Liu and Xu [33] propose a method for multiple-input multiple-output (MIMO) blind symbol equalization in which they take advantage of the discrete-alphabet property of the underlying digital communication signals by means of methods such as those described in [35] and [39]. Using the finite alphabet nature of the symbols, [35] and [39] show that the input vectors are identifiable, given sufficient data samples. Unfortunately, with the method in [35], the number of symbols needed for separating the users grows exponentially with the number of users, and the method developed in [39] is not guaranteed to produce the correct symbol estimates in the absence of noise.

Knowledge of the particular codeword associated with each user is a natural form of additional information, which, if properly exploited, can help resolve the remaining ambiguity. We next discuss one such algorithm for asynchronous CDMA systems which uses knowledge of the user signatures and their relative delays to extract the input symbol sequences [40]. The algorithm employs multiple antennas sampling at the chip rate and uses the resulting sequences to construct a ZF equalizer for each user.

3.3.2.1 Blind Equalization Using Multiple Antennas in CDMA Systems

As we have noted in the previous section, in MIMO blind-channel deconvolution, it is impossible to uniquely identify the multiple symbol sequences by using second-order statistics. With such methods, one can, at most, isolate the subspace of dimensionality P spanned by the symbol vectors $\mathbf{s}_i = [s_i[-L_c] \ \cdots \ s_i[N_s - 1] \ s_i[N_s]]^T$ for $i = 1, \cdots, P$. The remaining ambiguity can often be resolved if the receiver knows the user signature waveforms and their relative delays. Intuitively, one would use knowledge of any particular codeword to isolate the subspace of the P-dimensional subspace corresponding to the signal of the associated user.

In this section, we consider a representative blind ZF equalizer that can be used with asynchronous CDMA systems in the case that the user signatures are known to the receiver [40]. Specifically, consider the following modification of the modulator of Figure 3.2 where

$$x_i[n] = \sum_k s_i[k]c_i[n - kL - n_i],$$

where n_i is the relative delay of each user with respect to the receiver modulo L, i.e., $0 \leq n_i < L$. In this scenario, we assume that all relative delays are known at the receiver. This assumption implies that upon obtaining an estimate for $\sum_i x_i[n]$, the symbol sequence for each i can be recovered by despreading in the standard manner. We assume chip rate sampling at multiple ($M \geq 2$) receiver antennas. The received signals at the M antennas are described by the $M \times 1$ vector

$$\mathbf{y}[n] = \sum_{i=1}^{P} \sum_k \mathbf{h}_i[k]x_i[n - k] + \mathbf{w}[n]$$

where $R_{\mathbf{w}}[n, k] = \sigma_w^2 \delta[n - k]\mathbf{I}_{M \times M}$, and

$$\mathbf{h}_i[n] = [h_{1,i}[n] \ h_{2,i}[n] \ \cdots \ h_{M,i}[n]]^T$$

corresponds to the vector channel response of the ith user. We seek $M \times 1$ vector FIR equalizers of length $K + 1$ for combining the M received channel signals to obtain an estimate of $x_i[n]$ and, subsequently, $s_i[n]$. Specifically, we denote the equalizer for the ith user by the $M \times 1$ vector [40]

$$\mathbf{g}_i[n] = [g_{1,i}[n] \ g_{2,i}[n] \ \cdots \ g_{M,i}[n]]^T.$$

Assuming that the vector equalizer has finite support in $[0, K]$, it is convenient to collect all the equalizer taps for the ith user in the vector $\mathbf{g}_i$, i.e.,

$$\mathbf{g}_i = [\mathbf{g}_i^T[K] \ \mathbf{g}_i^T[K - 1] \ \cdots \ \mathbf{g}_i^T[0]]^T.$$

Then, the estimate of $x_i[n - K]$ will be given by

$$\hat{x}_i[n - K] = \sum_{k=0}^{K} g_i^\dagger[k]y[n - k] \tag{3.81}$$

$$= g_i^\dagger y_K[n],$$

where $y_K[n] = [y^T[n - K]\ \ y^T[n - K + 1]\ \ \cdots\ \ y^T[n]]^T$. By despreading, the estimate of the kth symbol of the ith user can be obtained, i.e.,

$$\hat{s}_i[k] = \sum_{\ell=0}^{L-1} \hat{x}_i[kL + n_i + \ell]c_i(\ell) \tag{3.82}$$

$$= g_i^\dagger X_i[k],$$

where the entries of the vector $X_i[k]$ are correlations of segments of the received sequence $y[n]$ with the signature of the ith user, $c_i[n]$. Specifically, we have $X_i[k] = Y_i[k]c_i$, where

$$Y_i[k] = [y_K[kL + n_i + K]\ \ y_K[kL + n_i + K + 1]\ \ \cdots\ \ y_K[kL + n_i + K + L - 1]],$$

and $c_i = [c_i[0]\ \ c_i[1]\ \ \cdots\ \ c_i[L - 1]]^T$. In the absence of noise, a ZF equalizer for the ith channel satisfies [40]

$$g_i^\dagger Y_i[k] = s_i[k]c_i^\dagger. \tag{3.83}$$

If N symbols are considered, i.e., if (3.83) is satisfied for $k = 1, 2, \cdots, N$, we can rewrite this set of equations in the following matrix form:

$$\begin{bmatrix} Y_i[1] & -c_i & 0 & \cdots & 0 \\ Y_i[2] & 0 & -c_i & \ddots & \vdots \\ \vdots & \vdots & \ddots & \ddots & 0 \\ Y_i[N] & 0 & \cdots & 0 & -c_i \end{bmatrix} \begin{bmatrix} g_i \\ s_i[1] \\ \vdots \\ s_i[N] \end{bmatrix} = 0. \tag{3.84}$$

For convenience, we refer to the matrix in (3.84) by the symbol Y. If the dimension of the null space of Y equals one, then the null space completely determines g_i and the ith symbol sequence. If the dimension of the null space is higher, there are multiple ZF solutions. In the presence of noise, a solution in the least-square sense can be obtained, i.e., one that minimizes the mean-square error $\sum_k \|g_i^\dagger Y_i[k] - s_i[k]c_i\|^2$. This solution is given by the least significant eigenvector of $Y^\dagger Y$. In [40], it is shown that $\hat{g}_{i,\text{ZF}}$ is the least significant eigenvector of $G_i = \hat{R}_{Y_i} - \hat{R}_{X_i}/L$, where $\hat{R}_{Y_i}$ and $\hat{R}_{X_i}$ are the sample covariances formed from $Y_i[k]$ and $X_i[k]$, respectively, for $k = 1, \cdots, N$. Consequently, the equalizer g_i is determined by first forming the sample-covariance matrices $\hat{R}_{Y_i}$ and $\hat{R}_{X_i}$ for the ith user, then the corresponding matrix G_i, and finally by identifying as $\hat{g}_{i,\text{ZF}}$ the least significant eigenvector of G_i. Provided that K satisfies $K \geq PL_c/(M - P)$, the ZF criterion can be satisfied, i.e., a ZF equalizer of length K exists [40].

We emphasize that this scheme is not blind in the true sense since the receiver knows both the signature of each user and the corresponding relative delay. In a sense, the signature of each user plays the role of a training/identifying sequence that enables the receiver to isolate a particular symbol stream based on a set of over-sampled observations. In the special case where $n_i = n_j$ for all i, j (synchronous transmission), we may note that for all i, j, we have $\hat{R}_{Y_i} = \hat{R}_{Y_j}$. Thus, knowledge of the codewords in this case is exploited in the form of $\hat{R}_{X_i}$ to separate the symbol streams. Furthermore, in a CDMA system where only one user is present, consider a receiver that samples at the chip rate and has no prior knowledge of the user signature. Such a receiver can obtain an estimate of the symbol sequence by means of one of the single-input, multiple-output blind strategies described in Section 3.3.1. However, the detector we considered here requires knowledge of the signature so that the sample covariance $\hat{R}_{X_i}$ can be formed, and from its use, superior and more robust symbol and channel estimates can be obtained.

One additional appealing aspect of this equalizer is that it does not require knowledge of the channel length. However, this advantage is implicitly due to the assumption that the receiver knows the relative delay of each of the users. A similar blind equalization scheme for CDMA systems is presented in [41]. More generally, many other strategies for the reverse link are also being pursued, as recent literature attests.

3.4 CONCLUDING REMARKS

In this chapter, we have focused on the problem of equalization in wireless channels in the context of multiple users. We first discussed the problem of equalization in the context of known wireless channels. This situation arises in systems where reliable estimates of the time-varying channels are first obtained via training and are subsequently used for equalization. We also discussed a class of equalization schemes that do not require a priori estimates of the wireless channels, which rely on signal reception via multiple unknown channels to obtain accurate channel and symbol estimates via short observation windows. Although in this chapter we have presented a number of potentially appealing strategies for multiuser equalization, equalizer design is a currently active area of research. In particular, many important and challenging issues remain in the development of practical bandwidth-efficient multiple-access systems which are robust to the fundamental impairments arising in multiple-access wireless scenarios. For the signal processing community, these are likely to constitute a rich set of research challenges for many years to come.

REFERENCES

[1] S. Verdú, "Demodulation in the presence of multiuser interference: progress and misconceptions," in *Intelligent Methods in Signal Processing and Communications* (D. Docampo, A. Figueiras-Vidal, and F. Perez-Gonzalez, eds.), pp. 15–44, Boston: Birkhauser, 1997.

[2] J. Proakis, *Digital Communications*. New York: McGraw-Hill, 1989.

[3] S. Verdú, "Minimum probability of error for asynchronous multiple-access channels," *IEEE Trans. Inform. Theory*, vol. 32, pp. 85–96, Jan. 1986.

[4] E. A. Lee and D. G. Messerschmitt, *Digital Communication*, 2nd ed. Boston: Kluwer, 1993.

[5] Z. Zvonar and D. Brady, "Optimum detection in asynchronous multiple-access multipath Rayleigh fading channels," in *Proc. Conf. Inform. Sci. Sys.*, pp. 826–831, Princeton U., 1992.

[6] A. Duel-Hallen, "Equalizers for multiple input/multiple output channels and PAM systems with cyclostationary input sequences," *IEEE J. Select. Areas Commun.*, vol. 10, pp. 630–639, Apr. 1992.

[7] R. Lupas and S. Verdú, "Linear multiuser detectors for synchronous code-division multiple-access channels," *IEEE Trans. Inform. Theory*, vol. 35, pp. 123–136, Jan. 1989.

[8] R. Price and P. E. Green, "A communication technique for multipath channels," *Proc. IRE*, vol. 46, pp. 555–570, Mar. 1958.

[9] R. Lupas and S. Verdú, "Near-far resistance of multiuser detectors in asynchronous channels," *IEEE Trans. Commun.*, vol. 38, pp. 496–508, Apr. 1990.

[10] A. Klein, G. K. Kaleh, and P. W. Baier, "Zero-forcing and minimum mean-square-error equalization for multiuser detection in code-division multiple-access channels," *IEEE Trans. Vehic. Technol.*, vol. 45, pp. 276–287, May 1996.

[11] M. K. Varanasi and B. Aazhang, "Near-optimum detection in synchronous code-division multiple-access systems," *IEEE Trans. Commun.*, vol. 39, pp. 725–736, May 1991.

[12] S. Verdú, "Multiuser detection," in *Advances in Statistical Signal Processing: Signal Detection* (H. V. Poor and J. B. Thomas, eds.), pp. 369–410. Greenwich, CT: JAI Press, 1993.

[13] Y. C. Yoon, R. Kohno, and H. Imai, "A spread-spectrum multiaccess system with cochannel interference cancellation for multipath fading channels," *IEEE J. Select. Areas Commun.*, vol. 11, pp. 1067–1075, Sept. 1993.

[14] P. Patel and J. Holtzman, "Analysis of a simple successive interference cancellation scheme in DS/CDMA systems," *IEEE J. Select. Areas Commun.*, vol. 12, pp. 796–807, June 1994.

[15] S. C. Park and J. F. Doherty, "Generalized projection algorithm for blind interference suppression in DS/CDMA communications," *IEEE Trans. Circuits and Systems*, vol. 44, no. 6, pp. 453–460, 1997.

[16] A. Duel-Hallen, "Decorrelating decision-feedback multiuser detector for synchronous code-division multiple-access channel," *IEEE Trans. Commun.*, vol. 41, pp. 285–294, Feb. 1993.

[17] S. Beheshti and G. W. Wornell, "Iterative interference cancellation and decoding for spread-signature CDMA systems," in *Proc. Vehic. Technol. Conf.*, vol. 1, pp. 26–30, May 1997.

[18] B. D. O. Anderson and J. B. Moore, *Optimal Filtering*. Englewood Cliffs, NJ: Prentice-Hall, 1979.

[19] U. Madhow and M. L. Honig, "MMSE interference suppression for direct-sequence spread-spectrum CDMA," *IEEE Trans. Commun.*, vol. 42, pp. 3178–3188, Dec. 1994.

[20] S. L. Miller, "An adaptive direct-sequence code-division multiple-access receiver for multiuser interference rejection," *IEEE Trans. Commun.*, vol. 43, pp. 1746–1755, Feb./Mar./Apr. 1995.

[21] S. L. Miller, "Training analysis of adaptive interference suppression for direct-sequence code-division multiple-access systems," *IEEE Trans. Commun.*, vol. 44, pp. 488–495, Apr. 1996.

[22] A. Duel-Hallen, "A family of multiuser decision-feedback detectors for asynchronous code-division multiple-access channels," *IEEE Trans. Commun.*, vol. 43, pp. 421–434, Feb./Mar./Apr. 1995.

[23] M. Kaverhard and J. Salz, "Cross-polarization cancellation and equalization in digital data transmission over dually polarized radio channels," *The Bell Sys. Tech. Jour.*, vol. 64, pp. 2211–2245, Dec. 1985.

[24] S. H. Isabelle and G. W. Wornell, "Recursive multiuser equalization for CDMA systems in fading environments," in *Proc. Allerton Conf. Commun. Contr. Signal Proc.*, pp. 613–622, 1996.

[25] Y. Sato, "A method for self recovering equalization," *IEEE Trans. Commun.*, vol. 23, pp. 679–682, June 1975.

[26] D. L. Donoho, "On minimum entropy deconvolution," in *Applied Time Series Analysis II* (D. F. Findley, ed.), pp. 565–608, Academic Press, 1980.

[27] A. Benveniste and M. Goursat, "Blind equalizers," *IEEE Trans. Commun.*, vol. 32, pp. 871–883, Aug. 1984.

[28] O. Shalvi and E. Weinstein, "Super-exponential methods for blind deconvolution," *IEEE Trans. Inform. Theory*, vol. 39, pp. 504–519, Mar. 1993.

[29] L. Tong, G. Xu, and T. Kailath, "Blind identification and equalization based on second-order statistics: a time domain approach," *IEEE Trans. Inform. Theory*, vol. 40, pp. 340–349, Mar. 1994.

[30] L. Tong, "Blind identification of channels with rational transfer functions," in *Proc. IEEE GLOBECOM*, no. 1, pp. 56–60, 1994.

[31] G. Xu, H. Liu, L. Tong, and T. Kailath, "A least-squares approach to blind channel identification," *IEEE Trans. Signal Processing*, vol. 43, pp. 2982–2993, Dec. 1995.

[32] E. Moulines, P. Duhamel, J.-F. Cardoso, and S. Mayrargue, "Subspace methods for the blind identification of multichannel FIR filters," *IEEE Trans. Signal Processing*, vol. 43, pp. 516–525, Feb. 1995.

[33] H. Liu and G. Xu, "Closed-form blind symbol estimation in digital communications," *IEEE Trans. Signal Processing*, vol. 43, pp. 2714–2723, Nov. 1995.

[34] H. Liu, G. Xu, and L. Tong, "A deterministic approach to blind equalization," in *Proc. Asilomar Conf. Signals, Sys., Computers*, (Pacific Grove, Calif.), pp. 751–755, 1993.

[35] D. Yellin and B. Porat, "Blind identification of FIR systems excited by discrete-alpha-bet inputs," *IEEE Trans. Signal Processing*, vol. 41, pp. 1331–1339, Mar. 1993.

[36] L. Tong, "Blind sequence estimation," *IEEE Trans. Commun.*, vol. 43, pp. 2986–2995, Dec. 1995.

[37] H. Liu, G. Xu, L. Tong, and T. Kailath, "Recent developments in blind channel equalization: From cyclostationarity to subspaces," *Signal Processing*, vol. 50, pp. 83–99, Jan. 1996.

[38] G. Giannakis and S. Halford, "Blind fractionally-spaced equalization of noisy FIR channels: adaptive and optimal solutions," in *Proc. Int. Conf. Acoust. Speech, Signal Processing*, vol. 4, (Detroit, Mich.), pp. 1972–1975, 1995.

[39] S. Talwar, M. Viberg, and A. Paulraj, "Blind estimation of multiple cochannel digital signals using an antenna array," *IEEE Sig. Proc. Lett.*, vol. 1, pp. 29–31, Feb. 1994.

[40] H. Liu and M. D. Zoltowski, "Blind equalization in antenna array CDMA systems," *IEEE Trans. Signal Processing*, vol. 45, pp. 161–172, Jan. 1997.

[41] M. Torlak and G. Xu, "Blind multiuser channel estimation in asynchronous CDMA systems," *IEEE Trans. Signal Processing*, vol. 45, pp. 137–147, Jan. 1997.

ACKNOWLEDGMENTS

The author would like to thank Gregory Wornell for a number of critical comments and valuable suggestions from which the presentation and content of this chapter have greatly benefited, and Steven Isabelle for a number of insightful remarks and discussions on the topic of multiuser equalization.

This work was supported, in part, by the Office of Naval Research under Grant No. N00014-96-1-0930, the National Science Foundation under Grant No. MIP-9502885, the Air Force Office of Scientific Research under Grant No. F49620-96-1-0072, and the Army Research Laboratories under Cooperative Agreement No. DAAL01-96-2-0002.

4

Blind Space-Time Signal Processing

Arogyaswami J. Paulraj
Constantinos B. Papadias
Vellenki U. Reddy
Alle-Jan van der Veen

As discussed in Chapters 1–3, signal processing functions in wireless communications include modulation/demodulation, channel coding/decoding, channel equalization and estimation of transmitted signals, and reduction of cochannel interference (CCI). One promising approach to improve signal processing performance is space-time (S-T) processing, which operates simultaneously on multiple antennas. A key leverage of this spatial dimension is CCI reduction. This reduction is possible since the CCI and the desired signal almost always arrive at the antenna array (even in complex multipath environments) with distinct and often well-separated spatial signatures, thus allowing the modem to exploit this difference to reduce the CCI. Likewise, the space-time transmit processing can use spatial selectivity to deliver signals to the desired mobile while minimizing the interference for other mobiles. Another leverage is the exploitation of blind methods. Use of training for equalization consumes bandwidth and is not efficient in rapidly time-varying channels. Therefore, blind channel equalization and estimation of multiple users' signals can improve network capacity and performance.

The spatial dimension can also be used to enhance other aspects of space-time modem performance. In reception, the antennas can be used to provide enhanced array gain, improve signal to thermal noise ratio, and enhance diversity gain, as discussed in Chapter 1. In transmission, the spatial dimension can enhance array gain, improve transmit diversity, as is also discussed in Chapter 1, and reduce delay spread.

The chapter focuses on the receiver S-T processing for nonspread modulation and is organized as follows. In Section 4.1, we summarize the propagation model. In Section 4.2, we develop a model for signals received at an antenna array and discuss the spatial and temporal structures in this model. In Section 4.3, we discuss a zero-forcing view of channel identifiability and equalizability and discuss the similarities and differences between intersymbol interference (ISI) and CCI cancellation. In Section 4.4, we present some recently proposed techniques for blind multiuser detection using block and recursive techniques. Section 4.5 concludes with a summary of the chapter.

4.1 THE WIRELESS PROPAGATION ENVIRONMENT

As discussed in earlier chapters, the propagation of radio signals on both the forward (base-station-to-subscriber unit) and reverse (subscriber unit-to-base-station) links is affected by a channel in several ways. Multipath propagation results in the spreading of the signal in three dimensions. These are the delay (or time) spread, Doppler (or frequency) spread and angle spread. These spreads (see Figure 4.1) have significant effects on the signal and are summarized below together with the terminology we will use in this chapter.

Doppler Spread: Time-Selective Fading. Doppler spread results from mobile motion and local scattering near the mobile. If one assumes uniformly dis-

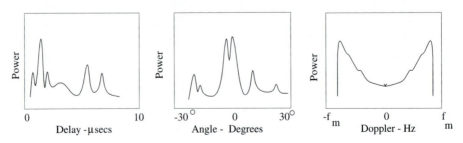

Figure 4.1   The three spreads of the wireless channel.

tributed local scatterers, then, as discussed in Chapter 2, the baseband power spectrum of the received vertical electrical field due to a continuous-wave tone has a U-shaped form, called the classical (or Jakes' model) spectrum [9].

If there is a direct line-of-sight path, the channel spectrum is modified by an additional line at a frequency corresponding to the relative velocity between the base and the mobile. Doppler spread causes time-selective fading and can be characterized by the *coherence time* of the channel. The larger the Doppler spread, the smaller the coherence time.

Delay Spread: Frequency-Selective Fading.   Due to the multipath propagation, several time-shifted and scaled versions of the transmitted signal will arrive at the receiver. Typically, a double negative exponential model is observed: the delay separation between paths increases negative exponentially with path delay, and the path amplitudes also fall off negative exponentially with delay. This spread of path delay is called *delay spread*. Delay spread causes frequency-selective fading and is also measured in terms of *coherence bandwidth*. The larger the delay spread, the smaller the coherence bandwidth.

Angle Spread: Space-Selective Fading.   Angle spread on receive refers to the spread of arrival angles of the multipaths at the antenna array. Likewise, angle spread in transmit refers to the spread of departure angles of the multipaths. The angle of arrival (or departure) of a path can be statistically related to the path delay. With a constant-delay, ellipse-scattering model, it can be shown that angle spread is proportional to delay spread and inversely proportional to the transmitter-receiver separation. Angle spread causes space-selective fading and is characterized by the *coherence distance*. The larger the angle spread, the shorter the coherence distance.

Multipath Propagation in Large Cells.   Multipath scattering underlies the three spreading effects described above and Doppler spread, which in addition requires subscriber unit motion. It is important to understand the types of scatterers and their contribution to channel behavior. These scatterers are illustrated in Figure 4.2.

Scatterers Local to Mobile.   Scattering local to the mobile is caused by buildings in the vicinity of the mobile (a few tens of meters). Mobile motion and local scattering give rise to Doppler spread, which causes time-selective fading. For a mobile traveling at 65 mph, the Doppler spread is about ± 200 Hz in the 1900 MHz band. Although local scattering contributes to Doppler spread, the delay

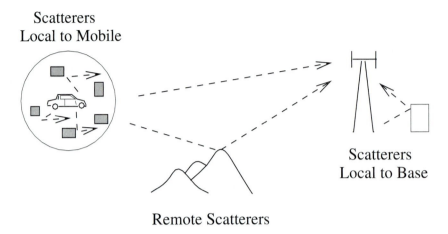

Scatterers
Local to Mobile

Scatterers
Local to Base

Remote Scatterers

Figure 4.2 Multipath propagation has three distinct scattering sources, each of which gives rise to different channel effects.

spread will usually be insignificant because of the small scattering radius. Likewise, the angle spread will also be small.

Remote Scatterers. The emerging wavefront from the local scatterers may then travel directly to the base or may be scattered toward the base by remote *dominant scatterers*, giving rise to specular multipath. These remote scatterers can be either terrain features or high-rise building complexes. Remote scattering can cause significant delay and angle spreads.

Scatterers Local to Base. Once these multiple wavefronts reach the base station, they may be scattered further by local structures such as buildings or other structures that are in the vicinity of the base. Such scattering will be more pronounced for low elevation and below-rooftop antennas. The scattering local to the base can cause severe angle spread, which in turn can cause space-selective fading. This fading is coherent, unlike the time-varying, space-selective fading caused by remote scattering.

Space-Time Channel Model. A multipath channel [9] is illustrated in Figure 4.3. Typical path amplitude, delay, and fading statistics can be obtained from published propagation models. The signal from the mobile travels through a number of paths, each with its own power fading and delay. The fading can be either Rayleigh or Rician and can have a Doppler spectrum that is flat or classical. These paths arrive at the receive antenna array with varying angles of arrival. The composite multipaths induce a different multipath channel at each antenna because of differences in relative phasing of the paths.

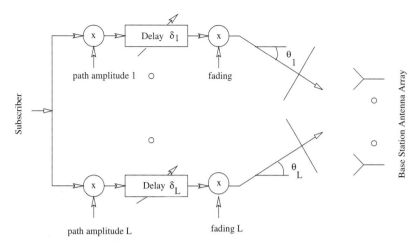

Figure 4.3   Multipath model.

A typical example of a macrocellular channel in a hilly terrain in the case of the European cellular system GSM (Groupe Speciale Mobile) is shown in Figure 4.4. We plot the frequency response at each antenna. Since the channel bandwidth is high (200 KHz), the channel is highly frequency-selective in a hilly terrain environment where delay spreads can reach 10 to 15 μs. Also, the large angle spread causes variations of the channel from antenna to antenna. The channel variation in time depends upon the Doppler spread. Note that since GSM uses a short time slot, the channel variation during the time slot is negligible.

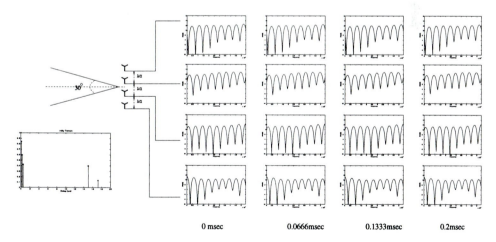

Figure 4.4   Channel frequency response at four different antennas for GSM.

4.2 SIGNAL MODEL AND STRUCTURE

In this section, we describe analytical models for the received signals at the antenna elements in a wireless system.

4.2.1 Signal Model

In this section, we develop suitable receive signal models for the single and multiple user cases shown in Figure 4.5. First, we develop the single-user (SU) model and later extend it to the multiple-user (MU) case. In both cases, on the reverse link, the subscriber uses a single antenna input (SI) and base station uses multiple antenna outputs (MO).

Let $c(t)$ denote the continuous-time impulse response of the multipath channel to an omnidirectional antenna (excluding that of transmitter and receiver filters), which we refer to as the physical channel impulse response. Assuming a specular multipath model of the type considered in earlier chapters, we can express $c(t)$ as

$$c(t) = \sum_{l=1}^{L} \alpha_l(t)\delta(t - \tau_l), \tag{4.1}$$

where $\alpha_l(t)$ and τ_l denote the complex path fading and the propagation delay of lth path, respectively, L is the number of multipaths, and $\delta(\cdot)$ is the Dirac delta function.

Let $u(t)$ denote the baseband equivalent of the transmitted signal that depends on the modulation waveform and the information data stream. In the IS-54 TDMA standard, $u(\cdot)$ is a $\pi/4$-shifted DQPSK (differential quadrature phase-shift keying), gray-coded signal that is modulated by a pulse with square-root raised

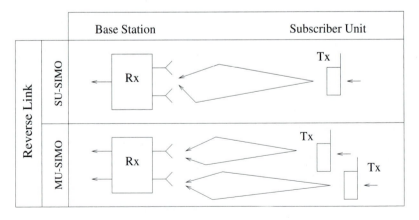

Figure 4.5   Single and multiple user receive configurations.

cosine spectrum with excess bandwidth of 0.35. In GSM, a Gaussian minimum shift keying (GMSK) modulation is used. See [4, 13, 33] for more details.

For a linear modulation (e.g., DQPSK), we can write

$$u(t) = \sum_k g(t - kT)s(k), \tag{4.2}$$

where $s(k)$ denote the transmitted symbols, T denotes the baud (or symbol) period, and $g(t)$ denotes the effective continuous-time pulse shape that includes the effects of the transmitting and receiving filters.

We now consider an m-element antenna array. We can express the noiseless baseband signal at the ith element of the array, $x_i(t)$, as

$$x_i(t) = \sum_{l=1}^{L} a_i(\theta_l)\alpha_l(t)u(t - \tau_l), \tag{4.3}$$

where $a_i(\theta_l)$ is the response of the ith sensor for an lth path from direction θ_l, and $\alpha_l(t)$ represents the complex path fading for the lth path. We can rewrite (4.3) as

$$x_i(t) = \sum_k \sum_{l=1}^{L} a_i(\theta_l)\alpha_l(t)g(t - \tau_l - kT)s(k) \tag{4.4}$$

In the above model, we have assumed that the inverse of the signal bandwidth is large compared to the travel time across the array (and that the channel fading bandwidth is assumed to be negligible compared to the signal bandwidth). This is the so-called narrowband assumption of the transmitted signal. Therefore, the signal complex envelope received by each antenna is identical except for phase (and perhaps amplitude) differences that depend on the path angle-of-arrival. This angle-of-arrival dependent phase shift along with any amplitude-phase response differences is included in $a_i(\theta_l)$ [19]. The complex reverse link signal fading amplitude $|\alpha_l(t)|$ is Rayleigh- or Rician-distributed.

The channel model described above uses physical path parameters such as path gain, delay, and angles-of-arrival, none of which are known nor are easily estimated. The noiseless baseband signal received at the ith antenna output can also be written as

$$x_i(t) = \sum_k h_i(t - kT)s(k), \tag{4.5}$$

where $h_i(t)$ represents the composite baseband impulse response of the channel from the user's transmitter to the output of the ith sensor and is the convolution of the physical channel impulse response and the pulse shaping function. Since $h_i(t) = c_i(t) * g(t)$, it follows that

$$h_i(t) = \sum_{l=1}^{L} a_i(\theta_l)\alpha_l(t)g(t - \tau_l) \qquad i = 1,\ldots,m. \tag{4.6}$$

This impulse response $h_i(t)$ has a finite duration called the channel length. In the following, we assume that the fading coefficients $\{\alpha_l\}$ remain constant over the time interval during which we collect the data and therefore treat $h_i(t)$ as time-invariant. Defining the vector impulse response[1] $\mathbf{h}(t) = [h_1(t) \ \cdots \ h_m(t)]^T$, (4.6) can be written as

$$\mathbf{h}(t) = \sum_{l=1}^{L} \mathbf{a}(\theta_l)\alpha_l(t)g(t - \tau_l), \tag{4.7}$$

where $\mathbf{a}(\theta) = [a_1(\theta) \ \cdots \ a_m(\theta)]^T$ is the array response vector.

Sampling $h_i(t)$ at the symbol rate, we get $h_i(n) = h_i(t)\big|_{t=t_0+nT}$, which is called the symbol-spaced channel, where t_0 is the initial sampling instant. The symbol-rate sampled received signal at the output of the ith antenna element can be expressed as

$$x_i(k) = \sum_{n=0}^{N-1} h_i(n)s(k - n), \qquad i = 1,\ldots, m, \tag{4.8}$$

where we have assumed, without loss of generality, the impulse response corresponding to each antenna to span N symbol periods (corresponding to a channel length of NT). Again, defining the array output $\mathbf{x}(k) = [x_1(k) \ \cdots \ x_m(k)]^T$, we can rewrite (4.8) as

$$\mathbf{x}(k) = \sum_{n=0}^{N-1} \mathbf{h}(n)s(k - n), \tag{4.9}$$

where $\mathbf{h}(n) = \mathbf{h}(t)\big|_{t=t_0+nT}$. If we define H as

$$H = [\mathbf{h}(0) \ \cdots \ \mathbf{h}(N - 1)], \tag{4.10}$$

(an $m \times N$ channel matrix), then (4.9) can be rewritten as

$$\mathbf{x}(k) = H\mathbf{s}(k), \tag{4.11}$$

where

$$\mathbf{s}(k) = \begin{bmatrix} s(k) \\ \vdots \\ s(k - N + 1) \end{bmatrix}. \tag{4.12}$$

In terms of the physical channel parameters and the pulse shaping function, H can be expressed as

$$H = \mathbf{A}(\theta)\Lambda\mathbf{G}(\tau), \tag{4.13}$$

where $A(\theta) = [\mathbf{a}(\theta_1) \ \mathbf{a}(\theta_2) \ \cdots \ \mathbf{a}(\theta_L)]$, $\Lambda = \mathrm{diag}\{\alpha_1, \cdots, \alpha_{L-1}, \alpha_L\}$ and

[1]The superscript ^{T} denotes the transpose operator.

$$G = \begin{bmatrix} g(t_0 - \tau_1) & g(t_0 + T - \tau_1) & \cdots & g(t_0 + NT - T - \tau_1) \\ \vdots & \vdots & \cdots & \vdots \\ g(t_0 - \tau_{L-1}) & g(t_0 + T - \tau_{L-1}) & \cdots & g(t_0 + NT - T - \tau_{L-1}) \\ g(t_0 - \tau_L) & g(t_0 + T - \tau_L) & \cdots & g(t_0 + NT - T - \tau_L) \end{bmatrix}. \tag{4.14}$$

In the following, we refer to the rows of H as the subchannel responses. From (4.13) we get the following expression for each coefficient of H in terms of the channel parameters [32]:

$$H_{ij} = \sum_{l=1}^{L} \alpha_l a_i(\theta_l) g(t_0 - \tau_l + (j-1)T). \tag{4.15}$$

As we will see later, there are advantages if $h_i(t)$ is sampled at a rate greater than the symbol rate. This oversampling can be easily incorporated in our signal model. Let the fractional sampling interval be $T_s = T/P$, where P is an integer. We define the pth phase, $p = 1, 2,..., P$, of the fractionally spaced response channel corresponding to ith antenna as

$$\mathbf{h}_p(n) = \mathbf{h}(t)\big|_{t=t_0+nT+\frac{(p-1)}{P}T}. \tag{4.16}$$

Now, the new $mP \times 1$ vector channel impulse response can be redefined as

$$\mathbf{h}(n) = \begin{bmatrix} \mathbf{h}_1(n) \\ \vdots \\ \mathbf{h}_P(n) \end{bmatrix}. \tag{4.17}$$

Similarly to (4.16), we define the pth phase of the vector received signal $\mathbf{x}_p(k)$ as

$$\mathbf{x}_p(n) = \sum_{n=0}^{N-1} \mathbf{h}_p(n) s(k - n), \tag{4.18}$$

where we have assumed again, for convenience, that the impulse response of each phase spans N symbol periods. H (now of dimension $mP \times N$) will correspond to a fractionally spaced vector channel with number of subchannels as mP and each channel length as N, and will still obey the definition (4.10) with $\mathbf{h}$ given in (4.17). The channel model is again

$$\mathbf{x}(k) = H\mathbf{s}(k), \tag{4.19}$$

with $\mathbf{x}(k) = [\mathbf{x}_1^T(k) \cdots \mathbf{x}_P^T(k)]^T$. Notice that with this definition of H, the factorization (4.13) is no longer valid (an alternative definition that would keep (4.13) valid would be to use H' of dimension $m \times PN$).

We now collect M vector samples of the received signal during M symbol periods. Stacking these samples (in increasing order) in a polyphase $mMP \times 1$ vector, i.e., $X_M(k) = [\mathbf{x}^T(k) \ \mathbf{x}^T(k+1) \ \cdots \ \mathbf{x}^T(k+M-1)]^T$, we obtain the factorization

$$X_M(k) = \mathcal{H}S(k), \tag{4.20}$$

where $\mathcal{H}$ is the $mMP \times (N + M - 1)$ matrix,

$$\mathcal{H} = \begin{bmatrix} 0 & \boxed{H} \\ & \ddots & \ddots \\ & \boxed{H} & \\ \boxed{H} & & 0 \end{bmatrix} \tag{4.21}$$

and

$$S(k) = \begin{bmatrix} s(k+M-1) \\ \vdots \\ s(k-N+1) \end{bmatrix}. \tag{4.22}$$

Notice the block-Hankel structure of $\mathcal{H}$ (an equivalent formulation of (4.20) with a decreasing time index in X_M would result in a block-Toeplitz instead of a block-Hankel matrix—see [12, 20]).

If $\mathcal{H}$ is of full column rank, then

$$\text{Column span } (X) = \text{Column span } (\mathcal{H}), \tag{4.23}$$

and full column rank of $\mathcal{H}$ is guaranteed if $mMP \geq (N + M - 1)$ and if the subchannel polynomials of H do not share a common root [12]. As will be seen later, the full column rank of $\mathcal{H}$ is an important property linked both to the identifiability and the equalizability of the channel.

Suppose we have M' polyphase vector samples of data and we wish to use these in blocks. We can then extend the data vector $X_M(k)$ to a block-Hankel matrix by left-shifting and stacking M' times:

$$\mathcal{X}_M = \begin{bmatrix} \mathbf{x}(k) & \mathbf{x}(k+1) & \iddots & \mathbf{x}(k+M'-M) \\ \mathbf{x}(k+1) & \mathbf{x}(k+2) & \iddots & \iddots \\ \iddots & \iddots & \iddots & \mathbf{x}(k+M'-2) \\ \mathbf{x}(k+M-1) & \iddots & \mathbf{x}(k+M'-2) & \mathbf{x}(k+M'-1) \end{bmatrix}, \tag{4.24}$$

which has size $mMP \times (M' - M + 1)$. In terms of H and the transmitted symbol matrix, this augmented matrix can be expressed as

$$\mathcal{X}_M = \mathcal{H}\mathcal{S} = \tag{4.25}$$

$$\begin{bmatrix} 0 & \boxed{H} \\ & \ddots \\ \boxed{H} & \\ \boxed{H} & 0 \end{bmatrix} \begin{bmatrix} s(k+M-1) & \ddots & s(k+M'-2) & s(k+M'-1) \\ & \ddots & \ddots & s(k+M'-2) \\ s(k-N+2) & s(k-N+3) & \ddots & \ddots \\ s(k-N+1) & s(k-N+2) & \ddots & s(k+M'-M-N+1) \end{bmatrix},$$

where $\mathcal{H}$ is $mMP \times (N+M-1)$ and $\mathcal{S}$ is $(N+M-1) \times (M'-M+1)$. Equation (4.25) shows that the augmented data matrix admits a factorization into two matrices, a block-Hankel space-time channel matrix and a block-Toeplitz transmitted symbol matrix. This observation, too, is crucial to the blind identification of $\mathcal{H}$ from the observations $\mathcal{X}_M$.

In the MU case, we assume that multiple subscribers transmit their information signals towards the antenna array at the same base station. The MU-SIMO model is a straightforward extension of the SU-SIMO model. Assuming Q users, the symbol-rate sampled, received signal at the antenna array is the sum of the signals from Q subscribers, and using (4.19) we have

$$\mathbf{x}(k) = \sum_{q=1}^{Q} H_q \mathbf{s}_q(k). \tag{4.26}$$

A data model for the MU case will have Q channels corresponding to Q users, and we assume, for ease of exposition, that each user's channel impulse response spans N symbol periods. We interleave the sampled impulse responses of the Q users, sample by sample, taking the size of H to $mP \times QN$. Defining the Q-tuple symbol vector as $\mathbf{s}(k) = [s^{(1)}(k)s^{(2)}(k)\ldots s^{(Q)}(k)]^T$, where $s^{(i)}(k)$ denotes the kth symbol of ith user, we get the following signal model:

$$\mathcal{X}_M = \tag{4.27}$$

$$\begin{bmatrix} 0 & \boxed{H} \\ & \ddots \\ \boxed{H} & \\ \boxed{H} & 0 \end{bmatrix} \begin{bmatrix} \mathbf{s}(k+M-1) & \ddots & \mathbf{s}(k+M'-2) & \mathbf{s}(k+M'-1) \\ & \ddots & \ddots & \mathbf{s}(k+M'-2) \\ \mathbf{s}(k-N+2) & \mathbf{s}(k-N+3) & \ddots & \ddots \\ \mathbf{s}(k-N+1) & \mathbf{s}(k-N+2) & \ddots & \mathbf{s}(k+M'-M+N+1) \end{bmatrix},$$

so that $\mathcal{H}$ is $mMP \times Q(N+M-1)$, and $\mathcal{S}$ is $Q(N+M-1) \times (M'-M+1)$, and where the M shifts of H to the left are now each over Q positions. Again, as in the single-user case, the blind identification and equalization of the channel will be affected by the size and conditioning of $\mathcal{H}$.

4.2.2 Spatial and Temporal Signal Structure

Given the signal model at (4.25), an important question is whether the unknown channel $\mathcal{H}$ and data $\mathcal{S}$ can be determined from the observations $\mathcal{X}_M$. This question leads us to examine the underlying constraints on $\mathcal{H}$ and $\mathcal{S}$, which we call *structure*.

4.2.2.1 Spatial Structure

The spatial structure of H is apparent from (4.15). The vector $\mathbf{a}(\theta_l)$ lies on the *array manifold* $\mathcal{A}$, which is the set of array response vectors indexed by θ

$$\mathcal{A} = \{\mathbf{a}(\theta) \mid \theta \in \Theta\}, \tag{4.28}$$

where the set Θ is the set of all possible values of θ. Knowledge of $\mathcal{A}$ helps determine $\mathbf{a}(\theta_l)$. $\mathcal{A}$ includes the effect of array geometry, element patterns, interelement coupling, scattering from support structures, and objects near the base station. $\mathcal{A}$, when measured at the receiver baseband after digitization, includes the effects by cable and receiver gain/phase response, in-phase/quadrature (I-Q) imbalance, and analog-to-digital (A/D) converter errors. $\mathcal{A}$ is frequency-dependent and needs to be calibrated at multiple points within the operating band.

4.2.2.2 Temporal Structure

The temporal structure relates to the properties of the signal $u(t)$ and includes modulation format, pulse-shaping function, and symbol constellation. Some typical temporal structures are described below.

Constant Modulus (CM). In many wireless applications, the transmitted waveform has a constant envelope (e.g., in FM modulation). A typical example of a constant envelope waveform is the GMSK modulation used in the GSM cellular system, which has the following general form:

$$u(t) = e^{j(\omega t + \varphi(t))},$$

where $\varphi(t)$ is a Gaussian-filtered phase output of a minimum shift keyed (MSK) signal [23].

Finite Alphabet (FA). Another important temporal structure in mobile communication signals is the *finite alphabet*. This structure underlies all digitally modulated schemes. The modulated signal is a linear or nonlinear map of an underlying finite alphabet. For example, the IS-54 signal is a $\pi/4$-shifted DQPSK signal given by

$$u(t) = \sum_p A_p g(t - pT) + j \sum_p B_p g(t - pT), \tag{4.29}$$

where

$$A_p = \cos(\varphi_p), \quad B_p = \sin(\varphi_p), \quad \varphi_p = \varphi_{p-1} + \Delta\varphi_p,$$

$g(\cdot)$ is the pulse shaping function (which is a square root raised cosine function in the case of IS-54), and $\Delta\varphi_p$ is chosen from a set of finite phase shifts $\left\{ \frac{5\pi}{4}, \frac{3\pi}{4}, \frac{\pi}{4}, \frac{7\pi}{4} \right\}$ depending on the data $s(\cdot)$. This finite set of phase shifts represents the FA structure.

Distance from Gaussianity. The distribution of digitally modulated signals is not Gaussian,[2] and this property can be exploited to estimate the channel from higher-order statistical quantities such as cumulants. See, e.g., [7] and [16]. Clearly CM signals are non-Gaussian. These higher-order statistics (HOS) based methods are usually slower converging than those based on second-order statistics.

Cyclostationarity. As discussed in Chapter 3, recent theoretical results [5, 12, 22, 25] suggest that exploiting the *cyclostationary* characteristic of the communication signal can lead to algorithms requiring only second-order statistics to identify the channel H and therefore are a more attractive approach than HOS techniques.

It can be shown [2] that the continuous-time stochastic process $x(t)$ defined in (4.3) (assuming the fade amplitudes α_ℓ are constant) is cyclostationary. Moreover, the discrete sequence $\{x_i\}$ obtained by sampling $x(t)$ at the symbol rate $\frac{1}{T}$ is wide-sense stationary, whereas the sequence obtained by temporal oversampling (i.e., at a rate higher than $1/T$) or spatial oversampling (multiple antenna elements) is cyclostationary. The cyclostationary signal consists of a number of *phases* (polyphase components), each of which is stationary. A phase corresponds to a shift in the sampling point in temporal oversampling and different antenna element in spatial oversampling. The duality between temporal and spatial oversampling is illustrated in Figure 4.6.

The cyclostationary property of sampled communication signals carries important information about the channel phase, and this information can be exploited in several ways to identify the channel. The cyclostationarity property can also be interpreted as a *finite duration* property. Put simply, the oversampling increases the number of samples in the signal $\mathbf{x}(t)$ and phases in the channel H but does not change the value of the data for the duration of the symbol period. This oversampling is what allows $\mathcal{H}$ to become tall (more rows than columns) and full column rank. Also, the stationarity of the channel makes $\mathcal{H}$ Hankel (or rather,

---

[2]The distribution may, however, approach a Gaussian when constellation shaping is used for spectral efficiency [34].

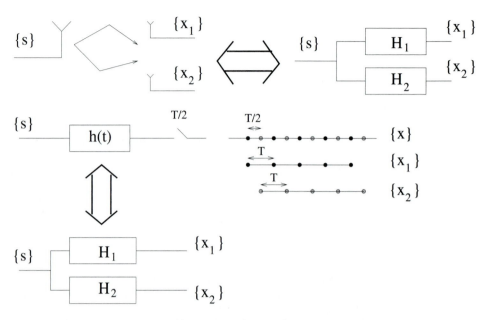

Figure 4.6 Antennas and/or oversampling result in polyphase SIMO channels.

block-Hankel). As indicated earlier, tallness and Hankel properties are key to the blind estimation of H.

The Temporal Manifold. Just as the array manifold captures spatial wave-front information, the *temporal manifold* captures the temporal pulse-shaping function information (see [32, 35]). We define the temporal manifold $\mathbf{k}(\tau)$ as the sampled response of a receiver to an incoming pulse with delay τ. The temporal manifold is a powerful structure for channel identification and tracking. Moreover, unlike the array manifold, it can be estimated with good accuracy since it depends only on our knowledge of the pulse-shaping function. Table 4.1 shows the duality between the array and the temporal manifold.

The different structures and properties inherent in the nature of the transmitted signals and the employed receivers in space-time processing are shown in Figure 4.7.

TABLE 4.1 The duality between the array and the time manifold

Manifold	Indexed by	Characterizes
Array	angle θ	antenna array response
Time	delay τ	transmitted pulse shape

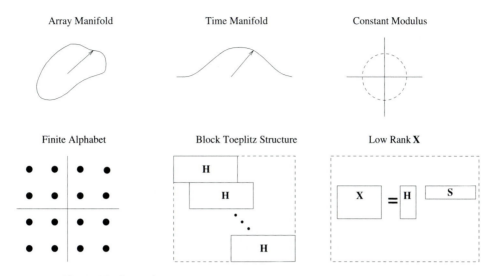

Figure 4.7   Space-time structures.

4.3 CHANNEL IDENTIFICATION AND EQUALIZATION

Given the model (4.25) or (4.27) and the received signal, the task of the receiver is to estimate the data S that were transmitted. This estimation is usually performed using one of two approaches. The first approach is to determine the channel H and then use a maximum-likelihood sequence estimator (MLSE) to find the data S. Another approach is to sidestep the channel estimation and invert or equalize the channel to reveal the data directly.

If we wish to equalize the channel, a key question is: What conditions on the channel make it invertible? On the other hand, if we wish to estimate the channel first, the corresponding question is under what conditions the received signal statistics alone can provide identification (this is called blind channel identification). Of course, the answers to these questions depend on the type of filters used for equalization and the type of signal statistics used for identification. We address these issues next, focusing on linear equalizers and blind identification based on second-order statistics (SOS).

4.3.1 Single-User Channels

We begin, in this section, with the equalization of single-user channels. The multi-user case is treated in the following section.

4.3.1.1 Conditions for Channel Equalization/Identification

We consider the problem of zero-forcing (ZF) equalization for an m-sensor receiver with a P-oversampled channel. Namely, we are interested in finding an equalizer of order M, $\mathbf{F}_M = [\mathbf{f}(0) \ \cdots \ \mathbf{f}(M-1)]^T$ (an $mMP \times 1$ vector), such that the following zero-forcing condition is satisfied for the equalizer output $y(k)$:

$$y(k) = \mathbf{F}_M^T X_M(k) \equiv s(k - \delta), \tag{4.30}$$

where $\delta \in [-M + 1, N - 1]$, which results according to (4.20) to

$$\mathbf{F}_M^T \mathcal{H} = [0 \cdots 0 \ 1 \ 0 \cdots 0], \tag{4.31}$$

where the position of the single non-zero element in the right-hand side of (4.31) depends of course on the choice of δ. To satisfy (4.31), the generalized Sylvester matrix $\mathcal{H}$ needs to be left-invertible, hence full column rank, which requires that it has at least as many rows as columns and that the different columns are linearly independent. This requirement results in the following two conditions:

(C1) $mMP \geq M + N - 1 \Rightarrow M \geq \underline{M} = \left\lceil \frac{N-1}{mP-1} \right\rceil$.

(C2) The polynomials $H_i(z)$, $i = 1,\ldots, mP - 1$ corresponding to the different rows of H must have no common roots.

Therefore, zero-forcing equalization is indeed possible with a finite-length equalizer, provided that the received signal is oversampled (or received with multiple antennas) so as to satisfy (C1) and that the mP channel phases have no common zeros. Notice that this is still possible if only oversampling ($m = 1$, $P > 1$) or only antennas ($P = 1$, $m > 1$) are used. This fact was noticed by Slock in [20].[3]

The same conditions that hold for perfect noiseless ZF equalization turn out to be necessary and sufficient for the blind identification of a polyphase channel with the use of second-order-only (cyclostationary) statistics. The cyclostationarity of the sampled received signal is crucial in obtaining this result. The result was first obtained by Tong et al. in [25] and was stated as follows.

Theorem I.   The channel transfer function $H(z)$ is uniquely determined (identified) by the cyclic spectrum of the oversampled channel output if and only if $H(z)$ does not have zeros uniformly spaced on a circle with separation of $2\pi/T$ radians.

In this theorem, $H(z) = \sum_{i=1}^{mP} z^{-i} H(z^{mP})$ represents the z-transform of the interleaved channel response. It is easy to show [28] that $H(z)$ has zeros equispaced on a circle if and only if the mP channel phase polynomials $H_i(z)$ have no common root (condition (C2) above).

[3]It turns out that similar conditions had appeared in a different context earlier (Massey and Sain [11]).

Hence, the common root condition is important for both the existence of finite-length linear equalizers and for SOS-based blind channel identification.

Whereas in theory it may be unlikely that all the subchannels share common roots, in practice they may have several roots that are very close to each other. This can make SOS-based methods ill conditioned. We show below that there exist some channel classes that will suffer from the common zeros problem and we present some approaches to overcome it.

4.3.1.2 Avoiding the Problem of Zeros in Common

According to Theorem I, SOS-based identifiability will fail when the phases of the different channels share common roots. Two questions are then of interest: how likely this is to happen and what can be done in practice to avoid it?

A partial answer to the first question was given in [28] and in [3]. In [28], it was observed that the following class of channels will always suffer from the zeros-in-common problem when oversampled in time:

- Class I: channels with delays that are all multiples of T

This case can be easily seen as follows. The impulse response of Class I channels will have the general form

$$h(t) = \sum_{l=0}^{L-1} \alpha_q g(t - lT); \tag{4.32}$$

hence, the ith phase $h_i(k) = h(t_0 + kT + (i-1)T/P$ will be given by

$$h_i(k) = \sum_{l=0}^{L-1} \alpha_q g_i(k-l), \tag{4.33}$$

where $g_i(k) = g(t_0 + kT + (i-1)/T)$. Taking the z- transform of (4.33) gives

$$H_i(z) = A(z)G_i(z), \tag{4.34}$$

where $A(z)$ is the z-transform of α_q. It is clear from (4.34) that the channel phases obtained from oversampling Class I channels have common zeros.

Some other channel classes that have the same identifiability problem when oversampled were reported in [3]. One of these classes is the following:

- Class II: band-limited channels with frequency nulls in $\left[-\frac{\pi(1-\beta)}{T}, \frac{\pi(1-\beta)}{T}\right]$, where β is the roll-off parameter

An example of such band-limited channels is shown in Figure 4.8.

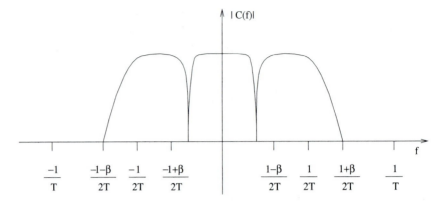

Figure 4.8   Typical example of a Class II band-limited channel.

The existence of the above channel classes suggests that the zeros-in-common identifiability problem is likely to arise in some practical cases. We now present some proposed approaches that avoid this problem.

Space-Time Oversampling.   We consider first Class I channels. Instead of using time oversampling to obtain a polyphase channel (4.33), we could use space oversampling: the received signal is sampled at the symbol rate and received by an array of m sensors. Assuming a uniform linear array (ULA), the continuous-time impulse response corresponding to the ith sensor of the array would then be given by

$$
\begin{aligned}
h_i(t) &= \left(\sum_{l=0}^{L-1} \alpha_l e^{-j\frac{2\pi d}{\lambda}(i-1)\sin\theta_l} \delta(t - lT) \right) * g(t) \\
&= \left(\sum_{l=0}^{L-1} \alpha_l e^{-j\frac{2\pi d}{\lambda}(i-1)\sin\theta_l} g(t - lT) \right),
\end{aligned}
\tag{4.35}
$$

where d is the interelement spacing, and λ is the wavelength of the carrier frequency. Defining the discrete impulse response of each antenna i as $h_i(k) = h_i(t)|_{t=t_0+kT}$ and the corresponding sampled pulse shaping function $g(k) = g(t)|_{t=t_0+kT}$, the z-transform of $h_i(k)$ is given by [18]

$$
H_i(z) = \sum_{l=0}^{L-1} \gamma_{il} z^{-l} G(z),
\tag{4.36}
$$

where $\gamma_{il} = \alpha_l e^{-j\frac{2\pi d}{\lambda}(i-1)\sin\theta_l}$. According to (4.36), in the multiple sensors case, the common factor among the subchannels comes from the pulse shape (while in the oversampling case it comes from the multipath channel—cf. (4.34)). Now, if $g(t)$ is a Nyquist pulse (e.g., a pulse with raised cosine spectrum) and perfect synchronization has been achieved (t_0 is a multiple of T), then $G(z) = 1$. In this case, according to (4.36), the m subchannels will have no common roots as long as the arrival angles of all the L multipaths are not the same or they do not correspond to array ambiguities. This reasoning establishes that for a Class I channel, the use of spatial instead of temporal oversampling helps avoid the identifiability problem. If synchronization is not perfect, then $G(z) \neq 1$; however, the identifiability problem can again be avoided if combined spatial and temporal oversampling is used [18].

We now consider the Class II channels, in which the identifiability problem comes from the fact that each frequency null in $[-\pi(1-\beta)/T, \pi(1-\beta)/T]$ gives rise, in the oversampled response, to a set of P roots that are located uniformly around the unit circle.

For example, we consider the class of multipath channels with two paths [3],

$$c(t) = \delta(t) + \delta\left(t - \frac{T}{(1-\eta)}\right), \eta > \beta, \tag{4.37}$$

which has a frequency null at $\omega = \frac{\pi(1-\eta)}{T}$ or $\omega = \frac{-\pi(1-\eta)}{T}$. We consider again the subchannels obtained from the m sensors of a uniform linear array. Following the steps similar to those used in arriving at (4.36), the ith antenna channel response is given in the z-domain by

$$H_i(z) = e^{-j\frac{2\pi d}{\lambda}(i-1)\sin\theta_0}(G(z) + \gamma_i G_\tau(z)), \tag{4.38}$$

where $g_\tau(k) = g(t-\tau)\big|_{t=t_0+kT}$. Observe that there is no common polynomial factor shared by the subchannels.[4] Thus, again, the use of an antenna array instead of oversampling can be used to allow the SOS identifiability of channels that would otherwise be unidentifiable.

Cyclostationarity Through Decision Feedback.   A different approach to obtain cyclostationarity at the receiver without using temporal or spatial oversampling is the use of a decision-feedback equalizer (DFE) receiver. As we show below, after sufficient opening of the channel eye by the receiver, decision feedback can provide a polyphase signal that does not suffer from the zeros-in-common problem [14].

Consider the receiver shown in Figure 4.9, where symbol-rate sampling has been assumed:

When the switch is at position 2, the setup corresponds to a standard symbol-rate decision-feedback receiver. Assuming that the channel eye has been sufficiently opened (as is typically assumed in the analysis of DFEs), the following condition holds:

$$\hat{a}(k) = a(k-j), \tag{4.39}$$

where j is some inherent delay. In this case, the DFE receiver corresponds to the setup of Figure 4.9, where the switch is at position 1. One may now notice that the

---

[4]For $\tau = kT$, however, $G_\tau(z) = G(z)z^{-k}$, and it corresponds to a special case of the Class I channels.

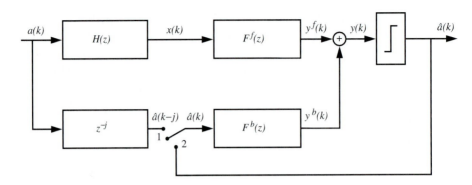

Figure 4.9   An equivalent DFE setup, assuming correct decisions.

vector input to the two filters $\tilde{\mathbf{x}}(k) = [x(k)\ \hat{a}(k)]^T$ can be seen as the output of the following single-input-two-output channel:

$$H_1(z) = H(z)$$
$$H_2(z) = z^{-j}. \tag{4.40}$$

Therefore, the following theorem holds:

Theorem II.   If (4.39) holds, the vector input $\tilde{\mathbf{x}}(k)$ to the feedforward and feedback filters is equal to the output of a single-input-two-output channel whose subchannels $H_1(z)$ and $H_2(z)$ have no common factor, except for a possible pure delay z^{-l}.

Based on the above theorem, the problem of zeros in common can be avoided with the use of decision feedback, provided that the channel eye has been sufficiently opened to provide correct decisions. A description of blind methods for DFE can be found in [14]. As compared to the above method of spatial oversampling, this approach requires less computational complexity since we need to compute only the values of a few coefficients in the feed-forward and feedback filters. On the other hand, the implementation of fully blind DFE techniques needs some care to guarantee cyclostationary structure (see [14]).

Transmission-Induced Cyclostationarity.   An alternative approach to avoid the channel conditioning problem is to design communication signals that are cyclostationary prior to transmission. This cyclostationarity can either be artificially introduced by redundancy in transmission or arise naturally through the cyclostationary character of the transmitted signal. For example, in [27] it is shown that if one uses repetitive interleaving of a factor 2 at the transmitter, one obtains a SIMO channel model that does not suffer from identifiability problems. Of course, this is done at the cost of reduced bandwidth efficiency (if no extra bandwidth is used, this approach will result in controlled ISI, whereas increasing the bandwidth will result in a repetition coding scheme). Similar results were presented in [1], where chip interleaving was used for the same reason prior to transmission in a CDMA system. The use of filter-bank-based precoding to induce cyclostationarity at the transmitter and avoid the zeros-in-common problem is also presented in [6]. Finally, in [8] it is shown that the SAT tone signal that is superimposed on the information bearing signal in the Advanced Mobile Phone Service (AMPS) analog cellular system used in North America also results in a cyclostationary transmitted signal.

4.3.2 Multiple-User Problem

Additional issues arise in multiuser problems, which we consider in the subsections that follow.

4.3.2.1 Joint ISI/CCI Cancellation

We consider again the general case of an m-antenna receiver with an oversampling factor of P. Assuming Q users, the ZF equalization condition is

$$\mathbf{F}_M^T X_M(k) = s^{(1)}(k - \delta), \tag{4.41}$$

where $\mathbf{F}_M^T$ and $X_M(k)$ are now $mMP \times 1$ and where we have assumed without loss of generality that we are interested in the recovery of the first user's signal (up to some delay). Using (4.27), this will give again

$$\mathbf{F}_M^T \mathcal{H} = [0 \;\; \cdots \;\; 0 \;\; 1 \;\; 0 \;\; \cdots \;\; 0] \qquad (1 \times Q(N + M - 1)). \tag{4.42}$$

Again, $\mathcal{H}$ needs to be left-invertible; hence, it is necessary that

$$M \geq \left\lceil \frac{Q(N - 1)}{mP - Q} \right\rceil. \tag{4.43}$$

Again, (4.43) will not be sufficient to achieve perfect ISI/CCI cancellation unless the row polynomials of H are guaranteed to share no common roots [15, 21].

We may notice from (4.43) that even if pure temporal processing is used ($m = 1$), it is still theoretically possible to cancel perfectly both the channel ISI and CCI if oversampling is used ($P > 1$). However, in practice, this type of performance will be limited by the channel conditioning, as we discuss below.

4.3.2.2 Channel Condition and ISI/CCI Cancellation

The preceding ZF analysis for ISI and CCI canceling with linear filters did not address the important aspect of performance deterioration of the ZF equalizer in the presence of noise. It is well known that in the presence of noise, zero-forcing equalizers can cause severe noise amplification, especially when the channel has deep spectral notches. This picture also holds, broadly, for the oversampled case.

We now discuss how path parameters affect ZF equalizer performance in the ISI and CCI channel cases. Based on the previous discussion on ISI and CCI canceling, one might infer that the ISI and CCI cancellation appear to be much the same. A ZF equalizer cannot distinguish which interference it combats, and only the differences in the two channels distinguish them from each other.

However, there are some critical differences in the way the ISI and CCI channels are influenced by the channel parameters. For example, consider the case of L paths arriving at the antenna array with near equal delays ($\tau_1 \simeq \cdots \simeq \tau_L \simeq \tau$). If all paths come from the same user, this corresponds to a low delay spread case: the channel eye is open (assuming synchronization is achieved), and no equalization is needed.

On the other hand, if the paths correspond to different users, the different user channels are similar to each other, making ZF CCI cancellation very ill conditioned ($\mathcal{H}$ is near singular). This ill-conditioning causes severe noise amplification. It is clear from the above that in this case the ISI and CCI cancellation are affected by the path parameters in opposite ways!

Denoting by $\mathbf{F}$ the fractionally spaced ZF equalizer that completely nulls interference in the absence of noise and assuming the noise at the equalizer input to be white of variance σ_v^2, the noise variance at the equalizer output is given by

$$\sigma_o^2 = \sigma_v^2 \|\mathbf{F}\|^2. \tag{4.44}$$

Hence, we can use the quantity $\|\mathbf{F}\|^2$ as a measure of noise amplification at the equalizer output.

To compare the performance between the ISI and the CCI case, we consider both an ISI and a CCI channel with two equipower signal paths with the interfering path arriving with a small delay equal to $\tau = T/20$. The received signal is of the form $x(t) = u(t) + u(t - T/20)$ in the ISI case and $x(t) = u_1(t) + u_2(t - T/20)$ in the CCI case. We have assumed linear modulation (see (4.2)) where $g(t)$ is a raised cosine pulse with roll-off parameter $\beta = 0.3$ (see [17]). The received signal is sampled with an offset of $\tau = T/20$ and oversampled by a factor of two. To evaluate the performance of zero-forcing equalization in this experiment, we have calculated for both cases the minimal-length zero-forcing equalizers that correspond to all possible cursor positions (δ in (4.30) or (4.41)).

In Figure 4.10, we have plotted the quantity $1/\|\mathbf{F}\|^2$ for each of the two cases (ISI and CCI). Notice from the figure both the effect on performance of the choice of δ and the dramatic noise amplification in the CCI case. As expected from the previous arguments, in the CCI case the two channels are very similar, leading to severe noise amplification, whereas in the ISI case the problem is well conditioned and good performance can be achieved.

We now consider a large delay-spread case: now the delay $\tau = 1.05T$ for both the ISI and the CCI channels. The performance in this case is shown in Figure 4.11.

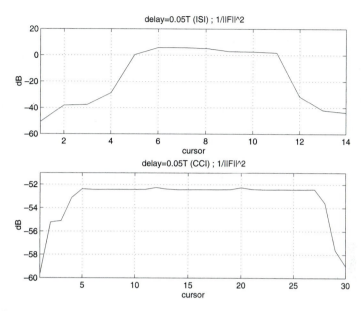

Figure 4.10   Performance comparison of ISI and CCI ZF equalizers (low delay spread).

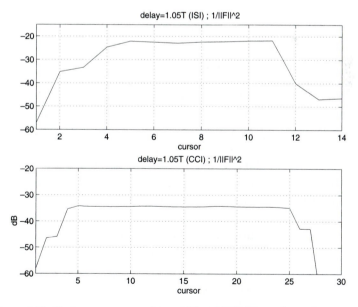

Figure 4.11   Performance comparison of ISI and CCI ZF equalizers (large delay spread).

Observe that whereas in the ISI case the performance is still superior, the gap between the two cases has reduced. Also, the performance of the ISI channel has considerably deteriorated. The first effect is due to the fact that now the two user channels in the CCI case are no longer similar; hence, the channel will be better conditioned. On the other hand, in the ISI case, the delay spread is now significant, leading to a performance reduction as compared to the earlier low delay spread.

These differences will vary depending on the channel and equalizer; e.g., if linear MMSE equalization is used instead of ZF equalization, the noise enhancement problem will be less pronounced. Also, if nonlinear equalization is used after the channels have been well identified, joint maximum-likelihood sequence detection will provide optimal performance irrespective of channel characteristics.

4.4 BLIND TECHNIQUES

To this point, we have studied channel identifiability, equalizability, and the ISI/CCI cancellation problem. However, we have not yet addressed the important question of *how* signal recovery can be achieved. In the following discussion, we present two approaches for blind signal recovery from the channel output data in the multiple-user channel case. The single-user problem can be seen as a special case of this problem and will not be addressed in this chapter.

As derived in Section 4.2, if the channel is FIR, then the oversampled output signal can be written as

$$\mathcal{X} = \mathcal{H}\mathcal{S}.$$

The objective is to blindly identify $\mathcal{S}$.

A number of properties of the signal can be used, as was listed in Section 4.2. In particular, in this section we use the following properties.

1. The *fixed symbol rate* of the signals (equivalent to the finite duration property mentioned in Section 4.2.2.2), which allows one to obtain independent linear combinations of the same symbols by using oversampling and/or multiple antennas (assuming linear modulation). This property gives rise to the Toeplitz structure of $\mathcal{S}$ and is due to the (assumed) *time invariant* nature of the channel.

2. The *constant modulus* (CM) of the signals or their *finite alphabet* (FA).

We begin with the presentation of a block technique that makes use of the finite duration and FA properties. Then, we present a recursive technique that relies on the CM property.

4.4.1 Block Methods

The method presented below consists of two steps. The first step is a straightforward extension from scalars to vectors of the blind single-user equalization algorithm proposed by Moulines et al. in [12] and by Slock in [20] and discussed in Chapter 3. At this point, the ISI caused by the channel is removed and the input signals are synchronized. However, the symbol sequences can be determined only up to a fixed linear combination of them. This problem can then be treated with the methods proposed in [10, 30, 31].

4.4.1.1 Linear Data Model

To describe the FIR-MIMO (multiple-input multiple-output) scenario, consider the linear data model as detailed in Section 4.2, which we repeat here for convenience. Assuming m antennas, P times oversampling, and an equalizer length of M symbols, the MP-dimensional complex-valued data vectors $\mathbf{x}_k$ received at the antenna array during M' symbol periods are collected in the $MmP \times (M' - M + 1)$ block-Hankel matrix

$$
\mathcal{X}_M = \begin{bmatrix} \mathbf{x}_0 & \mathbf{x}_1 & \cdot^{\cdot^{\cdot}} & & \mathbf{x}_{M'-M} \\ \mathbf{x}_1 & \mathbf{x}_2 & \cdot^{\cdot^{\cdot}} & & \cdot^{\cdot^{\cdot}} \\ \cdot^{\cdot^{\cdot}} & & \cdot^{\cdot^{\cdot}} & \cdot^{\cdot^{\cdot}} & \mathbf{x}_{M'-2} \\ \mathbf{x}_{M-1} & \cdot^{\cdot^{\cdot}} & \mathbf{x}_{M'-2} & \mathbf{x}_{M'-1} & \end{bmatrix} \tag{4.45}
$$

(see (4.24)). Let N_j be the channel length of the qth user. With Q users and a maximum channel length of $N = \max\limits_q N_q$ symbols per channel, $\mathcal{X}$ has a factorization (Section 4.2)

$$
\mathcal{X}_M = \mathcal{H}\mathcal{S} = \begin{bmatrix} \boxed{0\ \ \boxed{H}} \\ \cdot^{\cdot^{\cdot}} \\ \boxed{H} \\ \boxed{H}\ \ 0 \end{bmatrix} \begin{bmatrix} \mathbf{s}_{M-1} & & \mathbf{s}_{M'-2} & \mathbf{s}_{M'-1} \\ \cdot^{\cdot^{\cdot}} & \ddots & \cdot^{\cdot^{\cdot}} & \mathbf{s}_{M'-2} \\ & & \ddots & \cdot^{\cdot^{\cdot}} \\ \mathbf{s}_{-N+2} & \mathbf{s}_{-N+3} & \ddots & \\ \mathbf{s}_{-N+1} & \mathbf{s}_{-N+2} & \cdot^{\cdot^{\cdot}} & \mathbf{s}_{M'-M-N+1} \end{bmatrix}
$$

where we recall that $\mathcal{H}$ is an $MmP \times Q(N + M - 1)$ block-Hankel matrix, and $\mathcal{S}$ is a $Q(N + M - 1) \times (M' - M + 1)$ block-Toeplitz, finite alphabet matrix. The block H contains the impulse response of the channel, convolved with the modulating pulse shape function; $\mathbf{s}_k$ is a $Q \times 1$ vector containing the symbols transmitted by the Q users in the kth interval. For digital sources, the entries of $\mathbf{s}_k$ belong to a specific alphabet Ω, such as $\Omega = \{\pm1\}$ for BPSK signals.

If MmP is large enough and $\mathcal{H}$ has full column rank, then $\mathcal{X}$ is rank-deficient and is expected to have rank

$$
Q_{\mathcal{X}} = Q(N + M - 1). \tag{4.46}
$$

Our goal is to factor $\mathcal{X}_M$ into $\mathcal{H}$ and $\mathcal{S}$ with the indicated structures as above. The *necessary* conditions for $\mathcal{X}_M$ to have a unique factorization $\mathcal{X}_M = \mathcal{H}\mathcal{S}$ are that $\mathcal{H}$ is a "tall" matrix and $\mathcal{S}$ is a "wide" matrix, which for $N > 1$ leads to

$$mP > Q$$

$$M \geq \frac{QN - Q}{mP - Q} \tag{4.47}$$

$$M' > QN + (Q + 1)(M - 1).$$

The common root condition mentioned in the single-user case now extends to the condition that H is "irreducible and column reduced." (See also (4.43).) Given sufficient data, only $mP > Q$ poses a fundamental identification restriction since M and M' are usually large enough.

Note that these conditions are *not sufficient* for $\mathcal{H}$ and $\mathcal{S}$ to have full rank. One case where $\mathcal{H}$ does not have full rank is when the channels do not have equal lengths, in which case the rank of $\mathcal{X}$ is at most $\sum N_q + Q(M - 1)$.

4.4.1.2 Blind Multiuser Identification

The blind FIR-MIMO identification problem may be stated as a matrix factorization problem: given $\mathcal{X}$, find factors $\mathcal{H}$ and $\mathcal{S}$ with the indicated structure.

Suppose that the conditions (4.47) are satisfied and that $\mathcal{H}$ has full column rank $Q(N + M - 1)$. Then, $\text{row}(\mathcal{X}) = \text{row}(\mathcal{S})$, so that we can determine the row span of $\mathcal{S}$ from that of $\mathcal{X}$. The first step of the algorithm is to compute an orthonormal basis $\hat{V}$ of $\text{row}(\mathcal{X})$. The next step is to find linear combinations of the rows of $\hat{V}$ such that the result both belongs to the finite alphabet (FA) and has a Toeplitz structure.

Forcing the Toeplitz Property: Subspace Intersections. A standard procedure to find $\mathcal{S}$ as a block-Toeplitz matrix with $\text{row}(\mathcal{S}) = \text{row}(\mathcal{X})$ (but not forcing the FA property) is to rewrite this as

$$
\begin{aligned}
[\mathbf{s}_{M-1} \quad \mathbf{s}_M \quad \cdots \quad \mathbf{s}_{M'-1}] &\in \text{row}(\mathcal{X}) \\
[\mathbf{s}_{M-2} \quad \mathbf{s}_{M-1} \quad \cdots \quad \mathbf{s}_{M'-2}] &\in \text{row}(\mathcal{X}) \\
&\vdots \\
[\mathbf{s}_{-N+1} \quad \mathbf{s}_{-N+2} \quad \cdots \quad \mathbf{s}_{M'-N-M+1}] &\in \text{row}(\mathcal{X}).
\end{aligned} \tag{4.48}
$$

These conditions can be aligned to apply to the same block-vector in several ways. We choose to work with

$$S := [\mathbf{s}_{-N+1} \quad \mathbf{s}_{-N+2} \quad \cdots \quad \mathbf{s}_{M'-1}].$$

Let $\hat{V}$ be a basis for $\text{row}(\mathcal{X})$.

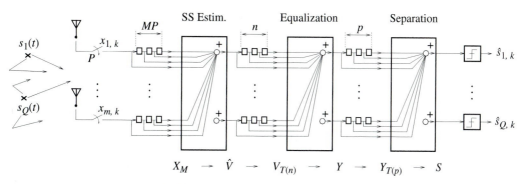

Figure 4.12   Multistage equalization/separation filter.

The conditions (4.48) can be transformed into

$$S \in \text{row}\, \hat{V}^{(1)}, \qquad \hat{V}^{(1)} := \begin{bmatrix} \hat{V} & \mathbf{0} \\ \mathbf{0} & I_{N+M-2} \end{bmatrix},$$

$$S \in \text{row}\, \hat{V}^{(2)}, \qquad \hat{V}^{(2)} := \begin{bmatrix} \mathbf{0} & \hat{V} & \mathbf{0} \\ 1 & \mathbf{0} & \mathbf{0} \\ \mathbf{0} & \mathbf{0} & I_{N+M-3} \end{bmatrix}, \qquad (4.49)$$

$$\vdots$$

$$S \in \text{row}\, \hat{V}^{(N+M-1)}, \quad \hat{V}^{(N+M-1)} := \begin{bmatrix} \mathbf{0} & \hat{V} \\ I_{N+M-2} & \mathbf{0} \end{bmatrix}.$$

Indeed, the identity matrices in each $\hat{V}^{(k)}$ reflect the fact that, at that point, there are no range conditions on the corresponding entries of **s**. Thus, S is in the intersection of the row spans of $\hat{V}^{(1)}$ until $\hat{V}^{(N+M-1)}$, and we have to determine a basis for the intersection of a set of given subspaces. If all channel lengths are equal, then we expect to have Q signal sequences in the intersection.

To compute the intersection, we can use the fact that, for *orthonormal* bases $\hat{V}^{(k)}$ as we have in (4.49), the subspace intersection is obtained by computing the singular-value decomposition (SVD) of a stacking of all the basis vectors or, more conveniently, (for n intersections, $n = N + M - 1$) by an SVD of

$$V_{T(n)} := \begin{bmatrix} \hat{V} & & & \mathbf{0} \\ & \hat{V} & & \\ & & \ddots & \\ \mathbf{0} & & \hat{V} & \\ J_1 & & \mathbf{0} & \\ \mathbf{0} & & & J_2 \end{bmatrix} \Biggl.\Biggr\} n , \qquad (4.50)$$

where the n copies of $\hat{V}$ are each shifted over one entry, and

$$J_1 = \begin{bmatrix} \sqrt{n-1} & & & 0 \\ & \ddots & & \\ & & \sqrt{2} & \\ 0 & & & 1 \end{bmatrix}, \quad J_2 = \begin{bmatrix} 1 & & & 0 \\ & \sqrt{2} & & \\ & & \ddots & \\ 0 & & & \sqrt{n-1} \end{bmatrix}.$$

The matrices J_1, J_2 summarize the identity matrices present in (4.49), which is possible because we are interested only in row spans.

The estimated basis for the intersection (hence, for S) is given by the right singular vectors of $V_{T(n)}$ that correspond to the *large* singular values of $V_{T(n)}$: those that are close to $\sqrt{n}$. This subspace intersection algorithm has complexity $\mathcal{O}(Q^2(N+M)^3 M')$ and is linear in M'.

Let Y be a matrix containing the estimated basis for the row span intersection. For $n = N + M - 1$, this matrix ideally consists of Q row vectors. To find S itself (hence, $\mathcal{S}$ as well), we have to determine which linear combination of the basis vectors gives a finite alphabet structure. Effectively, the subspace intersections perform a blind equalization jointly on all signals, but their separation is done based on the FA property.

Forcing the FA Property. For a given matrix Y, the iterated least-squares algorithms ILSP/E/F [24], and the constant modulus algorithm RACMA [29] (for BPSK or QPSK) solve the factorization

$$(Y = AS: A, S \text{ full rank}, [S]_{ij} \in \Omega), \tag{4.51}$$

where Ω is a prespecified finite alphabet, and A is any resulting nonsingular matrix.

Since the factorization $\mathcal{X} = \mathcal{H}S$ is of the form (4.51), we could, in principle, use the ILSP or RACMA algorithm directly on $\mathcal{X}$. However, $\mathcal{X}$ is generally a large matrix with many rows, limiting the performance of ILSP (mainly in the context of finding *all* independent signals) and giving an unacceptably large computational complexity in RACMA. A second problem is that the algorithm does not force the Toeplitz structure of $\mathcal{S}$. After finding a candidate $\mathcal{S}$, we must compare the rows and detect which rows are shifted copies (echos) of other rows.

Detection of Q and L. If $\mathcal{H}$ and $\mathcal{S}$ have full column rank and row rank, respectively, then the rank of $\mathcal{X} = \mathcal{X}_M$ is $Q_\mathcal{X} := Q(N + M - 1)$. (See (4.46).) In principle, the number of signals Q can be estimated by increasing the blocking factor M of $\mathcal{X}_M$ by one and looking at the increase in rank of $\mathcal{X}_M$. This property provides a useful detection mechanism even if the noise level is quite high since it is independent of the actual (observable) channel length $\hat{N}$. Furthermore, the property still holds if all channels do not have equal lengths. If they do, then N can be estimated from the estimated rank of $\mathcal{X}_M$, $\hat{Q}_x$, and the estimated number of signals, $\hat{Q}$, by $\hat{N} = \hat{Q}_x/\hat{Q} - M + 1$.

If the channels do not have equal length, but, say, lengths N_{ql}, then $\mathcal{H}$ is not full rank and a modification of the algorithm for estimating $\mathcal{S}$ is necessary. The approach in this case is to base the number of intersections on the *shortest* channel length among all sources. Doing so will equalize the corresponding channel and partially equalize the others. The remaining equalization is best carried out by means of the finite alphabet property. The details of this scheme are in [24]. Blind equalization is notoriously hard when channels have ill-conditioned and differing lengths.

4.4.2 Recursive Methods

We now summarize a recursive approach based on the CM property of the transmitted signals. Assuming Q users, we consider a linear $mM \times Q$ spatio-temporal equalization structure

$$\mathbf{F} = [F_1 \cdots F_Q],$$

where F_q, $q = 1, \ldots, Q$, denotes the filter corresponding to the qth signal. Then, the $Q \times 1$ equalizer output at time instant k can be written as

$$\mathbf{y}(k) = \mathbf{F}^T(k)X(k), \tag{4.52}$$

where $X(k) = [\mathbf{x}^T(k) \ \cdots \ \mathbf{x}^T(k + M - 1)]^T$, with $\mathbf{x}(k) = [x_1(k) \ \cdots \ x_m(k)]^T$. The MU-CM algorithm [15] is a simple blind technique to determine the coefficients of the spatio-temporal equalizer $\mathbf{W}$. We can set up a standard CM cost function and derive a set of coupled CM recursions that converge to the desired S-T equalizers. Global convergence to optimal settings can be guaranteed under most conditions.

The algorithm minimizes the following criterion:

$$\min_{\mathbf{F}} J(\mathbf{F}) = E \sum_{j=1}^{Q} \left(|y_j|^2 - 1 \right)^2 + 2 \sum_{l,n=1; l \neq n}^{Q} \sum_{\delta=\delta_1}^{\delta_2} |r_{ln}(\delta)|^2, \tag{4.53}$$

where $r_{ln}(\delta)$ is the cross-correlation function between users l and n defined as

$$r_{ln}(\delta) = E\left(y_l(k)y_n^*(k - \delta)\right), \tag{4.54}$$

and δ_1, δ_2 are integers that should be chosen in compliance with the channel delay spread in order to take into account all the achievable delays between different users. The cost function (4.53) is the sum of a CM term and a cross-correlation term:

the CM term penalizes the deviations of the equalized signals' magnitudes from a constant modulus, whereas the cross-correlation term penalizes the correlations between them. The corresponding stochastic-gradient algorithm has the form

$$\mathbf{F}(k+1) = \mathbf{F}(k) - \mu[\hat{\Delta}_1(k)\cdots\hat{\Delta}_Q(k)], \qquad (4.55)$$

where

$$\Delta_j(k) = 4E\{(\,|y_j(k)|\,^2 - 1)y_j(k)\mathbf{X}^*(k)\} + 4\sum_{l=1;l\neq j}^{Q}\sum_{\delta=\delta_1}^{\delta_2} r_{jl}(\delta)E\{y_l(k-\delta)\mathbf{X}^*(k)\}, \qquad (4.56)$$

and $\hat{\Delta}_j$ is an estimate of Δ_j based on instantaneous values or sample averaging. Equation (4.55) describes a stochastic gradient algorithm derived from the MIMO "constant modulus" criterion (4.53) and is suitable for the spatio-temporal equalization of multiple user signals in the presence of both ISI and CCI. Simulations and analysis have shown its MMSE behavior at steady state, as well as its robustness to the power imbalance of different users.

The parameters employed are the equalizer length M, the number of users Q, and the step-size parameter μ. The number as well as the weight of the autocorrelation functions in the criterion (4.53) can be made variable. Equation (4.55) has a low computational complexity (depending on the number of terms present in the criterion as well as the length of the averaging window). Notice that (4.55) may reduce to the standard CMA 2-2 algorithm [26] in the case of one user ($Q = 1$).

4.5 CONCLUDING REMARKS

Space-time processing is a rapidly growing field that is still in its infancy. In this chapter, we have surveyed several aspects of this important research area that integrate and extend many of the concepts developed in earlier chapters. We hope that the results presented in the chapter will stimulate fresh research in this fascinating field.

REFERENCES

[1] H. A. Cirpan and M. K. Tsatsanis. "Chip interleaving in direct sequence CDMA systems." *Proc. Int. Conf. Acoust., Speech, Signal Processing,* pages 3877–3880, Munich, Germany, April 1997.

[2] Z. Ding. "Blind channel identification and equalization using spectral correlation measurements, part I: frequency-domain analysis." In *Cyclostationarity in Communications and Signal Processing,* W. A. Gardner, ed., pages 417–436, New Jersey, USA, 1994.

[3] Z. Ding. "Characteristics of band-limited channels unidentifiable from second-order cyclostationary statistics." *IEEE Signal Processing Letters*, 3(5):150–152, May 1996.

[4] K. Feher. *Wireless Digital Communications*. Prentice Hall, Upper Saddle River, New Jersey, 1995.

[5] W. A. Gardner, editor. *Cyclostationarity in Communications and Signal Processing*. IEEE Press, New Jersey, USA, 1994.

[6] G. B. Giannakis. "Filter banks for blind channel identification and equalization." *IEEE Signal Processing Letters*, 4(6):181–183, June 1997.

[7] G. B. Giannakis and J. M. Mendel. "Identification of nonminimum phase systems using higher order statistics." *IEEE Trans. Acoust., Speech, Signal Processing*, 37(3):360–377, March 1989.

[8] R. He and J. H. Reed. "Spectrum correlation characterization of AMPS signal with its application to interference rejection." *Proc. IEEE MILCOM*, October 1994.

[9] William C. Jakes, ed. *Microwave Mobile Communications*. John Wiley, New York, 1974.

[10] H. Liu and G. Xu. "Closed-form blind symbol estimation in digital communications." *IEEE Transactions on Signal Processing*, SP-43(11):2714–2723, November 1995.

[11] J. L. Massey and M. K. Sain. "Inverse of linear sequential circuits." *IEEE Trans. Comput.*, pages 330–337, 1968.

[12] E. Moulines, P. Duhamel, J. F. Cardoso, and S. Mayrargue. "Subspace methods for the blind identification of multichannel FIR filters." *IEEE Trans. Signal Process.*, SP-43:516–525, 1995.

[13] Kaveh Pahlavan and Allen H. Levesque. *Wireless Information Networks*. John Wiley, New York, 1995.

[14] C. B. Papadias and A. Paulraj. "Decision-feedback equalization and identification of linear channels using blind algorithms of the Bussgang type." *Proc. Asilomar Conf. Signals, Systems, Computers*, Pacific Grove, California, October 1995, pages 335–340.

[15] C. B. Papadias and A. Paulraj. "A constant modulus algorithm for multi-user signal separation in presence of delay spread using antenna arrays." *IEEE Signal Processing Letters*, 4(6):178–181, June 1997.

[16] B. Porat and B. Friedlander. "Blind equalization of digital communication channels using higher order moments." *IEEE Trans. Acoust. Speech, Signal Processing*, SP-39(2):522–526, February 1991.

[17] J. G. Proakis. *Digital Communications*. McGraw-Hill, New York, 1983.

[18] V. U. Reddy, C. B. Papadias, and A. Paulraj. "Blind identifiability of certain classes of multipath channels from second-order statistics using antenna arrays." *IEEE Signal Processing Letters*, 4(5):138–141, May 1997.

[19] R. O. Schmidt. *A Signal Subspace Approach to Multiple Emitter Location and Spectral Estimation*. Ph.D. thesis, Stanford University, Stanford, California, 1981.

[20] D. T. M. Slock. "Blind fractionally-spaced equalization, perfect-reconstruction filter banks and multichannel linear prediction." *Proc. Int. Conf. Acoust., Speech, Signal Processing*, 4:585–588, Adelaide, Australia, April 1994.

[21] D. T. M. Slock. "Blind joint equalization of multiple synchronous mobile users using oversampling and/or multiple antennas." *Proc. Asilomar Conf. Signals, Systems, Computers*, Pacific Grove, California, October 1994.

[22] D. T. M. Slock and C. B. Papadias. "Further results on blind identification and equalization of multiple FIR channels." *Proc. Int. Conf. Acoust., Speech, Signal Processing,* 4:1964–1967, Detroit, Michigan, May 1995.

[23] R. Steele. *Mobile Radio Communications.* Pentech Press, 1992.

[24] S. Talwar, M. Viberg, and A. Paulraj. "Blind separation of synchronous co-channel digital signals using an antenna array. Part I. Algorithms." *IEEE Trans. Signal Process.,* 44(5):1184–1197, May 1996.

[25] L. Tong, G. Xu, and T. Kailath. "Blind identification and equalization of multipath channels: a time domain approach." *IEEE Trans. Inform. Theory,* 40(2):340–349, March 1994.

[26] J. R. Treichler and B. G. Agee. "A new approach to multipath correction of constant modulus signals." *IEEE Trans. Acoust. Speech, Signal Processing,* ASSP-31(2):459–472, April 1983.

[27] M. K. Tsatsanis and G. B. Giannakis. "Cyclostationarity in partial response signaling: a novel framework for blind equalization." *Proc. Int. Conf. Acoust., Speech, Signal Processing,* pages 3597–3600, Munich, Germany, April 1997.

[28] J. K. Tugnait. "On blind identifiability of multipath channels using fractional sampling and second-order cyclostationary statistics." *IEEE Trans. Inform. Theory,* 41:308–311, Jan. 1995.

[29] A. J. van der Veen. "Analytical method for blind binary signal separation." *IEEE Trans. Signal Processing,* 45(4):1078–1082, April 1997.

[30] A. J. van der Veen, S. Talwar, and A. Paulraj. "Blind estimation of multiple digital signals transmitted over FIR channels." *IEEE Signal Process. Letters,* 2(5):99–102, May 1995.

[31] A. J. van der Veen, S. Talwar, and A. Paulraj. "Blind identification of FIR channels carrying multiple finite alphabet signals." *Proc. Int. Conf. Acoust., Speech, Signal Processing,* 2:1213–1216, Detroit, 1995.

[32] M. C. Vanderveen, C. Papadias, and A. Paulraj. "Joint angle and delay estimation (JADE) for multipath signals arriving at an antenna array." *IEEE Commun. Letters,* 1(1):12–14, January 1997.

[33] M. D. Yacoub. *Foundations of Mobile Radio Engineering.* CRC Press, Boca Raton, Florida, 1993.

[34] E. Zervas, J. Proakis, and V. Eyuboglu. "Effects of constellation shaping on blind equalization." *Proc. SPIE,* 1565:178–187, July 1991.

[35] A. J. Paulraj and C. B. Papadias. "Space-time processing for wireless communications." *IEEE Signal Processing Magazine,* 14(6):49–83, November 1997.

Acknowledgment

This research was supported in part by the Department of the Army, Army Research Office, under Grant No. DAAH04-95-1-0249. The views and conclusions expressed in this document are those of the authors and should not be interpreted as necessarily representing the official policies or endorsements, either expressed or implied, of the Army Research Office or the U.S. Government.

5

Network Capacity, Power Control, and Effective Bandwidth

David N. C. Tse
Stephen V. Hanly

As the preceding chapters have emphasized, the mobile wireless environment provides several unique challenges to reliable communication not commonly found in wireline networks. These include scarce bandwidth, limited transmit power, interference between users, and time-varying channel conditions. A central problem in the design of wireless networks is how to use the limited resources most efficiently in such adverse environments, in order to meet the quality-of-service (QoS) requirements of applications as quantified in terms of bit rate and loss. The problem will become more acute for next-generation, integrated-services networks that aim to support a heterogeneous mix of high bandwidth media types with diverse QoS requirement and bursty traffic characteristics. As the demand for ubiquitous access to the backbone wireline network grows, the capacity of the wireless link will likely be a severe bottleneck.

To meet these challenges, there have been intense efforts in developing more sophisticated physical layer communication techniques, examples of which are described in preceding chapters. A significant thrust of work has been on developing *multiuser* receiver structures of the type described in Chapter 2, which mitigate the interference between users in spread spectrum systems. (See also, for example,

[10, 11, 12, 15, 16, 20, 23].) Recall that, unlike the conventional matched-filter receiver used in the IS-95 CDMA system, these techniques take into account the structure of the interference from other users when decoding a user. Another important line of work is the development of processing techniques in systems with antenna arrays, a class of which is described in Chapter 4. As discussed in Chapter 1 as well, while spread-spectrum techniques provide *frequency diversity* to the wireless system, antenna arrays provide *spatial diversity*, both of which are essentially *degrees of freedom* through which communication can take place.

Despite significant work done in the area, there is still much debate about the network capacity of the various approaches to deal with multiuser interference in spread-spectrum and multiple-antenna systems. One important reason is that the networking level problems of resource allocation and power control are less well understood in the context of multiuser techniques than with more traditional multiaccess schemes, such as TDMA, FDMA, and conventional-receiver CDMA systems. For example, in a TDMA or FDMA system, the network resource is shared among users via disjoint frequency and time slots, and this sharing provides a simple abstraction for resource allocation problems at the networking layer. Such clean separation between the networking and the physical layers does not exist when more sophisticated multiuser techniques are used. This in turn hampers the understanding of the capacity of networks with multiuser receivers and of the associated network-level resource allocation problems such as call admissions control, cell handoffs, and resource allocation for bursty traffic.

In this chapter, we show that under some conditions, a simple abstraction of the amount of resource consumed by a user is indeed possible for several important multiuser receivers. The specific scenario is a set of power-controlled mobile users communicating to a base-station in a single cell. Assuming that each user's QoS can be expressed in terms of a target signal-to-interference ratio (SIR), we show that a notion of *effective bandwidth* can be defined such that the QoS requirements of all the users can be met if and only if the sum of the effective bandwidths of the users is less than the total number of degrees of freedom in the system. These degrees of freedom can be provided by the processing gain in a spread-spectrum system or by the number of antenna elements in a system with an antenna array. These capacity characterizations are simple in that the effective bandwidth of a user depends only on its SIR requirement and nothing else. While these results are proved in an idealized model, they have the potential to provide a first step in bridging between resource allocation problems at the networking layer and multiuser techniques at the physical layer.

The effective bandwidth of a user depends on the multiuser receiver employed. Results for three receivers are obtained. They are the minimum mean-square error (MMSE) receiver [12, 15, 16, 23], the decorrelator [10, 11], and the conventional matched filter receiver. We show that the effective bandwidths are respectively $e_{mmse}(\beta) = \frac{\beta}{1+\beta}$, $e_{dec}(\beta) = 1$, and $e_{mf}(\beta) = \beta$, where β is the SIR requirement of the user. These effective bandwidth expressions also provide a succinct basis for performance comparison between different receiver structures. The MMSE receiver occupies a special place since it can be shown to lead to the minimum effective bandwidth among all linear receivers.

These effective bandwidth characterizations also illustrate the inherent flexibility in resource-sharing among users with heterogeneous SIR requirements in a CDMA system: the total degrees of freedoms can be divided arbitrarily according to each user's SIR. This case is in contrast to traditional FDMA or TDMA system where the resource allocation is much more rigid and coarse-grained. Such flexibility is supported by appropriate power control, and this philosophy is behind much of recent work in power control for conventional CDMA systems. (See, for example, [1, 3, 6, 25, 26, 27]). Our work can be viewed as an extension of this philosophy to more sophisticated multiuser receivers.

The outline of this chapter is as follows. In Section 5.1, we introduce our notation for the basic model of a multiple-access spread-spectrum system and the structure of the MMSE receiver. In Section 5.2, we present our key result: that in a large system with each user using random spreading sequences, the limiting interference effects under the MMSE receiver can be calculated as if they were additive; to each interferer can be ascribed a level of *effective interference* that it provides to the user to be decoded. In Sections 5.3 and 5.4, we apply this result to study the performance under power control and obtain a notion of *effective bandwidth*. In Section 5.5, we obtain analogous results for the decorrelating receiver. In Section 5.6, we show that similar ideas carry through for systems with antenna diversity. Section 5.7 contains some concluding remarks.

Proofs of results are not presented here but can be found in [18].

5.1 BASIC SPREAD-SPECTRUM MODEL AND THE MMSE RECEIVER

In a spread-spectrum system, each of the user's information or coded symbols is spread onto a much larger bandwidth via modulation by its own *signature* or *spreading sequence*. The following is a model for a symbol-synchronous, multiple-access, spread-spectrum system:

$$\mathbf{Y} = \sum_{m=1}^{M} X_m \mathbf{s}_m + \mathbf{W},$$

where X_m is a real scalar and $\mathbf{s}_m$ is a real L-dimensional vector that denote the transmitted symbol and signature spreading sequence of user m, respectively, and $\mathbf{W}$ is zero-mean, variance-σ^2 Gaussian background noise. The length of the signature sequences is L, which one can also think of as the number of degrees of freedom or diversity. The L-dimensional received vector is **Y**. We assume the X_m's are independent and identically distributed (i.i.d.) and that $E[X_m] = 0$ and $E[X_m^2] = P_m$, where P_m is the received power of user m.

Rather than looking at *symbol-by-symbol* detection, we are interested in the more general problem of *demodulation*, extracting good estimates of the (coded) symbols of each user as soft decisions to be used by the channel decoder [16]. From this point of view, the relevant performance measure is the SIR of the estimates.

We shall now focus on the demodulation of user 1, assuming that the receiver has already acquired the knowledge of the spreading sequences. The optimal linear demodulator that generates a soft decision $\hat{X}_1$, maximizing the SIR at the output of the demodulator, is the MMSE receiver [12, 15, 16].

As a comparison, note that the conventional CDMA approach simply matches the received vector to s_1, the signature sequence of user 1. This is indeed the optimal receiver when the interference from other users is white. However, in general, the multiple-access interference is not white and has structure as defined by $s_2, s_3, ..., s_M$, assumed to be known to the receiver. The MMSE receiver exploits the structure in this interference in maximizing the SIR of user 1.

The formulae for the MMSE demodulator and its performance are well known (cf. Chapter 2):[1]

$$X_{\mathrm{mmse}}(\mathbf{Y}) = \frac{1}{s_1^T (SDS^T + \sigma^2 I)^{-1} s_1} \, s_1^T (SDS^T + \sigma^2 I)^{-1} \mathbf{Y}, \qquad (5.1)$$

and the signal to interference ratio β_1 for user 1 is

$$\beta_1 = s_1^T (SDS^T + \sigma^2 I)^{-1} s_1 P_1, \qquad (5.2)$$

where $S = [s_2, ..., s_M]$ and $D = \mathrm{diag}(P_2, ..., P_M)$.

5.2 PERFORMANCE UNDER RANDOM SPREADING SEQUENCES

Equation (5.2) is a formula for the performance of the MMSE receiver, which one can compute for specific choice of signature sequences. However, it is not easy to obtain qualitative insights directly from this formula. For example, the effect of an individual interferer on the SIR for user 1 cannot be seen directly from this expression. In practice, it is often reasonable to assume that the spreading sequences are *randomly* and independently chosen (see, e.g., [13]). For example, they may be pseudorandom sequences, or the users may choose their sequences from a large set of available sequences as they are admitted into the network. In this case, the performance of the

---

[1]The superscript ^{T} denotes the transpose operator.

optimal demodulator can be modeled as a random variable since it is a function of the spreading sequences. In this section, we show that, unlike the deterministic case, there is a great deal of analytical information one can obtain about this random performance in a large network. In the development below, we assume that though the sequences are randomly chosen, they are known to the receiver once they are picked. In practice, this means that the change in the spreading sequences is at a much slower time scale than the symbol rate, so that the receiver has the time to acquire the sequences. (There are known adaptive algorithms for which this can even be done blindly; see [8].) However, the *performance* of the MMSE receiver depends on the initial choice of the sequences and, hence, is random.

As a model for random sequences, let $\mathbf{s}_m = \frac{1}{\sqrt{L}}(V_{1m}, \ldots, V_{Lm})^{\mathrm{T}}$, $m = 1, \ldots M$, where the random variables V_{km}'s are i.i.d., zero-mean, and unit-variance. The normalization by $\frac{1}{\sqrt{L}}$ ensures that $E[\|\mathbf{s}_m\|^2] = 1$. In practice, it is common that the entries of the spreading sequences are 1 or -1, but we want to keep the model general so that we can later apply our results to problems with other modes of diversity.

Our results are asymptotic in nature, for a large network. Thus, we consider the limiting regime where the number of users are large, i.e., $M \to \infty$. To support a large number of users, it is reasonable to scale up L as well, keeping the number of users per degree of freedom (equivalently, per unit bandwidth), $\alpha \equiv \frac{M}{L}$, fixed. We also assume that as we scale up the system, the empirical distribution of the powers of the users converges to a fixed distribution, say $F(P)$. The following is the main result of [18], giving the asymptotic information about the SIR for user 1. The proof of this result makes use of the theory of random matrices [14, 17].

Theorem 5.1   Let $\beta_1^{(L)}$ be the (random) SIR of the MMSE receiver for user 1 when the spreading length is L. Then, $\beta_1^{(L)}$ converges to β_1^* in probability as $L \to \infty$, where β_1^* is the unique solution to the equation

$$\beta_1^* = \frac{P_1}{\sigma^2 + \alpha \int_0^\infty I(P, P_1, \beta_1^*) dF(P)} \tag{5.3}$$

and

$$I(P, P_1, \beta_1^*) \equiv \frac{PP_1}{P_1 + P\beta_1^*} .$$

Heuristically, this means that in a large system, the SIR β_1 is deterministic and approximately satisfies

$$\beta_1 \approx \frac{P_1}{\sigma^2 + \frac{1}{L}\sum_{i=2}^M I(P_i, P_1, \beta_1)} , \tag{5.4}$$

where, as before, P_i is the received power of user i. This result yields an interesting interpretation of the effect of each of the interfering users on the SIR of user 1: for a large system, the total interference can be decoupled into a sum of the background noise and an interference term from each of the other users. (The factor $\frac{1}{L}$ results from the processing gain of user 1.) The interference term depends only on the received power of the interfering user, the received power of user 1, and the attained SIR. It does not depend on the other interfering users except through the attained SIR β_1.

One must be cautioned not to think that this result implies that the interfering effect of the other users on a particular user is additive across users. It is not: the interference term $I(P_i, P_1, \beta_1)$ from interferer i depends on the attained SIR which in turn is a function of the entire system. However, it can be shown that the equation:

$$x = \frac{P_1}{\sigma^2 + \frac{1}{L}\sum_{i=2}^{M} I(P_i, P_1, x)} \tag{5.5}$$

has a unique fixed point x^*, and moreover, the equation has the following monotonicity property: for any x, $x^* \geq x$ if and only if

$$\frac{P_1}{\sigma^2 + \frac{1}{L}\sum_{i=2}^{M} I(P_i, P_1, x)} \geq x. \tag{5.6}$$

It follows then that to check if the target for user 1's SIR, β_T, can be met for a given system of users, it suffices to check the following condition:

$$\frac{P_1}{\sigma^2 + \frac{1}{L}\sum_{i=2}^{M} I(P_i, P_1, \beta_T)} \geq \beta_T.$$

Based on this interpretation, it is natural to refer to the term $I(P_i, P_1, \beta_T)$ as the *effective interference* of user i on user 1, at a target SIR of β_T.

To gain further insight into this concept of effective interference, it is helpful to compare the above situation with that when the conventional matched filter s_1 is used for the demodulation. For that case, it can be shown that if $\beta_{1,mf}$ is the (random) SIR of the conventional matched-filter receiver for user 1, then for large processing gain L, $\beta_{1,mf}$ converges in probability to

$$\beta_{1,mf}^* = \frac{P_1}{\sigma^2 + \alpha\int_0^\infty P dF(P)}, \tag{5.7}$$

where, as before, F is the limiting distribution of the powers of the users and α is the number of users per degree of freedom. Hence, for large L, the performance of the matched receiver is approximately

$$\beta_{1,\,\mathrm{mf}} \approx \frac{P_1}{\sigma^2 + \dfrac{1}{L}\sum_{i=2}^{M} P_i}. \tag{5.8}$$

Comparing this expression with (5.4), we see that the interference due to user i is simply P_i in place of $I(P_i, P_1, \beta_1)$. Since the matched-filter receiver is independent of the signature sequences of the other users, it is not surprising that the interference is linear in the received powers of the interferers. In the case of the MMSE receiver, the filter does depend on the signature sequences of the interferers, thus resulting in the interference being a nonlinear function of the received power of the interferer. Also, observe that $I(P_i, P_1, \beta_1) < P_i$, which is expected since the MMSE receiver maximizes the SIR among all linear receivers. But more importantly, we see that while for the conventional receiver the interference grows without bound as the received power of the interferer increases, for the MMSE receiver, the effective interference from user i is bounded and approaches $\frac{P_1}{\beta_1}$ as P_i goes to infinity. Thus, while the SIR of the matched-filter receiver goes to zero for large interferers' powers, the SIR of the MMSE receiver does not. This is the well-known *near-far resistance* property of the MMSE receiver discussed in Chapter 2 [see also [12]]. The intuition is that as the power of an interferer grows to infinity, the MMSE receiver will null out its signal. While the near-far resistance property has been reported by previous authors, Theorem 5.1 goes beyond that as it not only quantifies the worst-case performance (i.e., large interferer's power) but also the performance for all finite values of the interference. This quantification is useful for example in situations when power control is exercised, as we turn to in the next section.

In general, we have no explicit solution for the SIR β_1^* in (5.3). However, for the special case when the received powers of all users are the same, the equation is quadratic in β_1^* and a simple solution is obtained:

$$\beta_1^* = \frac{(1-\alpha)P}{2\sigma^2} - \frac{1}{2} + \sqrt{\frac{(1-\alpha)^2 P^2}{4\sigma^4} + \frac{(1+\alpha)P}{2\sigma^2} + \frac{1}{4}}.$$

We see that the β_1^* is positive for all values of α and approaches 0 as α, the number of users per degree of freedom, goes to infinity.

Two performance measures commonly used in the literature for multiuser receivers (and discussed in Chapter 2) are their *efficiency* and their *asymptotic efficiency* [21]. In the context of linear receivers, the efficiency for user 1 is defined to be the ratio of the achieved SIR to the SIR when there is no interferer and only background noise. For the MMSE receiver with random spreading sequences and equal received power for all users, the efficiency is given by:

$$\frac{\beta_1^* \sigma^2}{P},$$

where β_1^* is given by the above expression. Recall from Chapter 2 that the asymptotic efficiency η_1 is the limiting efficiency as the background noise level goes to zero. If $\alpha \leq 1$, this asymptote is given by

$$\eta_1 := \lim_{\sigma \to 0} \frac{\beta_1^* \sigma^2}{P} = 1 - \alpha.$$

For $\alpha > 1$, the limiting SIR is positive but bounded:

$$\lim_{\sigma \to 0} \beta_1^* = \frac{1}{\alpha - 1}, \tag{5.9}$$

and so the asymptotic efficiency is zero.

5.3 CAPACITY AND PERFORMANCE UNDER POWER CONTROL

We observed in Section 5.2 that in the conventional receiver case, the interference of a user is proportional to its power, and hence a strong interferer can completely overcome a weaker signal. This is the so-called near-far problem, and a well-known consequence is that the conventional receiver can only avoid this problem via tight power control. We also observed that the MMSE receiver does not suffer arbitrarily poorly from the near-far problem, and indeed, this is one of the key motivations for the original work on multiuser detection [20]. Nevertheless, a MMSE receiver still suffers interference from other users, and it follows that capacity can be increased and power consumption reduced if power control is employed.

In this section, we consider the case in which all users require an SIR of exactly β^*, given a processing gain of L degrees of freedom per symbol. For a given number of users, we compute the minimum power consumption required to achieve β^* for all users and then look at the maximum number of users per degree of freedom supportable for a given power constraint under power control. Of particular interest is the maximum number without power constraint, which we define to be the *capacity* of the system (in terms of number of users per degree of freedom.) This coincides with the definition of capacity taken in [5]; "capacity" is then the point at which saturation occurs as we put in so many users that we drive the required power level to infinity. We show that this capacity is different but finite for both the conventional and the MMSE receivers; thus, both are interference-limited systems. As before, our results are asymptotic as the the processing gain L goes to infinity.

Let us focus first on the conventional receiver. With the matched filter receiver, (5.7) tells us that, asymptotically, users receive the same level of interference and hence must be received at the same power level to get the same SIR β^*. It is easy to compute that with $L\alpha$ users, and a processing gain of L, the common received power required for the conventional receiver is given asymptotically as $L \rightarrow \infty$ by

$$P_{\mathrm{mf}}(\beta^*) = \frac{\beta^* \sigma^2}{1 - \alpha \beta^*} .$$
(5.10)

For a given constraint P on the received power, the maximum number of users supportable is then

$$\alpha_{\mathrm{max}} = \frac{1}{\beta^*} - \frac{\sigma^2}{P} \text{ users/degree of freedom.}$$

The capacity of the conventional receiver when $P = \infty$ is then

$$C_{\mathrm{mf}}(\beta^*) = \frac{1}{\beta^*} \text{ users/degree of freedom.}$$
(5.11)

Phrased differently, as $\alpha \rightarrow \frac{1}{\beta^*}$, the system saturates and the required power level goes to infinity.

Now, let us turn to the MMSE receiver. To satisfy given target SIR requirements for each user, [9, 19] showed that there is an optimal solution for which the received power of *every user* is minimized; moreover, they gave an iterative algorithm to compute it. However, here we can give an explicit solution and characterize the resulting system capacity.

To begin, we fix the number of users per degree of freedom at α. As in the conventional receiver case, it turns out that the system saturates if α is too high, so we first obtain a necessary and sufficient condition for feasibility. It can be shown, from the monotonicity property of (5.6), that in the limit of a large number of degrees of freedom, the system is feasible if and only if the SIR can be met with equal received powers for all users. Setting the received powers of all users to be equal in (5.6) tells us that a given target SIR requirement β^* can be met if and only if

$$\alpha < \frac{1 + \beta^*}{\beta^*} .$$

If this condition is satisfied, it can further be shown that the minimum power solution is given by having the received powers of all users be

$$P_{\text{mmse}}(\beta^*) = \frac{\beta^* \sigma^2}{1 - \alpha \dfrac{\beta^*}{1 + \beta^*}} . \tag{5.12}$$

Hence, the capacity of the system under MMSE receiver is

$$C_{\text{mmse}}(\beta^*) = \frac{1 + \beta^*}{\beta^*} \text{ users/degree of freedom.} \tag{5.13}$$

Moreover, for a given received power constraint P, the maximum number of users that can be supported is attained by assigning each user the same received power, and that number is given by

$$\alpha_{\text{max}} = \frac{1 + \beta^*}{\beta^*} \left(1 - \frac{\beta^* \sigma^2}{P} \right) \text{ users/degree of freedom.}$$

Contrasting (5.10) and (5.11) with (5.12) and (5.13), we note that if α is feasible for both types of receiver, then the power consumption of the MMSE receiver system is less than that of the matched-filter system, and the MMSE system has potentially much greater capacity. Indeed, if $\alpha < 1$, then we can accommodate arbitrarily large β^* without saturating the MMSE receiver, whereas the the conventional receiver saturates as β^* approaches $\frac{1}{\alpha}$. For fixed β^*, we also note that the MMSE receiver system saturates at a higher value of α, yielding a capacity of precisely 1 more user per degree of freedom than the system with a conventional receiver. On the other hand, the relative gain of the MMSE receiver system is not so large for small values of β^*.

5.4 MULTIPLE CLASSES, MAXIMUM POWER CONSTRAINTS, AND EFFECTIVE BANDWIDTHS

It is straightforward to generalize these results to the case in which there are J classes, with all class j users requiring an SIR of β_j. We denote the number of users of class j by $\alpha_j L$ and again consider the limiting regime $L \to \infty$.

The conventional matched-filter receiver results generalize very easily to

$$P_{\text{mf}}(j) = \frac{\beta_j \sigma^2}{1 - \sum_{j=1}^{J} \alpha_j \beta_j} ,$$

where $P_{\text{mf}}(j)$ denotes the common, received power level of all users of class j (see [7]). Thus, the capacity constraint on feasible values of $(\alpha_1, \ldots, \alpha_J)$ is the linear constraint $\sum_{j=1}^{J} \alpha_j \beta_j < 1$. Furthermore, if class j users have a maximum power constraint that $P_{\text{mf}}(j) \leq \overline{P}_j$ for each j, then the tighter capacity constraint

$$\sum_{j=1}^{J} \alpha_j \beta_j \leq \min_{1 \leq i \leq J} \left[1 - \frac{\beta_i \sigma^2}{\overline{P}_i} \right]$$

emerges [5]. It is convenient to refer to β_j as the *bandwidth* of class j users, in degrees of freedom per class j user. Let us denote this bandwidth by

$$e_{\text{mf}}(\beta_j) \equiv \beta_j \text{ degrees of freedom per class } j \text{ user.}$$

We now show that the MMSE receiver results generalize in a similar manner. It is clear in this case also that the minimal power solution consists of the same received power for each class: let all users in class j be received at power P_j. Then, the power control equations become

$$\frac{P_i}{\sigma^2 + \sum_{j=1}^{J} \alpha_j I(P_j, P_i, \beta_i)} = \beta_i, \qquad i = 1, 2, \ldots, J, \tag{5.14}$$

where, as in Theorem 5.1, $I(P_i, P_j, \beta_j) \triangleq \frac{P_i P_j}{P_j + P_i \beta_j}$. But (5.14) implies that $\frac{\beta_j}{P_j}$ is a constant, which allows us to simplify (5.14) down to

$$P_{\text{mmse}}(i) = \frac{\beta_i \sigma^2}{1 - \sum_{j=1}^{J} \alpha_j \frac{\beta_j}{1 + \beta_j}} \qquad i = 1, 2, \ldots, J. \tag{5.15}$$

The capacity constraint for the MMSE receiver with J classes is therefore given by

$$\sum_{j=1}^{J} \alpha_j \frac{\beta_j}{1 + \beta_j} < 1, \tag{5.16}$$

which is linear in $\alpha_1, \ldots, \alpha_J$.

As above, maximum power constraints provide tighter capacity constraints, and in this context we note that (5.15) implies that

$$\sum_{j=1}^{J} \alpha_j \frac{\beta_j}{1 + \beta_j} = 1 - \frac{\beta_i \sigma^2}{P_{\text{mmse}}(i)}, \qquad i = 1, 2, \ldots, J.$$

Thus, if $P_{\text{mmse}}(i) \leq \overline{P}_i$ is a maximum power constraint on class i, then the linear constraint

$$\sum_{j=1}^{J} \alpha_j \frac{\beta_j}{1 + \beta_j} \leq \min_{1 \leq i \leq J} \left[1 - \frac{\beta_i \sigma^2}{\overline{P}_i} \right], \qquad i = 1, 2, \ldots, J$$

defines the restricted capacity region of the system. It is natural to define the effective bandwidth of class j users as $e_{\text{mmse}}(\beta_j)$ degrees of freedom per user, where

$$e_{\text{mmse}}(\beta_j) \equiv \frac{\beta_j}{1 + \beta_j} \ .$$

Linearity in the matched-filter case is a straightforward consequence of the fact that powers add. However, our MMSE effective bandwidth results are rather surprising, as the receiver itself depends on the signature sequences and the received powers of the users. Another interesting observation is that no matter how high β is, the MMSE effective bandwidth of a user is upper bounded by unity. We will gain further insight into why this is so in the next section.

5.5 THE DECORRELATOR

To this point, we have contrasted the performance of the MMSE receiver with that of the conventional matched-filter receiver. It is also illuminating to compare the performance of the MMSE receiver with that of the decorrelator. The decorrelator was in fact the first linear "multiuser detector" described by Lupas and Verdu [10]. As discussed in Chapter 3, this receiver is known to have optimal *near-far resistance* [11], as measured by the worst-case performance over all choices of interferers' powers and in the limit of vanishing background noise power. Here, we focus on the SIR performance for finite noise power and random sequences and obtain simple answers. It can be shown that in a large system with α users per degree of freedom, the (random) SIR under the decorrelating receiver for user 1 converges in probability to β_1^*, given by

$$\beta_1^* = \begin{cases} \dfrac{P_1(1 - \alpha)}{\sigma^2} & \alpha < 1 \\[2mm] 0 & \alpha \geq 1 \end{cases} . \tag{5.17}$$

We observe that as the number of users α per degree of freedom approaches 1, the SIR goes to zero. Geometrically, as the dimensionality of the orthogonal complement to the span of the interference decreases to zero, the length of the projection of the desired signal onto this orthogonal complement tends to zero, and so in the limit the projected signal is lost in the background noise. This behavior is the high price paid for ignoring the background noise. In contrast, the MMSE receiver can support more users than the number of degrees of freedom because it takes both the interference and the background noise into account.

By comparing (5.17) and (5.3), one can see that the effective interference for an interferer on user 1 under the decorrelator is $\frac{P_1}{\beta_1}$, which does not depend on the power of the interferers. Equation (5.17) further implies that the capacity constraint on the system is $\alpha < 1$.

We also observe that if all users require an SIR of β and employ power control, then it is sufficient for each user to be received with power at least $\beta\sigma^2/(1 - \alpha)$. Thus, for a given received power constraint $\overline{P}$, the maximum number of users with SIR requirement β supportable is $1 - \beta\sigma^2/\overline{P}$ (per degree of freedom). Similarly, for multiple classes of users with SIR requirement β_j and power constraint $\overline{P}_j$ for the jth class, the system can support α_j users (per degree of freedom) from each class if

$$\sum_{j=1}^{J} \alpha_j \leq \min_{1 \leq j \leq J} \left[1 - \frac{\beta_j \sigma^2}{\overline{P}_j} \right].$$

Thus, the capacity region under the decorrelator is given by

$$\sum_{j=1}^{J} \alpha_j \leq 1 \tag{5.18}$$

when there are no power constraints or, equivalently, when the background noise power σ^2 goes to zero. So, each user occupies an effective bandwidth of 1 degree of freedom, independent of the value of β.

From (5.17), it can be immediately inferred that the efficiency of a decorrelator in a large system with random spreading sequences is $1 - \alpha$ if α, the number of users per degree of freedom, is less than 1, and is zero otherwise. Since this efficiency does not depend on the background noise power σ^2, it is also the asymptotic efficiency.

It is well known [12] that the MMSE receiver has the same asymptotic efficiency as the decorrelator, and hence the decorrelator is optimal in this sense among all linear receivers. However, comparing (5.16) and (5.18), one can see that the capacity region under the MMSE receiver is strictly larger than that under the decorrelator, even as the background noise goes to zero. In particular, the MMSE receiver can in general accommodate more users than the number of available degrees of freedom, while the decorrelator cannot. This apparent paradox can be resolved by noting that when $\alpha > 1$, the SIR attained by the decorrelator is zero (5.17) while the attained SIR by the MMSE receiver is strictly positive but bounded as the noise power σ^2 goes to zero. Since the asymptotic efficiency measures only the *rate* at which the SIR goes to infinity as σ^2 goes to zero, they are the same (zero) for both receivers. On the other hand, the capacity region quantifies the number of users with *fixed* SIR requirements a receiver can accommodate; hence the difference between the decorrelator and the MMSE receiver is reflected. In practice, users have target SIR requirements and hence the capacity region characterization seems to be a more natural performance measure than the asymptotic efficiency. In this context, the decorrelator remains suboptimal even as the noise power σ^2 approaches zero.

5.6 ANTENNA DIVERSITY

In spread-spectrum systems, diversity gain is obtained by spreading over a wider bandwidth. However, there are other ways to obtain diversity benefits in a wireless system. A technique, particularly effective for combating multipath fading, is the use of an *adaptive antenna array* at the receiver. Multipath fading can be very detrimental because the received signal power can drop dramatically due to destructive interference between different paths of the transmitted signal. By placing the antenna elements greater than half the carrier wavelength apart, one can ensure that the received signal fades more or less independently at the different antenna elements. By appropriately weighing, delaying and combining the received signals at the different antenna elements, one can obtain a much more reliable estimate of the transmitted signal than with a single antenna. Such antenna arrays are said to be *adaptive* since the combining depends on the strengths of the received signals at the various antenna elements. This signal strength in turn depends on the location of the users. Moreover, the combining weights will be different for different users, allowing the array to focus on specific users while mitigating the interference from other users. This process is called *beamforming*. From our previous results, it turns out that the capacity of such an antenna array system can again be characterized by effective bandwidths.

The following is a model for a synchronous, multiple-access antenna-array system:

$$\mathbf{Y} = \sum_{m=1}^{M} X_m \mathbf{h}_m + \mathbf{W}.$$

Here, X_m is the transmitted symbol of the mth user, and $\mathbf{Y}$ is an L-dimensional vector of received symbols at the L antenna elements of the array. The vector $\mathbf{h}_m$ represents the fading of the mth user at each of the antenna array elements. The entries are complex to incorporate both phase and magnitude information. The vector $\mathbf{W}$ is zero-mean, variance-σ^2 Gaussian background noise.

The fading is time-varying, as the mobile users move, but usually at a much longer time scale than the symbol rate of the system. Assuming then that the channel fading of the users can be measured and tracked perfectly at the receiver, we would like to combine the vector of received symbols appropriately to maximize the SIR of the estimates of the transmitted symbols of the users. The optimal linear receiver is clearly the MMSE. Assuming that the fading of each user is independent and identically distributed from antenna element to antenna element, we are essentially in the same setup as for spread-spectrum systems. Thus, for a system with a large number of antenna elements and large number of users, we can treat each of the interfering users as contributing an additive *effective interference*. Under perfect power control, the system capacity is characterized by sharing the L degree of freedom among the users according to their *effective bandwidths* given by the previous expressions for the different receivers. The only difference here is that the L degrees of freedom are obtained by spatial rather than frequency diversity.

These results should be compared with that of Winters et al. [22], which showed that for a flat Rayleigh fading channel, a combiner that attempts to null out all of the interferers costs one degree of freedom per interferer. This combiner is, of course, the suboptimal decorrelator, which we have shown earlier to be very wasteful of degrees of freedom if interferers are weak. While Winters' result holds for the Rayleigh model and any number of antennas, our results hold for *any* fading distribution but are asymptotic in the number of antennas.

5.7 CONCLUDING REMARKS

It is illuminating to compare the effective interference and effective bandwidths of the users in the three cases: the conventional matched filter, the MMSE filter, and the decorrelating filter. This comparison is shown in Figures 5.1 and 5.2. The

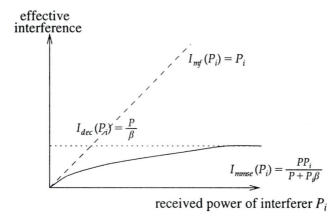

Figure 5.1   Effective interference for the three receivers as a function of interferer's received power P_i. Here, P is the received power of the user to be demodulated, and β is the achieved SIR.

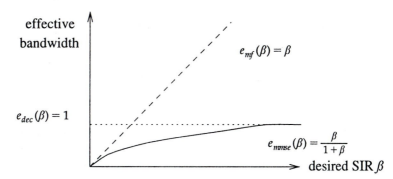

Figure 5.2   Effective bandwidths for three receivers as a function of SIR.

effective interference under MMSE is nonlinear and depends on the received power P of the user to be demodulated as well as on the achieved SIR β. The effective interference under the conventional matched filter is simply P_i, the received power of the interferer. Under the decorrelator, the effective interference is $\frac{P}{\beta}$, independent of the actual power of the interferer. The intuition here is that the decorrelator completely nulls out the interferer, no matter how strong or weak it is.

Assuming perfect power control, we can define effective bandwidths that characterize the amount of network resources a user consumes for a given target SIR. The effective bandwidths under the conventional, MMSE, and decorrelating receivers are β, $\frac{\beta}{1+\beta}$, and 1, respectively. We note that the conventional receiver is more efficient than the decorrelator when β is small and far less efficient when β is large. Intuitively, at high SIR requirements, a user has to transmit at high power, thus causing a lot of interference to other users under the conventional receiver. Not surprisingly, since it is by definition optimal, the MMSE filter is the most efficient in all cases. When β is small, the MMSE filter operates more like the conventional receiver, allowing many users per degree of freedom; but when β is large, each user is decorrelated from the rest, much as in the decorrelator receiver, and therefore the interferers can still occupy no more than one degree of freedom per interferer.

The effective bandwidth concept for the MMSE receiver is valid only in the perfectly power-controlled case. By contrast, the concept of effective interference applies with or without perfect power control and may prove more useful in the multicell context.

While these results provide much insight into the performance of these filters, we must emphasize that they pertain only to a single cell, without fading, and in the time-synchronous case. It remains to be seen how these filters perform in more realistic scenarios.

REFERENCES

[1] Chen, S. C., N. Bambos, and G. J. Pottie, "On distributed power control for radio networks," *Proc. Int'l Conf. Commun.*, Vol. 3, pp. 1281–1285, 1994.

[2] Cimini, L. J., and G. Froschini, "Distributed algorithms for dynamic channel allocations in microcellular systems," *Proc. IEEE Vehic. Tech. Conf.*, Vol. 2, pp. 641–644, 1992.

[3] Foschini, G. J., and Z. Miljanic. "A simple distributed autonomous power control algorithm and its convergence," *IEEE Trans. Vehic. Techn.*, Vol. 42, pp. 641–646, Nov. 1993.

[4] Gilhousen, K. S., I. M. Jacobs, R. Padovani, A. J. Viterbi, L. A. Weaver, and C. E. Wheatley, "On the capacity of a cellular CDMA system," *IEEE Trans. Vehic. Techn.*, Vol. 40, pp. 303–312, May 1991.

[5] Hanly, S. V. "Information Capacity of Radio Networks," Ph.D. Thesis, Cambridge University, Aug. 1993.

[6] Hanly, S. V. "An algorithm for combined cell-site selection and power control to maximize cellular spread spectrum capacity," *IEEE Select. Areas Commun.*, issue on the fundamentals of networking, Vol. 13, Sept., 1995.

[7] Hanly, S. V. "Capacity and power control in spread spectrum macrodiversity radio networks," *IEEE Trans. Commun.*, Vol. 44, pp. 247–256, Feb. 1996.

[8] Honig, M., U. Madhow, and S. Verdu, "Blind adaptive multiuser detection," *IEEE Trans. Inform. Theory*, Vol. 41, pp. 944–960, July 1995.

[9] Kumar, P., and J. Holtzman, "Power control for a spread-spectrum system with multi-user receivers," *Proc. IEEE PIMRC*, Vol. 3, pp. 955–959, Sept. 1995.

[10] Lupas, R., and S. Verdu, "Linear multiuser detectors for synchronous code-division multiple access," *IEEE Trans. Inform. Theory*, Vol. IT-35, pp. 123–136, Jan. 1989.

[11] Lupas, R., and S. Verdu,"Near-far resistance of multiuser detectors in asynchronous channels," *IEEE Trans. Commun.*, Vol. COM-38, pp. 496–508, Apr. 1990.

[12] Madhow, U., and M. Honig, "MMSE interference suppression for direct-sequence spread-spectrum CDMA, *IEEE Trans. Commun.*, Vol. 42, pp. 3178–3188, Dec. 1994.

[13] Madhow, U., and M. Honig, "MMSE detection of direct-sequence CDMA signals: performance analysis for random signature sequences," submitted to *IEEE Trans. Inform. Theory*, 1997.

[14] Marcenko, V. A., and L. A. Pastur, "Distribution of eigenvalues for some sets of random matrices," *Math. USSR-Sb*, pp. 457–483, 1967.

[15] Rapajic, P., and B. Vucetic, "Adaptive receiver structures for asynchronous CDMA systems," *IEEE J. Select. Areas Commun.*, Vol. 12, pp. 685–697, May 1994.

[16] Rupf, M., F. Tarkoy, and J. Massey, "User-separating demodulation for code-division multiple access systems," *IEEE J. Select. Areas Commun.*, Vol. 12, pp. 786–795, June 1994.

[17] Silverstein, J. W., and Z. D. Bai, "On the empirical distribution of eigenvalues of a class of large dimensional random matrices," *J. Multivariate Analysis*, Vol. 54, pp. 175–192, 1995.

[18] Tse, D., and S. V. Hanly, "Linear multiuser receivers: effective interference, effective bandwidth and capacity," submitted to *IEEE Trans. Inform. Theory*, 1997.

[19] Ulukus, S., and R. D. Yates, "Adaptive power control and MMSE interference suppression," to appear in *ACM Wireless Networks*.

[20] Verdu, S., "Minimum probability of error for asynchronous Gaussian channels," *IEEE Trans. Inform. Theory*, Vol. IT-32, pp. 85–96, Jan. 1986.

[21] Verdu, S., "Optimum multiuser asymptotic efficiency," *IEEE Trans. Commun.*, Vol. COM-34, pp. 890–897, Sept. 1996.

[22] Winters, J. H., J. Salz, and R. Gitlin. "The impact of antenna diversity on the capacity of wireless communications systems," *IEEE Trans. Commun.*, Vol. 42, pp. 1740–1751, 1994.

[23] Xie, Z., R. Short, and C. Rushforth, "A family of suboptimum detectors for coherent multi-user communications," *IEEE J. Select. Areas Commun.*, Vol. 8, pp. 683–690, May 1990.

[24] Yates, R. D., and C. Y. Huang, "Integrated power control and base station assignment," *IEEE Trans. Vehic. Techn.*, Vol. 44, pp. 638–644, Aug. 1995.

[25] Yates, R. "A framework for uplink power control in cellular radio systems," *IEEE J. Select. Areas Commun.*, issue on the fundamentals of networking, Vol. 13, pp. 1341–1347, Sept. 1995.

[26] Yun, L. C., and D. G. Messerschmitt, "Power control for variable QoS on a CDMA channel," *Proc. IEEE MILCOM*, Vol. 1, pp. 178–182, Oct. 1994.

[27] Zander, Z. "Performance of optimum transmitter power control in cellular radio systems," *IEEE Trans. Vehic. Techn.*, Vol. 41, pp. 57–62, 1992.

ACKNOWLEDGMENTS

This research was supported by the Air Force Office of Scientific Research under Grant No. F49620-96-1-0199, and by the Australian Research Council.

6

Architectural Principles for Multimedia Networks

Paul Haskell
David G. Messerschmitt
Louis Yun

Many treatments of wireless communications focus on a wireless link as an isolated entity. Our concern here is with networks that support all multimedia services (including data, graphics, audio, images, and video) for tetherless (not physically wired to the network), nomadic (able to access the network from many locations), and mobile (accessing the network while moving) users. Such a network is termed an *integrated-services multimedia network with wireless access*. Wireless access is a key component of tetherless and mobile access in particular. Important components of such a multimedia network include a (typically broadband) backbone network, wireless access links to that backbone, terminals associated with each user (where some terminals are tetherless and others are not), and centralized data and computational servers. It is expected that an integrated-services multimedia network will serve a large and heterogeneous mix of applications. Overall, in this large and complex system, the fact that there is wireless access should have broad implications to all the system components, not just to the wireless access link. Conversely, design issues in the remainder of the system impact the wireless access link design. A primary objective of this chapter is to identify these cross-cutting issues. As such, the issues explored in this chapter are complementary to many of those discussed in the preceding chapters of this volume.

Most of our attention is focused on the *signal processing* technologies in a multimedia network, including *compression, modulation, forward error-correction coding,* and *encryption*, as well as limited attention to other elements that interact with signal processing (such as protocols). Additional compression and coding issues are explored in detail in Chapter 7. From a networking perspective, we define as signal processing those functions that modify or hide basic syntactical and semantic components of a bit stream payload, as opposed to those functions that are oblivious to the payload bits (such as protocols, routing, etc.).

In treating these issues, it is important to identify the objectives that are to be achieved. We can list those objectives relevant to this chapter as follows:

- For continuous-media services like audio and video, as well as graphics, the relevant "quality" criterion is subjective.
- As discussed further below, the most critical objective performance criteria are low interactive delay and high traffic capacity for wireless access links.
- Privacy by end-to-end encryption will be important for some users; it also is one aspect of intellectual property protection and authorized access control.
- Applications, terminal capabilities (from desktop to various flavors of portable terminals with different processing and resolution), and transport media (especially wireless and fiber backbones) will be heterogeneous. It is important to support both desktop and portable terminals with a common set of applications. It is also important that applications deploy seamlessly to the network without the application developer needing to deal explicitly with a diversity of transport and terminal capabilities.
- Not only point-to-point connections, but also point-to-multipoint and multipoint-to-point connections will exist [1, 2].
- Propagation characteristics will vary widely, depending on assumptions about carrier frequency, bandwidth, propagation characteristics (especially in-building vs. wide-area networks), terminal velocity, etc. This chapter is primarily focused on broadband in-building wireless networks, where we expect relatively slow terminal velocity and hence relatively slowly varying channel characteristics. However, most considerations discussed in this chapter apply to more general situations, and we will mention the impact of less ideal channel characteristics.

It is quite challenging to meet all these objectives simultaneously. To have any hope requires a carefully crafted architecture. One clear conclusion is that the wireless access link typically will be the limiting factor in achieving good subjective quality, as it is inherently unreliable and typically has limited bandwidth resources relative to backbone networks. Thus, any architectural constructs should first and foremost be aimed at achieving the best subjective quality as limited by the wireless access link, making compromises in other components (terminals, servers, and backbone network) as necessary. The wireless access link should not be considered just an "add on" to an existing backbone infrastructure, as is unfortunately the most common design philosophy today.

Another important consideration is complexity management [4]. The internet protocols have managed to contain complexity by partitioning most functionality within the terminals and keeping the internet layer relatively simple and state-free. As a result, a rich suite of applications has been deployed swiftly. In contrast, the public telephone network, with a relatively limited set of services based on 64 kb/s circuits and a centralized control model, is straining at the limits of complexity within the switching-node software. Because the central-control telephone model is not extensible to achieving the flexibility required in future multimedia networks, distributed "intelligent networking" approaches to control are being deployed [5] and even more sophisticated approaches are being considered [6, 7]. The internet model also becomes considerably more complicated when extended to continuous media (CM) services, due to the need for resource reservations and multicast connections [2]. Careful attention should be paid to complexity management from the outset.

In this chapter, we propose some architectural principles and discuss their implications to the constituent signal processing technologies listed above. These principles suggest many new requirements for the signal processing and present opportunities to signal processing researchers and developers for years to come. A more general perspective on the convergence of communications and computing as embodied in multimedia networking is available [8, 9], as is a treatise on some of the societal impacts [10].

6.1 BASIC CONSIDERATIONS

In this section, we describe some of the constituent technologies and their relationship to multimedia networking.

6.1.1 High-Level Network Architecture

One group has made a proposal for an architecture for the future Global Information Infrastructure (GII), and for consistency, we draw upon their architecture [3].[1] As in [9], we use the terminology shown in Figure 6.1, which differs slightly from that in [3]. *Applications* draw upon the *services layer* (termed transport services in [3]), which calls upon the *bitway layer* (called bearer services in [3]). The bitway layer establishes

---

[1]Actually, [3] adds a fourth layer, middleware services, which we delete here because it is generally unrelated to signal processing functions of concern in this chapter.

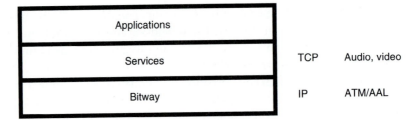

Figure 6.1 An architecture for the GII, including both CM and data services. Each layer may be subdivided into appropriate sublayers.

connections between endpoints, carries data between the endpoints, and monitors its own performance. The services layer provides a set of common generic capabilities that are available to all applications; examples include reliable streams (supporting applications like file transfer), reliable transactions, electronic payments, directory services, and audio or video transport (for multimedia applications). One of the functionalities in the services layers is the conditioning of data for the bitway (for example, the compression of audio or video) and compensating for impairments in the bitway (for example, resequencing of packets, as in the Transport Control Protocol (TCP), or retransmission of lost packets, as in TCP, or resynchronization of audio and video streams, as in the MPEG-2 transport stream [11, 12]).

A concrete example of the functional groupings of these layers is shown in Figure 6.2 for a video application. The application presents a stream of raw

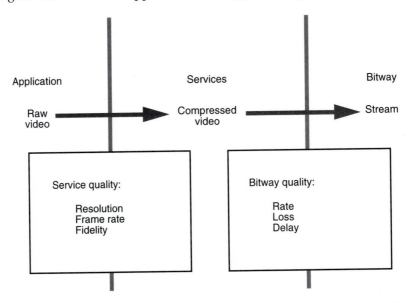

Figure 6.2 Illustration of the three layers in a video service and an incomplete list of service and bitway quality attributes.

(uncompressed) video frames to the services layer, where the quality attributes describing the video service to the application include resolution, frame rate, pixel depth, compression fidelity, etc. The services layer does video compression, as well as perhaps encryption, and presents to the bitway a compressed video stream (to save bandwidth on the transport). The service is described to the bitway by its rate attributes (average and peak rate), and the bitway to the service by its quality-of-service (QoS) attributes including loss, corruption, and delay characteristics.

6.1.2 Signal Processing Functions and Constraints

In this section, we discuss briefly and qualitatively some of the interactions between signal processing functions and the CM systems and network architecture within which they are embedded.

Compression removes signal redundancy as well as signal components that are subjectively unimportant, so as to increase the traffic-carrying capacity of transmission links within the bitway layer. Compression typically offers a trade-off between a signal's decoded fidelity and its transmitted bandwidth and often has the side effect of increasing the reliability requirements (loss and corruption) for an acceptable subjective quality. Compression can be divided into two classes: *signal semantics based* (such as conventional video and audio compression), and *lossless*, which processes a bit stream without cognizance of the underlying signal semantics. The compression typically has to make some assumptions about the bitway characteristics, such as the relative importance of rate and reliability (see Section 6.1.5).

Encryption reversibly transforms one bit stream into another such that a reasonable facsimile of the original bit stream is unavailable to a receiving terminal without knowledge of appropriate keys [13]. Encryption is one component of a conditional access system, with which a service provider can choose whether and when any individual receiver can access the provided service, and is also useful in ensuring privacy. It precludes any processing of a bit stream because it hides the underlying syntactical and semantic components, except in a secure server that has keys available to it. It also increases susceptibility to bit errors and synchronization failures, as discussed in Section 6.3.1.

Forward error-correction coding (FEC) adds a controlled redundancy so that transmission impairments such as packet loss or bit errors can be reversed. We distinguish *binary* FEC techniques from *signal-space* techniques. Binary FEC is applied to a bit stream and produces a bit stream; examples include Reed-Solomon coding and convolutional coding. Binary FEC has the virtue of flexibility, as it can be applied on a network end-to-end basis in a manner transparent to the individual links. FEC can be combined with other techniques, such as interleaving (to change the temporal pattern of errors) and retransmission (to repeat lost or corrupted information), if the temporal characteristics or error mechanisms are known. Signal-space coding, on the other hand, is tightly coupled with the *modulation* method (used to encode bits onto waveforms) and the physical characteristics of the medium and is often accompanied by soft decoding. Examples include lattice and trellis coding. It is usually custom tailored to the physical characteristics of each link and, as a result, can offer significantly higher performance. Wireless link modulation methods often incorporate power control and temporal and spatial diversity reception as well, as discussed in earlier chapters.

Figure 6.3 illustrates some fundamental syntactical constraints that we should keep in mind while designing a network architecture for CM services:

- Compression must precede encryption, and decryption must precede decompression. Encryption would hide basic statistical characteristics of an uncompressed audio or video signal, such as spatial and temporal correlations, that are heavily exploited by compression algorithms.
- Compression must precede FEC, and decompression must follow FEC since there is no point to "correcting" the benign and desired changes in a bit stream due to compression and decompression.

The relationship between encryption and error-correction coding is more complicated. Since encryption, like binary coding, transforms one bit stream into another, it can precede or follow binary error-correction coding. However, since a

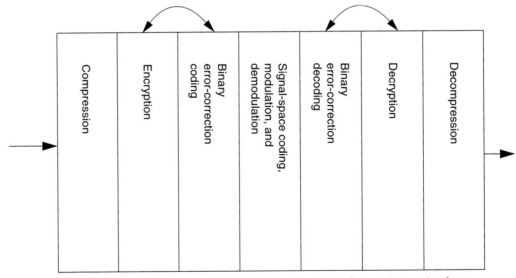

Figure 6.3   Illustration of some fundamental syntactical constraints on signal processing functions.

signal-space code generates an output in the real-number field, it cannot precede encryption and signal-space decoding cannot follow decryption. The purpose of binary coding before encryption is to attempt to correct postdecryption errors. The purpose of binary or signal-space coding after encryption is to prevent errors in the transport of the encrypted bit stream, which will indirectly prevent postdecryption errors.

6.1.3 Bitway Architecture

A bitway is the layer of a CM system responsible for transmitting data bits from one place (and time, for storage applications) to another. As part of this task, the bitway commonly carries out:

- Routing, establishment of a path from one communication endpoint to another;
- QoS establishment, the negotiation of service bit-rate characteristics and bitway impairment characteristics;
- Resource reservation, the assignment of resources to a particular connection to ensure compliance with the "QoS contract"; and
- Monitoring of network component status, source rate behavior, and QoS of active connections.

CM services commonly rely on a combination of several underlying transmission sublinks. This is especially true for wireless-based services, e.g., paging and cellular telephony. Heterogeneous sublinks certainly complicate the implementation of the bitway functions listed above. However, they also complicate the design and configuration of the signal processing functions described in Section 6.1.2.

The best trade-off between signal fidelity and bandwidth appropriate for a high-speed wired sublink may be very unreasonable for a wireless link. Frequently, this motivates system designs that include transcoders within the network, as with the IS-54 digital cellular system described in Section 6.1.5. Of course, if decompression and recompression are performed at transcoders within the network, there is a requirement for a secure server to perform decryption and encryption.

However, it is with FEC that the existence of multiple heterogeneous subnetworks complicates CM system design the most. It would be simplest to provision a single end-to-end FEC system across a heterogeneous network; however, the efficiencies of closely coupled error-correction coding, suitably designed modulation, and specific subnet characteristics may be too significant to pass up in some cases. Section 6.1.5 and Section 6.4.1 discuss this further.

6.1.4 Corruption, Loss, and Delay Effects

Packet-based communications networks inevitably introduce three types of impairments. There is packet *loss* (failure to arrive), packet *corruption* (bit errors occurring within the payload), and packet *delay*. Packet loss can occur due to several mechanisms, such as bit errors in the header, or buffer overflow during periods of network congestion.

Data networks do not make a distinction between loss and corruption since a packet that is corrupted is useless and hence is discarded. CM services can tolerate some level of loss and corruption without undue subjective impairment, especially if appropriate masking is built into the signal decoders. This is fortunate, since absolute reliability such as afforded by data networks, requires retransmission mechanisms, which can introduce indeterminate delay, often excessive to interactive applications like telephony and video conferencing. Another distinct characteristic of CM services is that loss and corruption are different effects. Lost data must be masked, for example, in video by repeating information from a previous frame or in audio substituting a zero-level signal. Under some circumstances, it is possible to make good use of corrupted information, for example, by displaying it as if it were correct. The resulting subjective impairment may be less severe than if the corrupted data were discarded and masked.

Some CM compression standards, generally those presuming a reliable transport mechanism (such as MPEG video [14, 15, 16, 17]) discard corrupted data and attempt to mask the discarded information. Other standards—those designed for a very unreliable transport (such as the voice compression in digital cellular telephony [18] and video compression designed for multiple access wireless applications [19])—use corrupted data as if it were error free and minimize the subjective impact of the errors. An important research agenda for the future is audio and video coding algorithms that are robust to loss and corruption introduced by wireless networks, recognizing that these effects are more severe than in backbone networks.

CM services are real-time, meaning that they require transport-delay bounds. However, there is a wide variation in delay tolerance, depending on the application. For example, a video-on-demand application will be relatively tolerant of delay, whereas it is critical that transport delay be very small (on the order of 50 ms or so) for a multimedia editing or video conferencing application. Much recent attention is focused on achieving bounded delay through appropriate resource reservation protocols [2, 20, 21]. Given this wide range of delay tolerance, it is clear that the highest traffic capacity can be obtained only by segmenting services by delay, coupled with delay-cognizant scheduling algorithms within the bitway statistical multiplexers.

Audio and video services are usually considered to be synchronous, implying that network transport jitter is removed by buffering before reconstruction of the audio or video. For the special case of voice, it is possible to change the temporal relationship of talkspurts somewhat without any noticeable effect, but video display is organized into periodic frames (at 24, 25, or 30 frames/s), and all information destined for a frame must arrive before it can be displayed. (We make an alternative proposal in Section 6.4.3.)

Packets arriving after some prescribed delay bound are usually considered to be lost, as if they did not arrive at all.[2] This illustrates another important characteristic of CM services: the existence of *stale information* that will be discarded by the receiver if it does not arrive in timely fashion. As another example, the bitway may be working feverishly to deliver a pause-frame video when the motion suddenly resumes. Any state residing in the bitway relevant to the pause-frame will not be used at the receiver. The purging of stale information within the bitway layer will increase traffic capacity.

6.1.5 Joint Source/Channel Coding

Joint source/channel coding (JSCC) is a way to increase the traffic capacity of a network, subject to a subjective quality objective. While a classic "separation theorem" of Shannon states that it is possible to separate the source and channel coding without loss of performance, his result requires conditions (on channel memory and time variation) not usually satisfied on wireless channels [22, 23] and, further, takes no account of delay or complexity. In fact, substantial gains can be achieved in traffic capacity for a given subjective quality using JSCC on wireless channels, for three reasons:

- Since wireless channels typically do not satisfy the assumptions of the separation theorem, that theorem does not rule out greater performance through better source-channel coordination.
- For some interactive applications like video conferencing, we are particularly concerned about minimizing delay, which is outside the scope of the separation theorem.

---

[2]We argue in Section 6.4.3 that this model for the reconstruction of video may not be the best in case of packet networks with substantial delay jitter.

- For applications like audio, video, and graphics, the only meaningful criterion of quality is subjective. This implies that we must improve the system
through experimentation; subjective quality falls outside the scope of the separation (or any other) theorem.

We can divide JSCC roughly into two classes: *tightly coupled* and *loosely coupled*. Tightly coupled JSCC, which predominates in the literature, designs the
source coding, modulation, and channel coding jointly, assuming therefore that
the channel coding and modulation are cognizant of the full details of the source
coding, and vice versa [24, 25, 26, 27]. This approach is applicable when designing
a stand-alone system, such as wireless transmission of high-definition television
(HDTV) [27].

If JSCC is to be applied to an integrated services multimedia network, we
have to deal with complications like the fact that a single source coder must be
able to deal with a variety of transport links (broadband backbone and wireless in
particular), the concatenation of heterogeneous transport links, and multicast connections with common source representations flowing over heterogeneous links
in parallel. In this environment, it is appropriate to consider loosely coupled JSCC,
which is the only variety we pursue in this chapter. Loosely coupled JSCC
attempts to abstract those attributes of the source and the channel that are most
relevant to the other and to make those attributes generic; that is, broadly applicable to all sources and channels and not tightly coupled to the specific type.

Loosely coupled JSCC is thus viewed differently from the perspective of the
"source" and the "channel," where channel is usually taken to mean a given phys-
ical-layer medium, but which we take here to mean the entire bitway network.
From the perspective of the bitway, JSCC ideally adjusts the allocation of network
resources (buffer space, bandwidth, power, etc.) to maximize the network traffic
capacity subject to a subjective quality objective. From the perspective of the
source, JSCC ideally processes the signal in such a way that bitway network
impairments have minimal subjective effect, subject to maximizing the network's
traffic capacity. This suggests that the source coding must take account of how the
bitway allocates resources, and the effect this allocation has on end-to-end impairments as well as on traffic capacity, and conversely the bitway needs to know the
source coding strategy and the subjective impact of the bitway resource allocations. However, to embed such common knowledge would be a violation of the
loosely coupled assumption, creating an unfortunate coupling of source and channel that precludes further evolution of each. Rather, we propose a model in which
the source is abstracted in terms of its bitrate attributes only, and the bitway is
abstracted in terms of its QoS attributes only. The benefits of JSCC can still be
achieved with this limited knowledge, but only if the source and channel are

allowed to *negotiate* at session establishment. During negotiation, each source or channel is fully cognizant of its internal characteristics and can influence the other only through a give-and-take in establishing the rate and QoS attributes, taking into account some measure of cost. The substream architecture discussed in Section 6.2 will increase the effectiveness of loosely coupled JSCC.

A simple example of JSCC is compression [33]. The classical goal of compression is to minimize bit rate, which is intended to maximize the traffic capacity of the network without harming the subjective quality appreciably. However, minimizing the bit rate (say, in the average sense) is simplistic because traffic capacity typically depends on more than average bit rate. To cite several examples:

- The statistical multiplexing advantage in congestion-dominated subnets depends on the peakiness of the offered bit streams, at least for a constant loss and delay objective, and the manner in which the bit rate varies with time is an important factor in the traffic capacity.

- A side effect of compression, at least at a relatively constant subjective quality, is usually to generate a variable bit rate, and exploiting that variable bit rate through statistical multiplexing often results in packet losses under high traffic loads. This, in turn, causes subjective impairment.

- Compression normally results in an increase in the susceptibility to bit errors. On interference-dominated subnets, such as cellular radio wireless access links, it is expensive (in terms of traffic capacity) to provision consistently low error probability since doing so requires large transmitted power and hence increased interference to other users.[3] Thus, the traffic capacity of such a subnet depends strongly on the reliability requirement, as well as the bit rate, and it is not automatically the case that minimizing the bit rate is equivalent to maximizing the traffic capacity [19, 29, 30, 31, 32].

Having stated our objective for JSCC in multimedia networks, let us now examine some current examples of tightly coupled JSCC and point out their shortcomings for an integrated-services network. Some systems effectively ignore the benefits of JSCC by focusing on a limited set of environments. Even standards such as MPEG targeted at widespread use commonly make specific limiting assumptions about the transport. The MPEG designers assume that uncorrected errors are infrequent enough that blocks of data with errors can be discarded and masked, with the resulting artifacts propagating until the next intraframe coded video frame. This results in error rate requirements on the order of 10^{-9} to 10^{-12}

---

[3]For example, for wireless CDMA, the traffic capacity is related to the product of average bit rate and a monotonic function of bit error rate [28].

(depending on the application) [34, 35]. While this rate is feasible in storage, fiber, and broadcast wireless applications (such as terrestrial HDTV [33]), this is likely not feasible in multiple-access wireless applications.[4] (Voice standards intended for multiple-access channels and mobile receivers with fading generally assume a worst-case error rate in the range of 10^{-2} to 10^{-3}, which is more representative on these types of channels during deep fades [18].) MPEG illustrates the difficulty in designing compression standards with sufficient flexibility and scalability to accommodate a variety of transport scenarios.

MPEG-1 is limited not just to low-error-rate bitways but to low-delay-jitter bitways as well. Fortunately, this limitation was addressed during the design of MPEG-2. The MPEG-2 Real-Time Interface (RTI) permits system designers to choose the maximum delay jitter expected in their systems; given this value, the RTI specifies how decoders can handle the specified jitter. The generic nature of the RTI came about specifically because the MPEG-2 designers wanted to handle delay jitter in a variety of bitways: satellite, terrestrial, fiberoptic, cable, etc. This is an example of transport characteristics influencing compression design. See Section 6.4.2 for further discussion of MPEG.

The critical role of traffic capacity in wireless access subnets typically results in systems with intricate but inflexible schemes for JSCC, as can be illustrated by a couple of concrete recent examples. These examples also illustrate some of the pitfalls of the coupling of the CM service and the network, and they point to some opportunities to reduce this coupling.

One example is the IS-54 digital cellular telephony standard. This standard uses radio time-division multiple-access (TDMA) transport, which due to vehicular velocity is subject to rapid fading. Due to fading and also in an effort to increase traffic capacity by an aggressive cellular frequency reuse pattern, worst-case error rates on the order of 10^{-3} are tolerated (the error rate could be reduced at the expense of traffic capacity, of course). The speech is aggressively compressed, and as a result, the error susceptibility is increased, particularly for a subset of the bits. Therefore, the speech coder bits are divided into two groups, one of which is protected by a convolutional code and the other left unprotected. Interleaving is used to spread out errors (which are otherwise grouped at the demodulator output). What do we consider undesirable about this system design? At least a few things:

- The close coupling between the speech compression algorithm (which determines which bits require more FEC protection) and the transport (FEC and modulation) makes it impossible to modify one without the other. This situation will not be acceptable in a heterogeneous bitway.

[4]While FEC may be able to achieve such error rates, countering the worst-case error rate environment during deep fades will require very high levels of redundancy, which, because it is present even during favorable channel conditions, will severely restrict the traffic capacity [19].

- The bitway (including FEC) is designed with a particular CM service in mind, namely, speech. It would be better if the transport were more flexible so that other services could easily be introduced.
- The transcoder (which converts from 64 kb/s speech to a compressed speech) introduces a substantial delay of about 80 ms. Two subscribers conversing via two digital cellular telephones encounter two tandem transcoders (where neither is necessary) and a round-trip delay on the order of 320 ms.
- The pulse code modulation (PCM) landline network connecting the cellular base station to another subscriber is carrying a much higher bit rate than necessary. While this bit rate does not affect the traffic capacity of the circuit-switched telephone network, a more flexible network would have a reduced traffic capacity (expressed in terms of simultaneous telephone calls) due to this mismatch of resources.

These issues are addressed in a more general context in Section 6.3.

A final example illustrates an architecture that begins to redress some of these problems. The Advanced Television Research Consortium (ATRC) proposal for terrestrial broadcast TV [36] attempts to separate the design of the video compression from the transport subsystem by defining an intermediate packet interface with fixed-length packets (cells). Above this interface is an adaptation layer that converts the video compression output byte stream into cells, and below this interface the cells are transported by error-correction coding and radio frequency (RF) modulation. Much more enlightening is the way in which a modicum of JSCC is achieved. First, the compression algorithm splits its output into two substreams, where, roughly speaking, the more subjectively important information is separated from the less subjectively important (and a reasonable rendition of the video can be obtained from the first substream).[5] This separation is maintained across the packet interface and is thus visible to the bitway. The bitway transmits these two substreams via separate modulators on separate RF carriers, where the first substream is transmitted at a higher power level. The motivation for doing so illustrates another important role of JSCC on wireless access links; namely, achieving graceful degradation in quality as the transmission environment deteriorates. In this case, in the fringe reception area the quality will deteriorate because the second substream is received unreliably, but a useful picture, based on the first substream, is still available. This system illustrates some elements of an architecture that will be proposed later. Chapter 7 focuses on these and other aspects of JSCC and also proposes some specific JSCC structures.

---

[5]Reference [36], and other work on packet video, use the term *priority* to distinguish between the substreams. We avoid that term here because it is usually applied to control the order of arrival or discard in congestion-dominated packet networks, and can be misleading when applied in more general contexts.

6.2 MODULARITY OF SERVICES AND BITWAY LAYERS

To allow different transmission media to work with the same source coding, and different source coders to work with different transmission media, it is especially important that we logically separate the design of source coders (in the services layer) from the transmission (in the bitway layer) as much as possible. (As discussed in Section 6.3, this separation is even more advantageous in heterogeneous transport environments.) This separation requires a careful partitioning of functionality between these layers and appropriate abstractions at their interface. This section concentrates on this partitioning and interface and describes a basic bitway model appropriate for multimedia services. See [31] for a description of the video compression problem in this heterogeneous environment.

6.2.1 Partitioning of Functionality

While [3] does not attempt a detailed partitioning of functions between services and bitway layers, we make a proposal here specifically with respect to signal processing functions, as shown in Figure 6.4. FEC has been placed in the bitway layer, and compression and encryption in the services layer, where we have termed the interface between these two layers the *medley gateway* [45]; the term *gateway* refers to the connection between layers, and *medley* refers to the heterogeneous substream structure we envision at this gateway, as we discuss later.

Compression is inherently a "conditioning for transport" function and hence belongs in the services layer. We explicitly avoid compression, or transcoding (converting from one compression to another), within the bitway layer. The reasons for this are elaborated further in Section 6.3.

The reasons that we include encryption within the services layer are more subtle:

- Encryption must follow compression (and precede decompression) and hence cannot reside in the application layer.

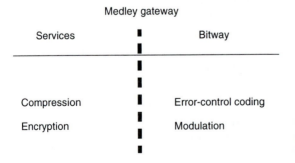

Figure 6.4 Partitioning of signal-processing functions between the service and bitway layers.

- There may be two or more bitway layers in a given connection in a heterogeneous environment (Section 6.3). Including encryption in the services layer opens the possibility of doing encryption on an end-to-end basis, with resulting simplified key management and higher level of security. Encryption in the bitway layer could result in two or more encryption/decryption operations, with complications to key management and reduced security due to "in the clear" signals available at intermediate points. Also, encryption would not be under the control of the user, but rather the service provider, dramatically reducing the security from the perspective of the user.

- The proposed architecture eliminates the increased burden of error multiplication due to encryption (see Section 6.3.1) on the FEC algorithms since FEC decoding occurs prior to decryption.

The reasons we have placed FEC in the bitway layer include the following:

- The most unreliable transmission media, wireless, are also the most critical with respect to spectral efficiency. On such media, signal-space coding techniques (for example, trellis coding and multidimensional signal constellations [46]) are tightly integrated into the modulation system and hence are inherently localized to each transmission link in the connection.

- There are many error correction techniques available, such as retransmission, FEC, interleaving, etc. It is most efficient for these techniques to be tightly coupled to the transmission environment. For example, the temporal characteristics of wireless access links depend heavily on the level of mobility, and the level of interleaving (to counter error-correlation effects) and the coding techniques are best coordinated with that mobility.

- Achieving high traffic capacity on time-varying media (such as wireless channels in the presence of terminal motion) requires techniques that take account of the state of the channel, so that parameters such as FEC redundancy, transmit power, etc., are varied with time. This important class of techniques is practical to implement only within the bitway because of the close coupling to the physical layer and the need for low-latency feedback between modulation and coding and the physical layer.

- As discussed in Section 6.3.3.2, performing FEC on an end-to-end basis implies codes that deal with a variety of different loss and corruption mechanisms, such as packet loss due to congestion (erasure codes), independent errors, and correlated errors due to interference in wireless access links (interleaving). In practice, this implies that different codes would have to be concatenated to deal with every possible contingency, and the resulting multiple layers of redundancy would be carried by every link with a resultant traffic penalty.

- End-to-end FEC would require sufficient redundancy for the worst-case link, resulting in a rate penalty on links with less severe impairments. In the absence of adaptive configuration, the redundancy has to be adjusted for the global worst case.
- End-to-end acknowledgment and repetition protocols will generally impose too large a delay for critical interactive CM services like video conferencing.

While we propose that the primary responsibility for error correction fall to the bitway, there is no reason to dogmatically preclude the involvement of the service, as discussed further in Section 6.2.5. For example, in "best effort" data services without delay guarantees, services retransmission protocols (as in TCP) may be acceptable. As another example, a subset of the data in a CM service may require extraordinary reliability but be relatively insensitive to delay, as, for example, coder configuration and state information. In the latter case, relying on a reliable transport protocol may be a better solution than imposing a high reliability requirement on the bitway layer. More generally, experience has shown that:

- Turning a poor reliability channel into one with moderate reliability is best done within the physical layer and utilizing signal space or binary coding techniques with soft decoding.
- Turning a modest reliability channel into one with almost complete reliability is best done with acknowledgment and retransmission protocols. These protocols are best done on an end-to-end basis, rather than embedded into each link, because of the delay that they introduce.

Thus, the best approach depends on circumstances, but very high reliability streams will involve a combination of FEC in the bitway layer and retransmission in the services layer. This is yet another reason to place encryption in the services layer—so as to perform decryption on the most reliable representation of the bit stream and thus minimize error multiplication effects.

6.2.2 Abstracted View of the Bitway

To maintain flexibility and contain complexity, it is important that abstractions of both services and bitway be defined at the medley gateway. These abstractions should retain information that is relevant and critical, while hiding unnecessary details. One of our major goals is to separate, insofar as is possible, the design of the service from the bitway. Not only is this an important complexity management technique, but it is critical to our ability to deal with complex bitway entities such as concatenated heterogeneous links and multicast connections.

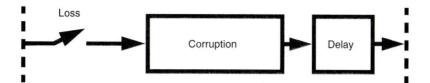

Figure 6.5   Abstracted model of the bitway from the perspective of the service, for a single substream.

Since the bitway core function is to transport packets, the abstract view should focus on the fundamental packet transport impairments of corruption, loss, and delay. A basic model incorporating these three elements is shown schematically in Figure 6.5. Often, the service will be interested in the temporal properties of these impairments; that is, a characterization of whether impairments like losses, corruption, or excessive delays are likely to be bunched together, or if they are statistically spread out in time. This issue is discussed further later.

The description of the properties of the connection that the bitway provides to the service is called a *flowspec* [2]. The most relevant of these properties are:

- *Rate* attributes, such as average rate, peak rate, and a characterization of the temporal characteristics of the rate.
- *QoS* attributes, including loss, corruption, and delay, and the temporal characteristics of these impairments. Other QoS attributes also may be specified explicitly, for example, whether a connection guarantees to deliver packets in sequence.

Note what information is *not* included in the bitway model. We deliberately exclude knowledge of the detailed transmission and switching structure within the bitway. For example, we hide from the service any knowledge of whether loss and delay is caused by congestion or by FEC and interleaving techniques, etc. Similarly, knowledge of whether corruption is caused by thermal noise, or interference, or is affected by time-varying mechanisms like Ricean or Rayleigh fading, is omitted. This strategy places on the bitway modeling the burden of specifying fundamental impairments with sufficient detail that the transmission characteristics are sufficiently characterized for purposes of the service.

6.2.3 Abstracted View of the Service

In considering the abstraction of the service as seen by the bitway, a primary objective is to allow JSCC, in spite of our careful separation of the design of the two layers. To this end, we include in the services layer abstraction the substream structure

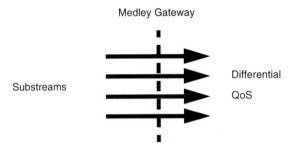

Medley Gateway

Substreams

Differential

QoS

Figure 6.6   Abstracted view of the service from the perspective of the bitway.

shown in Figure 6.6. The stream of packets is logically divided into *substreams*, which are visible to the bitway. The integrity of substreams is maintained across multiple links (see Section 6.3). Each substream is associated with distinct QoS and rate attributes established by negotiation with the application. The QoS attributes are aggregated values from the individual links, so that each substream on each link has a potentially different QoS objective. Thus, within the bitway, each packet is identified as to its substream, which implicitly specifies the QoS objective for that packet. JSCC then takes a specific form: each source coder segments its packets according to QoS objectives and then associates that packet with the appropriate substream. The system is also cognizant of the traffic it has generated for each substream.

For example, the two-level priority schemes in video coding can be thought of as associating high-importance packets with one substream and low-importance packets with another substream. The higher-importance substream would have a QoS requirement associated with a lower loss probability than the lower-importance substream. The bitway can exploit the relaxed QoS requirement of the lower-importance substream to achieve a higher traffic capacity.

More generally, the service, knowing the QoS to be expected on the substreams, can associate packets with substreams in a way that results in acceptable subjective quality. The bitway, knowing the QoS expectations and rates, can allocate its internal resources, such as buffer capacity, power, etc., in a way that maximizes the traffic capacity. In the absence of the substream structure, the bitway would have to provide the tightest or most expensive QoS requirements to the entire stream in order to achieve the same overall subjective quality.

Fortunately, the substream model is consistent with the most important existing protocols. Substreams have been proposed in ST-II, the second-generation Internet Stream Protocol [47]. Version 6 of the Internet Protocol (IP) includes the concept of a *flow*, which is similar to our substream, by including a *flow label* in the packet header [48]. Asynchronous transfer mode (ATM) networks incorporate virtual circuits (VC), and associate QoS classifications with those VCs, where nothing precludes a single application from using multiple VCs. The notion of separating packets into (usually two) priority classes is often proposed for video [49, 50, 51], usually with the view toward congestion networks. In particular, a two-level priority for video paired with different classes of service in the transmission has been proposed for broadcast HDTV [36]. We believe that substreams should be the universal paradigm for interconnection of services and bitways for a number of reasons elucidated below, and especially the support of wireless access and encryption. By attaching the name "medley gateway" to such an interface, we are not implying that a totally new gateway function is required. Rather, we propose this name as a common terminology applying to these disparate examples of a similar concept.

The distinction between a stream composed of a set of *substreams* and a *set of streams* with different QoS requirements is that a stream composed of substreams can have the rate and QoS descriptions of the substreams "linked together." For example, a service could specify that the temporal rate characteristics of all of its substreams are highly correlated (or that two substreams' rates are very negatively correlated).[6] Also, a service could request "loss priorities" from a bitway by explicitly specifying that packets on one substream should not be discarded while packets on another substream are delivered successfully. Another example is a service that requests one substream be given a higher "delay priority" than another substream to ensure that packets on the first substream experience less delay than packets on the second.

Combining the bitway and service abstractions, the overall situation is illustrated in Figure 6.7. Each of a set of substreams receives different QoS attributes

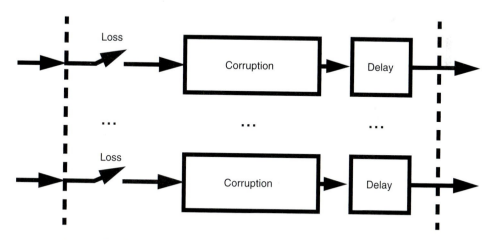

Figure 6.7   The abstracted bitway for a set of substreams.

[6]A little thought confirms that correlations can be expected among substreams emanating from a single source. For example, in video, high- and low-motion information will typically have negatively correlated rates attributes.

and hence a quantitatively different bitway model. As discussed in Section 6.4.4, this bitway abstraction opens up some interesting new possibilities in the design of services.

6.2.4 Loosely Coupled Joint Source/Channel Coding

The abstractions introduced in the bitway model make opportunities in loosely coupled JSCC more transparent. The JSCC functionality is now divided between the services layer and the bitway layer. The bitway, in an effort to maximize its traffic-carrying capacity, does the following:

- Affords each packet a loss or corruption probability lower than required by the QoS specified for the substream with which it is associated.
- Takes maximum advantage of the delay flexibility afforded by the QoS on a per packet basis. This is a new opportunity in JSCC not anticipated in previous approaches and is discussed further in Section 6.4.3.

Simultaneously, the service attempts to maximize the subjective quality afforded to the application or user within the constraints of the agreed flowspec. For example, packets less sensitive to delay are associated with a substream with a relaxed delay specification.

In the absence of the substream structure, the bitway would have to provide the tightest or most expensive QoS requirements to the entire stream in order to achieve the same overall subjective quality (and the QoS needs of different packets may vary over several orders of magnitude, e.g., for MPEG video headers vs. chrominance coefficients). Thus, the bitway has the option of exploiting the substream structure to achieve more efficient resource use through JSCC. Critically, substreams are generic and not associated with any particular service (for example, audio or video or a specific audio or video coding standard).

The medley gateway model does impose one limitation on JSCC. It does not include a feedback mechanism by which information on the current conditions in the bitway layer can be fed back to affect the services layer. Nor does it allow the flowspec to be time-dependent. One can envision scenarios under which this "closed loop" feedback would be useful. One example is flow control, in which compression algorithms are adjusted to the current information-carrying capacity of a time-varying channel. Another is an adjustment of compression algorithms to the varying bit-error rate due to time-varying noise or interference effects. We do not include these capabilities because we question their practicality in the general situation outlined in Section 6.3, where the services layer implementation may be geographically separated from the bitway entity in question, implying an unacceptably high delay in the feedback path. This does raise questions of how to deal with time-varying wireless channels. In this case, we do not preclude feedback within a bitway link, adapting various functions like FEC and power control in an attempt to maintain a fixed QoS.

6.2.5 Substream-Based Transport Protocols

Within the services layer, there is typically a transport protocol, the purpose of which is to serve as a "translation" between the characteristics of the bitway layer and the differentiated needs of the applications. An example in the Internet would be TCP, which adds, among other things, retransmission and acknowledgment to ensure reliable and in-sequence delivery of packets for data applications. TCP adds significant delay and hence may not be appropriate for critical interactive CM services, especially those that do not require reliable delivery, as discussed in Section 6.1.4. The question then arises, what is the appropriate transport protocol? Since the transport protocol by definition impacts the QoS as seen by the application, of course constrained by the QoS provisioned by the bitway, any consideration of QoS and JSCC must incorporate the transport protocols.

The multiple substream model of the medley gateway has several characteristics that may particularly require a transport protocol:

- Each substream is simply a packet delivery mechanism. There are configured QoS attributes relative to packet loss, corruption, and delay, but no guarantee that a particular packet is actually delivered, nor any guarantee that packets are delivered in the same order in which they were transmitted.
- The substreams are asynchronous at the receiver, implying that there is no predictable temporal relationship between their delivery. This asynchrony is quite deliberate, since the substreams may have different delay QoS attributes.

Should the application desire more control, for example, guaranteed packet delivery, guaranteed order of delivery, or synchronization of the substreams at the receiver, an appropriate transport protocol can be invoked. A general architecture for such a protocol in the context of a medley bitway is shown in Figure 6.8. The medley transport protocol presents a service with N substreams to the application and makes use of M medley bitway substreams. While it would be likely that $M = N$, that is not necessarily the case, as will be illustrated by a concrete example in Section 6.4.4. The general purpose of the transport protocol is to modify the semantics of the bitway so as to ensure ordered delivery or synchronization among

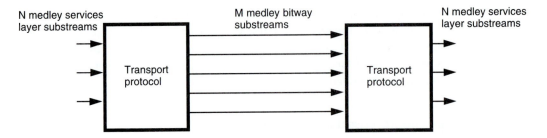

Figure 6.8   Abstract medley transport protocol making use of a medley bitway.

substreams. The transport protocol may require a feedback stream not shown, for example, carrying acknowledgments or requests for retransmission. Generally, QoS attributes such as reliability and delay will be substantially affected by the transport protocol. For example, ordered delivery will add delay since it will be necessary to buffer packets arriving before one or more of their predecessors, and synchronization of substreams will make all substreams suffer the worst-case bitway delay.

Thus far, to our knowledge there have been no proposals for substream-based transport protocols, although of course an existing transport protocol such as User Datagram Protocol (UDP) could be used independently on each substream. Medley transport protocols should be a profitable area for research.

6.2.6 Scalability and Configurability Issues

Requiring services and bitway to be mixed and matched arbitrarily puts a much greater burden on each. A service entity that is designed to utilize any bitway entity must exhibit scalability to deal, for example, with both a broadband backbone bitway and a wireless access bitway. Similarly, the bitway must be prepared to allocate its resources differently for different rate attributes and QoS requirements, for example, to provision both an audio and a video service.

In the loosely coupled JSCC model, we envision a connection establishment *flowspec negotiation* between service source and sink and bitway. These three entities can iterate through multiple sets of flowspec attributes to find a set that balances service performance and connection cost goals well. For example:

- The service entity, based on subjective quality criteria requested by the application, requests a flowspec of the bitway. However, since the bitway can conceivably be anything between a broadband backbone and a wireless access link, this request may be wildly unrealistic or too expensive.

- The bitway entity determines the feasibility of the flowspec, and if feasible, passes back to the service a cost[7] associated with that flowspec.
- The service and the bitway exchange sets of flowspecs, choosing flowspecs that improve service performance or reduce cost. This process results in a final agreed-to flowspec.
- Both the service and the bitway configure themselves. This action implies appropriate resource allocation by the bitway to guarantee that the agreed flowspec will be achieved. This also implies that the service chooses signal processing operations and a substream decomposition to conform to the rate attributes in the flowspec to maximize subjective quality subject to the agreed-to flowspec.

During the negotiation, the bitway entity must aggregate QoS impairments and costs for all sublinks in a connection. Suitable modeling of these impairments, their costs, and their aggregation will be a big challenge.

Unfortunately, an establishment negotiation in this form is not advisable for multicast connections because it is not scalable and is likely to be overly complex. The service source would have to negotiate with an unknown number of service sinks and associated bitway entities—potentially, thousands. Further, sinks will typically be joining and leaving the multicast connection during the session, and it is not reasonable to expect that the source will reconfigure (e.g., new compression algorithm or substream decomposition) on each of these events, especially if doing so requires all other sinks to reconfigure as well. Mobility of receiving terminals raises similar issues.

To avoid this problem, we can envision a different form of configuration for multicast groups, with some likely compromise in performance, inspired by the multicast backbone (MBone) [52] and the resource reservation protocol (RSVP) [2]. The service source generates a substream decomposition that is designed to support a variety of bitway scenarios, unfortunately without knowing in advance their details. It also indicates to the bitway (and potential service sinks) information as to the trade-offs between QoS and subjective quality for each substream. Each new sink joining the multicast group subscribes to this static set of substreams, based on resources and subjective quality objectives, and this subscription would be propagated to the nearest feasible splitting point. The QoS up to this splitting point would be predetermined, but possibly configurable downstream to the new sink. The resulting compromise—the bitway QoS to each new sink would

---

[7]In a commercial context, cost is likely to be expressed in monetary terms, or in other contexts, it may be expressed in other terms. In any case, an important component of the cost will be the traffic capacity implications of the requested flowspec.

be constrained by the QoS to the splitting point established by other sinks—could be mitigated by allowing a sink to request the addition of bitway resources upstream from the splitting point.

For wireless links, the ability to configure QoS depends on assumptions about the propagation environment and terminal speed. For well-controlled, indoor, wireless local-area networks, it may be relatively easy to configure reproducible QoS attributes because low terminal speeds will result in a slowly varying propagation condition due to fading. In that case, the media-access layer may be able to adaptively maintain a reasonably constant QoS over time. In contrast, in wide-area wireless networks with high terminal velocities and high carrier frequencies, fading and shadowing effects may make it extremely difficult to adaptively maintain QoS. In this case, it may be more appropriate to view the configured QoS as an objective rather than as a guarantee and to assume that there is an outage probability (possibly configurable but at least provided to the application); that is, probability that the QoS objective is violated. Many intermediate situations are surely possible.

6.3 Edge vs. Link Architecture for Service Layer

In Section 6.2, we addressed the problem of separating the designs of the service from the bitway while leaving open most possibilities for JSCC. Our motivation was to allow the flexibility to substitute freely the service or bitway realizations. In this section, we consider a related set of issues in the provision of CM services through two or more heterogeneous subnets. Many of the issues addressed in Section 6.2 become more important.

Consider two basic architectures, illustrated in Figure 6.9, for concatenated *links*, where each link corresponds to one homogeneous bitway subnet. For example, in wireless access to a broadband network, the wireless subnet would constitute one bitway link, and the broadband subnet would constitute the second link. The distinction between the *link architecture* and the *edge architecture* is whether or not a services layer is included within each subnet.[8] The back-to-back services layers in the link architecture include, for CM services like audio and video, a decompression signal processing function followed by a compression function. These functions together constitute a *transcoder*.

A transcoder is functionally equivalent to introducing an analog link in the network by converting from one compressed digital representation to analog (by decompressing and D/A converting) and then converting from analog to a different compression standard (by A/D converting with a synchronous sampling clock and compressing). This virtual analog link circumvents many interoperability

---

[8]The term *edge* denotes the entry point to the first bitway link in the network.

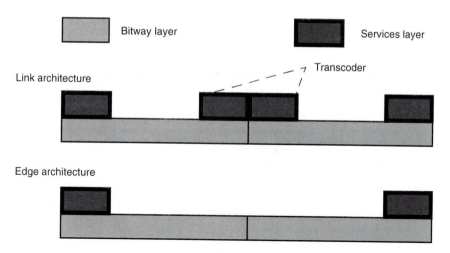

Figure 6.9 Contrast of link and edge architectures for concatenated heterogeneous subnets, where the former includes a transcoder function at the gateway between the two subnets.

issues, such as ensuring an allocation of the same bit rate on all network links. In some situations, such as introducing new technology into a legacy system, transcoding may be unavoidable. For example, in the telephone network, in a call from a wired to a digital cellular telephone, one voice coding technique (8 kHz sampled PCM) is used on the wired network and another vector-sum-excited linear prediction (VSELP) coding, in the case of the North American IS-54 standard) is used on the digital cellular subnet [18]. This approach is for valid and important technical reasons; namely, the desire for spectral efficiency on the digital cellular subnet, resulting in more aggressive compression (traded off against implementation cost and reduced subjective quality) and the need for JSCC between speech coder and wireless link.

In the Internet, services layers like TCP or UDP are realized at the edges. That is, the Internet uses today the edge architecture. In extensions to the Internet architecture for realizing CM services, under some limited circumstances transcoders are proposed to be included within the network[9]; thus, the Internet is currently proposed to move (at least to a minor extent) in the direction of the link architecture.

In designing a new infrastructure, it should be possible to avoid transcoders, and we believe very desirable as well. We argue that the edge architecture is superior and should be adopted for the future. The resulting architecture is structured as in Figure 6.10. Compression, encryption, and QoS negotiation occur in the services layer, at the network edge. FEC, modulation, and resource reservation occur in the

---

[9]Specifically, in multicast CM services, bridges incorporating transcoder functionality are allowed at the nodes of the multicast spanning tree as a method of accommodating heterogeneous downstream terminals [53]. We later propose an alternative method to solve this problem.

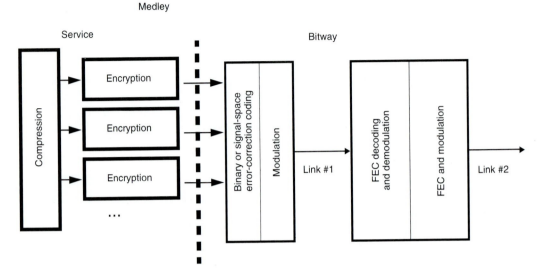

Figure 6.10 A proposed architecture including compression, encryption, and error-correction encoding. Encryption is performed independently on each substream so that the QoS after decryption can be controlled within the bitway.

bitway layer, at each network link. In favor of this architecture, we mention five factors:

- *Privacy and security.* The link architecture is incapable of providing privacy by end-to-end encryption under user control since an encrypted signal cannot be transcoded. The best that can done is encryption on a link basis by the service provider(s), with no ability for the user to verify that encryption has been performed. In our opinion, this problem alone should preclude serious consideration of the link architecture.[10]

- *Openness to change.* The edge architecture is open to substitution of different services layers at the network edge (user terminal or access point). This flexibility leads to an economically viable method to upgrade services over time, as well as to introduce new ones, as discussed in Section 6.3.2.

- *Performance.* The link architecture suffers from the accumulation of delay and subjective impairment through tandem compressions and decompressions of the CM signal. This problem has already become serious in digital cellular telephony, where each transcoder introduces on the order of 80 ms of delay. This delay is inherent to the compression in transcoding since compression

---

[10]This issue is discussed in [54], where it is pointed out that end-to-end encryption alone allows routing information to be intercepted internal to the network. It is argued that a combination of both end-to-end and link-by-link encryption is the most secure option.

is at its heart a time-averaging process. In more complicated heterogeneous scenarios, delay could become unacceptable for delay-sensitive interactive applications.

- *Complexity.* The edge architecture has a number of challenges, as discussed later, but overall we believe it substantially reduces the complexity of establishment and configuration.

- *Mobility.* The link architecture embeds considerably more state within the network associated with the realization of a CM service, creating additional requirements for migration of state when terminals are mobile (requiring the movement or the disestablishment/establishment of multipoint connection spanning trees).

The impairment accumulation and mobility considerations are relatively straightforward; the following subsections discuss the other factors.

6.3.1 Privacy and Security

Encryption is an important requirement for privacy and for preventing unauthorized interception in intellectual property protection schemes. Of course, encryption is accompanied by a host of other issues, such as key management and distribution, that are beyond the scope of this chapter. Not all services will require encryption, but the network architecture has to accommodate it for those cases where it is required. One issue with encryption is whether it is applied end-to-end or only on selected links of the network (especially the wireless link). End-to-end encryption affords much greater protection to the user than does link-by-link encryption because keys are known only to the user. Since encryption deliberately hides the syntactical and semantic components of the signal, no compression can be incorporated into the network where streams may be encrypted, including the conversion from one compression standard to another.

Encryption techniques can be divided into two classes [13]. In the *binary additive stream cipher*, which is used, for example, to encrypt the speech signal in the Groupe Speciale Mobile (GSM) digital cellular system [41], the data is exclusive-or'ed with the same random-looking *running key generator (RKG)* bit sequence at the transmitter and receiver. The RKG depends on a secret key known to both encryption and decryption [42]. The stream cipher has the advantage of no error multiplication and propagation effects; however, the loss of synchronization of the RKG will be catastrophic. A *block cipher* algorithm applies a functional transformation to a block of data plus a secret key to yield the encrypted block, and an inverse function at the receiver can recover the data if the key is available. For example, the Data Encryption Standard (DES) applies its transformation to blocks of 64 bits by means of a 56-bit key [43]. In fact, error propagation within the block is considered a desirable property of the cryptosystem; that is, block ciphers should on average modify an unpredictable half of the plaintext bits whenever a single ciphertext bit is changed (this behavior is called the "strict avalanche property" [44]). There are variations on block ciphers with feedforward and feedback of delayed ciphertext blocks that cause error propagation beyond a single block.

Another important issue is the impact of encryption on QoS. In a general, integrated-services multimedia network, encryption techniques with error propagation should not be used for CM services since this use will preclude strategies designed to tolerate errors rather than mask them.

Neither a stream nor block cipher is ideal: the stream cipher introduces serious synchronization issues in a packet network, while the block cipher has severe error propagation. This is a serious issue for wireless multimedia networks that should be addressed by additional research.

6.3.2 Openness to Change

The history of signal processing operations like compression is one of relentless improvement in performance parameters like compression ratio, subjective quality, and delay. Algorithm improvements are usually accompanied by increasing processing requirements, but fortuitously the cost/performance of electronics also advances relentlessly. It would, given this history, be unfortunate to "freeze" existing performance attributes through an architecture that discourages or precludes change.

In this regard, the argument in favor of the edge architecture is economic: it allows the latest technologies to be introduced into the network in an economically viable way. New signal processing technologies are initially more expensive than older technologies since innovation and engineering costs must be recovered and because such technologies usually require more processing power. In the edge architecture, the services signal processing is realized within the user terminal or at a user access point; that is, it is provisioned specifically for the user. Only users who are willing to pay the cost penalty of the latest technology need upgrade, and only services desired for that user need be provisioned.

In contrast, in the link architecture, service signal processing elements are embedded widely throughout the network. At each point, it is necessary to deploy all services, including the latest and highest performance. The practical result is that for *any* users to benefit from a new technology, a global upgrade throughout the network is required. If only a relatively few users are initially willing to pay the incremental cost of new technology, there is no business case for this upgrade. There is also the question of who provisions and pays for the substantial infrastructure that would be required to support transcoding in or near base stations.

Further, the link architecture also requires that, for N different performance flavors of a given service, internal nodes in the network be prepared to implement $N(N-1)$ distinct transcoders, and that nodes be prepared to implement all distinct services. These nodes must also implement all feasible encryption algorithms and must be cognizant of encryption keys. In contrast, in the edge architecture, the edge nodes need implement only those services desired by the local application/user and only the flavor with the highest desired performance (as well as fallback to lower-performance flavors).

Past examples of these phenomena are easy to identify. The voiceband data modem, realized on an end-to-end basis, has advanced through two orders of magnitude in performance while simultaneously coming down in price. Users desiring state-of-the-art performance must pay a cost increment, but other users need not upgrade. If a higher-performance modem encounters a less capable modem, it falls back to that mode. Realizing the older modem standards introduces only a tiny cost increment since the design costs have been amortized and the lower performance standard requires *less* processing power. This example provides a useful model of how a service can be incrementally upgraded over time in the edge architecture. It illustrates that each terminal does *not* have to implement a full suite of standards, but rather needs to include only those services and the highest performance desired by the local application or user, as well as fallback modes to all lower speed standards.[11] The fallback modes, which are the only concession to interoperability with other terminals, do not add appreciable cost—the lower performance standards require *less* processing power, and the design costs of the older standards have been previously amortized.[12] The total end-to-end performance will be dictated by the lowest performance at the edges.

Contrast this behavior with the circuit-switched telephone network, where the same voice coding has been entrenched since the dawn of digital transmission. This voice coding standard is heavily embedded in the network, which was originally envisioned as a voice network. Today, it would be feasible to provide a much-improved voice quality (especially in terms of bandwidth) at the same bit rate, but there is no economically viable way to introduce this technology into the network.

The ability for users or third-party vendors to add new or improved services, even without the involvement of the network provider, is perceived as one of the key features of the Internet, leading to the rapid deployment of new capabilities such as the World Wide Web (WWW). In the link architecture, the necessary involvement of network service providers in services is undoubtedly a major barrier to innovation within the services domain of functionality, such as signal compression.

---

[11]This style of progressive improvement in a standard is already evident in MPEG, where MPEG-2 decoders are required to also be MPEG-1 compliant. Numerous examples of this methodology exist in other domains, such as microprocessor architectures.

[12]This argument is valid for software-defined standards, and is valid today in audio applications and will be increasingly valid in video as well.

6.3.3 Performance and Efficiency

In this section we examine the relative performance and efficiency of the architectures discussed above.

6.3.3.1 Modulation

Packet loss, corruption, and delay are especially problematic in wireless communication, which is limited by low bandwidth, time-varying multipath fading and interference. Moving to higher radio frequencies may alleviate spectrum congestion, but this solution is attended by a host of other difficulties, including susceptibility to atmospheric attenuation from fog and rain. Thus, the application of physical layer signal processing to combat impairments is more important in a wireless context than in a wireline backbone network.

We can distinguish between two categories of physical layer signal processing. As discussed in earlier chapters, transmit waveform shaping, spatial and temporal diversity-combining, and equalization are commonly employed wireless physical layer techniques that strive to unilaterally improve the reliability of *all* information bits. These methods trade off reliability for signal processing overhead (hardware cost), delay, and reduced traffic capacity. In contrast, power control and signal-space codes (such as trellis-coded modulation and shell mapping) form a class of methods with an additional dimension: given a fixed amount of resources—transmit power in the case of the former, hardware complexity and radio spectrum in the case of the latter—these strategies can selectively allocate impairments to different information bits, thereby controlling QoS. The ability to match transmit power to loss and corruption requirements is essential for maximizing capacity in wireless cellular networks, where excessive power creates unnecessary interference to other users.

The substream abstraction (Section 6.2.3) enables this matching. As shown in Figure 6.11, each bitway link is obligated to maintain the structure of the medley gateway at its output. That is, the medley gateway is the interface between service

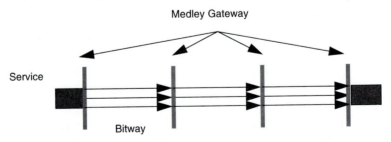

Figure 6.11 Each bitway link maintains the structural integrity of the medley gateway, making the structure available to downstream bitway links.

and bitway layers and *also* the interface between distinct bitway entities. This is why we call it a *gateway*—since it serves as a common protocol interface between heterogeneous bitway subnets. The substream structure is visible to each bitway link, which is able to allocate resources and to tailor its modulation efficiently in accordance with JSCC.

6.3.3.2 Forward Error-Correcting Coding

With end-to-end FEC, the transport may provision reliability by applying binary FEC on an end-to-end basis. The FEC-encoded information bitstream may then transparently pass through multiple transport links to be FEC decoded (again at the network edge) by the sink. Priority encoding transmission (PET) [37] is an example of the end-to-end FEC approach. The goal of PET is to provide reliable transmission of compressed video over wired packet networks. The primary error mechanism in these networks is congestion, leading to excessively delayed packets or buffer overflow. PET combats congestion losses by using a form of binary FEC known as erasure coding: B packets are encoded into N packets, such that all B packets can be recovered from any B out of N packets successfully received. This approach is appealing for its simplicity and, in fact, can be efficient for a homogeneous wired transport whose links have very similar characteristics.

An alternative architecture is to provision reliability by applying physical layer signal processing on a link-by-link basis: each link is made cognizant of the loss and corruption requirements of an application, then applies its own specific physical layer processing to meet these requirements. This is the architecture we prefer, for reasons we now elaborate.

The link-by-link approach to providing reliability necessitates a mechanism for QoS negotiation on each link so that it can configure itself in accordance with the requirements of a particular source stream. With binary FEC, we can do away with QoS negotiation altogether, which is certainly an advantage. However, consider the reliability requirements of the wireless link. Since the wireless link has no knowledge of the source requirements, it must be designed for a homogeneous QoS across all streams. There are two options. First, the designer can adjust the reliability for the *most* stringent—or most demanding—source. This conservative design approach will, for less stringent source requirements, overprovision resources such as bandwidth and power and overly restrict interference, thus reducing traffic capacity. While we don't expect this to be a major issue on backbone networks, it may severely decrease the capacity of the bottleneck wireless access network if there is a wide variation in source QoS needs.

The second option is to design the wireless access link to be suitable for the *least* stringent source requirement and compensate by FEC on an end-to-end basis, as in PET. This option introduces several sources of inefficiency for heterogeneous networks with wireless access. As noted earlier, binary erasure codes are very efficient in combatting congestion-based losses in a wired packet network. Their performance is significantly poorer in a wireless environment, where packets are likely to be corrupted due to the inherently high bit-error rate (BER). For the 10^{-2} uncoded BER typical of high mobility wireless and a packet size of $M = 120$ bits, the packet loss rate after applying an ($N = 8, B = 2$) erasure code is:

$$\Pr[\text{packet error}] = \sum_{i=N-B+1}^{N} p^i(1-p)^{N-i} = 0.26, \tag{6.1}$$

where

$$p = 1 - (1 - \text{BER}_{\text{uncoded}})^M. \tag{6.2}$$

This performance is a modest improvement over the uncoded packet error rate of 70% but was achieved by quadrupling the bandwidth. A better way of lowering losses is to attempt to reduce the corruption rate instead, for example, by using a convolutional code. For the same bandwidth expansion as an (8,2) erasure code, a convolutional code can lower the BER by two orders of magnitude [38, 39], thereby lowering the packet error rate to 1%. On a wireless link with rapid fading, this code will usually be accompanied by interleaving to turn correlated errors into quasi-independent errors. While convolutional coding may be attractive for wireless links, it is largely ineffective in a wired network, where losses are congestion-derived. Thus, it will be necessary with end-to-end FEC to concatenate different codes and interleaving designed to combat all anticipated error mechanisms, implying that the wireless link traffic will be penalized by redundancy intended for the other links in the network as well as its own.

The link-by-link architecture also permits us to apply physical layer signal processing techniques not possible in end-to-end binary FEC. In end-to-end binary FEC, one has no choice but to perform a *hard* decision on the information bits as they cross from one link to another. On a wireless link, we have control over the modulation and demodulation process and thus can apply *soft* decoding to the information bits. Hard decisions made prior to the final decoding result in an irreversible loss of information. This loss is equivalent to a 2 dB drop in the signal-to-noise ratio (SNR) [40], and the effect on loss and corruption is cumulative across multiple links. In addition, we can consider making the FEC and interleaving adaptive to the local traffic and propagation conditions on the wireless link.

Overall, active configuration of the QoS on a wireless link based on individual source requirements will substantially increase traffic capacity. The price to be paid is an infrastructure for QoS negotiation and configuration and the need to provision variable QoS in a wireless network; the latter issue is addressed further in Section 6.4.1. Fully quantifying this benefit requires further research since it depends on the characteristics and requirements of the source traffic, as well as on the benefits of variable QoS.

6.3.4 Complexity and Resource Allocation

Both the link and the edge architectures raise important issues in resource allocation in session establishment. In both cases, for CM services the overriding objective is to obtain acceptable and controllable subjective quality in the audio or video service. Subjective quality is measured objectively by attributes such as frame rate and resolution (for video), bandwidth (for audio), and delay (for both video and audio). It is also measured by other factors more difficult to characterize, such as the perceptual impact of artifacts introduced in the process of decompression by information corrupted or discarded in the service (i.e., in the compression) and in the bitway (packet losses), and also artifacts introduced by corruption in the bitway.

Inherently, resources belong to individual links, not to end-to-end connections. However, the QoS negotiation between the services and bitway layers that establishes each link's resource use can be done end-to-end or link-by-link.

In the link architecture, overall subjective quality objectives must be referenced back to the individual links, since each link will contribute artifacts that impair subjective quality (such as quantization, blocking effects, error masking effects, etc.). These artifacts will accumulate across links in a very complicated and difficult-to-characterize way. (For example, how is a blocking or masking artifact represented in the next compression/decompression stage?) It is relatively straightforward to partition objective impairments like delay among the links. Other objective attributes like frame rate, bandwidth, and resolution will be dictated by the worst-case link and are thus also straightforward to characterize. Subjective impairments due to loss and corruption artifacts will, however, be very difficult, if not impossible, to characterize in a heterogeneous bitway environment. Simple objective measures like mean-square error are fairly meaningless in the face of complex impairments like the masking of bitway losses. Thus, as a practical matter, it will be very difficult to predict and control end-to-end subjective quality.

The situation in the edge architecture is quite different. The first step is to generate an aggregated bitway model for all the concatenated bitway links. That is, the loss models for the individual links must be referenced to a loss model for the overall connection, and similarly for corruption and delay. There are no doubt serious complications in this aggregation, for example, correlations of loss mechanisms in successive links due to common traffic. Nevertheless, this is a relatively straightforward task susceptible to analytical modeling. Once this analysis is done, the aggregate bitway model must be related back to service subjective quality, much in the fashion of a *single* link in the link architecture. There is no need to characterize the accumulation of artifacts in multiple compression/decompression stages. Accurate prediction and control of subjective quality in the edge architecture should be feasible, and this is an additional advantage over the link architecture.

6.3.5 Multicast Connections

The problem of multicast connections is illustrated in Figure 6.12. With heterogeneous receiving terminals, or heterogeneous subnets, we may need different representations (say, with different bandwidth or resolution) of the CM service after a splitting bridge, but to conserve bitway resources we want to share a common stream before the bridge. An obstacle to this is encryption, which will hide the syntax of the originating stream. One solution is to locate transcoding at the bridge, preceded by decryption and followed by encryption, but this solution introduces all the disadvantages of the link architecture.[13] The medley gateway provides a framework for the solution to this problem, as shown in Figure 6.13. At the point

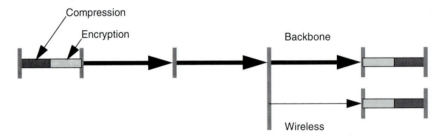

Figure 6.12   Illustration of a multicast connection with heterogeneous receiving terminals.

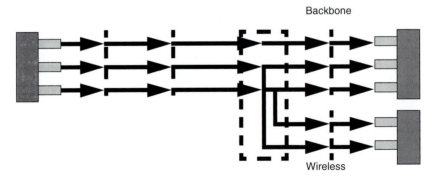

Figure 6.13   The medley gateway substream structure allows multicast bridging to be performed within the bitway layer, without interference from encryption.

[13]A transcoding approach to multicast splitting is currently envisioned as part of the future Internet architecture [53].

where two representations are split, a (not necessarily proper) subset of the medley substreams is extracted for each downstream branch. From the perspective of the bitway, different endpoint terminals receive different subsets of substreams, with the great simplification that the bridging function can be accomplished entirely within the bitway layer. If each substream is independently encrypted, encryption does not interfere with this bridging function. Substreams in this context play a similar role to multicast groups in the MBone [52].

Support for heterogeneous terminals in the edge architecture presents to the service a well-defined design problem: perform a layered compression, such that a subset of the substreams embodies a minimal representation of the source, and the additional substreams provide additional information (higher resolution, higher sampling rate, etc.) to terminals with greater capabilities. Thus, in the edge architecture, the substream structure is used for three distinct but complementary purposes:

- *JSCC*. It allows the service to present to each bitway entity, in a generic fashion separated from particular service standards, the differing QoS requirements of different packets, thus allowing the bitway to efficiently allocate its resources.
- *Layered coding*. It allows the service to decompose its layered encoding in a way that is also generic and visible to the bitway layer, so that the splitting function required in multicast connections with heterogeneous terminals can be performed entirely within the bitway.
- *Privacy and security*. Independent encryption of the substreams allows the privacy and security of end-to-end encryption without interfering with either JSCC or multicast splitting.

6.4 DESIGN EXAMPLES

JSCC for the medley gateway model has serious implications to the design of the wireless bitway, source coding, and services. In this section, we illustrate this by a few design examples.

6.4.1 Variable QoS in Wireless Bitways

In Section 6.2, we discussed two design philosophies for multimedia networks: homogeneous QoS in the network with end-to-end unequal error protection (UEP), and active configuration of QoS within the individual links of the network. In the latter case, the approach is to adjust the QoS, and hence resources, of individual links in accordance with the requirements of each constituent stream. As we believe the latter is a superior approach for wireless bitway design, we now discuss the provisioning of *variable QoS*. Our development complements that of Chapter 5. For generality, we consider a medley bitway (with substreams), although these results would apply equally well to the more restrictive case of QoS provisioned on a stream rather than substream granularity.

A variable QoS medley bitway has two design challenges: provide flexibility in loss/corruption/delay attributes with a substream granularity, and exploit the configured characteristics to maximize the traffic capacity. We illustrate the design issues for a wireless direct-sequence code-division multiple-access (CDMA) system. We focus on achieving variable reliability and ignore the issue of variable delay discussed elsewhere [60].

There are two handles for controlling reliability in CDMA: FEC and power control. Achieving variable reliability with FEC would require UEP. Many forms of UEP coding have been developed, notably algebraic codes for UEP and embedding asymmetric constellations in trellis-coded modulation [61]. However, the number of different levels of reliability provided by these techniques is limited, and it is difficult to apply them to hierarchical UEP. Variable rate (VR) convolutional codes have also been suggested for UEP coding of speech [62]. By adopting UEP, we can increase the reliability of a substream by adjusting the coding rate, at the expense of bandwidth expansion from the redundancy. Signal space codes such as trellis-coded modulation are particularly attractive for wireless networks because they provide redundancy without increasing bandwidth. However, it is difficult to generate (by varying the constellation size) the trellis equivalent of a variable rate convolutional code since, as Ungerboeck has shown, virtually all of the coding gain is attained by doubling the alphabet size [40].

As developed in Chapter 5, an alternative mechanism to control reliability QoS would be to adjust the signal-to-interference + noise ratio (SINR) by adjusting the transmitted power, taking into account the interference from other user's traffic being simultaneously transmitted on different spreading codes. Of course, it is beneficial for overall traffic capacity to minimize the transmitted power for any given user in order to minimize the interference to other users. Hence, overall traffic capacity is maximized by achieving, for each packet, no greater SINR than is necessary to meet the QoS objective. If we use power control only, then in a CDMA system we will be transmitting to a particular user at less than 100% duty cycle whenever the bit rate required by that user is less than the peak rate enabled by the chip rate and processing gain. Power control has some important advantages:

- It is easy to achieve variable reliability over a wide range of bit error rates by changing the power level. The power level can be dynamically varied to track time-varying channel conditions and maintain relatively constant reliability for any substream.

- Power control is transparent to the receiver, requiring no special processing.
- For CDMA, using a fixed-rate code to provide maximum coding gain without bandwidth expansion is easy, since the spectrum has already spread; i.e., channel coding and spreading can be combined to provide redundancy without bandwidth expansion [63].

The first question is whether, ignoring implementation issues, it is most advantageous to use UEP or power control. To address this issue, let us examine UEP and power control from an information theory perspective, using the following elementary calculation. In a CDMA system, focus on a single user's data and approximate the total interference as white Gaussian noise with power spectrum N_0, and let the bandwidth be B. Let P be the average transmitted power for this particular user, and transmit with a duty cycle $\gamma \leq 1$. Then, the transmitted power during that duty cycle is P/γ. The channel capacity using this duty cycle is γ times the capacity if we transmitted at this same power level at 100% duty cycle, where the latter is $B \cdot \log\left(1 + \frac{P/\gamma}{2N_0B}\right)$. Thus, the overall channel capacity is

$$C = \gamma B \cdot \log\left(1 + \frac{P}{2N_0\gamma B}\right), \tag{6.3}$$

which is precisely the same as the capacity of a channel with bandwidth γB with 100% duty cycle transmission and transmitted power P. Since this capacity is maximum for $\gamma = 1$, we conclude that it is advantageous to transmit with 100% duty cycle in order to minimize the average transmitted power P for a fixed bit rate C. To minimize P, and hence the interference to other users, we should always transmit at 100% duty cycle by adding channel coding redundancy as necessary. Intuitively, it is advantageous to use coding to increase the duty cycle of transmission to 100%, regardless of the required bit rate, and take advantage of the coding gain to reduce the average transmit power.

Thus, information theory teaches us that in an interference-dominated wireless channel such as CDMA, it is best to use *coordinated* UEP and power control. Either UEP or power control in isolation is suboptimum at the fundamental limits. If the bit rate for a given CDMA spreading code is low, coding redundancy should be added and the transmitted power simultaneously reduced. The bitway coding and power control layers in a wireless cellular bitway design are illustrated in Figure 6.14. A set of substreams is applied to a coding layer that is cognizant of the propagation characteristics of the channel and that configures itself to provide the

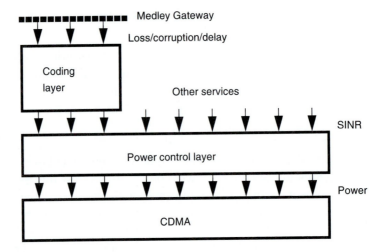

Figure 6.14   Internal architecture of a CDMA bitway.

negotiated QoS contract.[14] Based on the coding selected, each substream is associated with a required SINR. The power control layer then associates a transmitted power with each substream, taking into account a maximum power requirement, the SINR requirement for each substream, and which substreams currently have packets awaiting transmission.[15]

Finally, let us quantify the capacity gain in a wireless CDMA system attainable by using joint power and error control for variable QoS. The traffic capacity is the amount of traffic that the system can support, subject to the (possibly distinct) QoS demands of the traffic. Let M be the number of users and K_m be the number of substreams of user m. Specify a user (CDMA spreading code) by subscript m and a substream by subscript k. The SINR experienced by the kth substream of user m on the uplink is

$$\text{SINR}_{\text{experienced}} = \frac{G_m x_{k,m}}{\sigma^2 + I_m^{\text{intra}}} ,$$ (6.4)

where G_m is the path loss from user m to the base station; $x_{k,m}$ is the transmit power assigned to substream k; I_m^{intra} is the intracell interference experienced by user m, and σ^2 is the lump sum of background noise and intercell interference experienced at the base station.

[14]For example, in the design of the trellis code, using a metric different from the Euclidean metric typically used for the additive Gaussian noise channel is advantageous for Rayleigh fading channels [64].

[15]In practice, we would also like to schedule the most opportune time for a packet transmission to take advantage of allowed delay jitter. This problem is considered elsewhere [60].

The intracell interference experienced by a substream of user m is

$$I_m^{\text{intra}} = \sum_{\substack{n=1 \\ n \neq m}}^{M} f_{m,n} G_n \sum_{k=1}^{K_n} \beta_{k,n} x_{k,n}. \tag{6.5}$$

where $\beta_{k,m}$ is an indicator function, equaling one if substream k of user m is currently active, zero otherwise. Each user's traffic can be decomposed into multiple substreams, but the substreams are statistically multiplexed together onto one user stream so that only one of the user's substreams is active at any time. In (6.5), $f_{m,n}$ is the partial correlation coefficient (or degree of nonorthogonality) between channels of users m and n: because signals from different mobile users travel through different multipath channels to reach the base station, perfect orthogonality between user codes may be lost and $f_{m,n}$ may be non-zero. Uplink transmission is inherently asynchronous, so $f_{m,n}$ is well modeled by f, the correlation between *random* signature sequences, with $E[f] = 2/3$ [65].

The indicator function of a substream as it evolves over time $\beta_{k,m}(t), t \in [0, \infty)$ is a random process. Let $\overline{\beta}_{k,m}$ denote the long-term time average of $\beta_{k,m}(t)$; e.g., $\overline{\beta}_{k,m} = 1/4$ if the average bit rate of substream k is 500 kbps and it belongs to a 2-Mbps user stream. We assume ergodicity in the mean, so that $E[\beta_{k,m}(t)] = \overline{\beta}_{k,m}$. This assumption is based on the intuition that at any given time slot, the probability that you receive a packet from substream k equals the average rate of that substream, divided by the aggregate rate of the user stream to which it belongs. The expected value of the total power is then

$$E[P] = E\left[\sum_{m=1}^{M} \sum_{k=1}^{K_m} \beta_{k,m}(t) x_{k,m}\right] = \sum_{m=1}^{M} \sum_{k=1}^{K_n} \overline{\beta}_{k,m} x_{k,m}. \tag{6.6}$$

Our objective is to minimize the average overall power $E[P]$ while promising each substream that the *expected value* of the SINR it experiences will meet or exceed the desired SINR:

$$\text{minimize } E[P] = \sum_{m=1}^{M} \sum_{k=1}^{K_n} \overline{\beta}_{k,m} x_{k,m} \quad \text{such that} \tag{6.7}$$

$$\forall k, m, \qquad \frac{G_m x_{k,m}}{\sigma^2 + E[f] \sum_{\substack{n=1 \\ n \neq m}}^{M} G_n \sum_{j=1}^{K_n} \overline{\beta}_{j,n} x_{j,n}} \geq \text{SINR}_{k,m}, \tag{6.8}$$

$$x_{k,m} \geq 0, \text{SINR}_{k,m} > 0$$

In (6.8), $\text{SINR}_{k,m}$ is the SINR requested by substream k of user m, and the inequality in 6.8 implies that the expected value of the SINR achieved at the receiver must equal or exceed the desired SINR.

It can be shown [66] that the feasible capacity region of a CDMA system is given by

$$\gamma < 1, \tag{6.9}$$

where

$$\gamma = \sum_{m=1}^{M} \gamma_m, \tag{6.10}$$

$$\gamma_m = E[f] \sum_{k=1}^{K_m} \alpha_{k,m}, \text{ and} \tag{6.11}$$

$$\alpha_{k,m} = \frac{\text{SINR}_{k,m} \overline{\beta}_{k,m}}{1 + E[f] \sum_{j=1}^{K_m} \overline{\beta}_{j,m} \text{SINR}_{j,m}} \tag{6.12}$$

$$k = 1,\ldots, K_m, \quad m = 1,\ldots, M.$$

The left hand side of (6.9), γ, represents the load of the system. The closer γ is to unity, the closer the system is to violating the QoS requirements for all users and substreams. If the QoS requirements are too stringent, then the interference will be too great and no solution exists, regardless of the transmit power. Examining (6.12) we see that for a cellular wireless system whose capacity is interference-limited, the "cost" of transmitting an information substream is the product of its reliability requirement (specified by an SINR) and bandwidth requirement (specified by its average rate $\overline{\beta}$).

As noted earlier, information theory suggests that the application of VR coding to adapt a variable-rate information source to a bandlimited channel is necessary in order to maximize capacity. In a CDMA system, a user is associated with a code, and the bandwidth afforded by the code is shared by the user's substreams. Each substream is allocated a time-averaged fraction of the bandwidth, $\overline{\beta}$. To ensure optimal capacity in the information theoretic system, the bandwidth associated with the code should be used at 100% duty cycle; i.e., in (6.9)–(6.12), $\{\overline{\beta}_{j,n}\}$ should satisfy

$$\sum_{j=1}^{K_n} \overline{\beta}_{j,n} = 1 \quad \text{for all users } j, \tag{6.13}$$

with the SINR requirements of all substreams adjusted for the resulting coding gain. We note that the interference-limited capacity result as stated in (6.9)–(6.12) is quite general and can also be applied toward suboptimal systems that do not achieve 100% utilization of the channel bandwidth.

We can apply these results to find the capacity gain of power control for variable QoS with substreams over power control without substreams. In the absence of substreams, each stream's reliability requirement would be equal to the reliability need of its worst-case (most error-sensitive) information component. A fine-grained substream architecture therefore achieves a capacity gain of

$$
\text{capacity gain} = \frac{\displaystyle\sum_{\text{user } m}\ \sum_{\text{stream } k}\ \sum_{\text{substream } i} \alpha_{k,m}^{(\text{max})}\overline{\beta}_{i,k,m}}{\displaystyle\sum_{\text{user } m}\ \sum_{\text{stream } k}\ \sum_{\text{substream } i} \alpha_{i,k,m}\overline{\beta}_{i,k,m}}\ , \tag{6.14}
$$

where $\alpha_{k,m}^{(\text{max})}$ corresponds to the maximum SINR requirement among the substreams of stream k.

6.4.2 MPEG-2 Compression

The International Organization for Standardization's Moving Pictures Experts' Group (ISO/MPEG) has developed several well-known audiovisual compression standards:

- MPEG-1 is designed for VCR-quality audio and video compression and for delivery via reliable media such as CD-ROMs [16].
- MPEG-2 is designed for high-quality broadcast applications, including entertainment, remote learning, electronic publishing, and more [17].
- MPEG-3 was intended to address high-definition television, but this effort was folded into MPEG-2.
- MPEG-4, just beginning development, is intended to address wireless interactive multimedia coding and transmission [67]. MPEG-4 currently borrows some technology from the ITU-T Study Group 15, for example, SG15's wireless multiplexing protocol, H.245.

MPEG-2 is not designed for wireless multiaccess but rather for wireless and wired broadcast; as such, the decisions made in the design of MPEG-2 are quite different from those that would be made for a wireless multiaccess system. Broadcast channels differ from wireless multiaccess channels in that they are noise rather than interference-limited and thus can be provisioned to deliver lower bit-error rates and very much lower burst-error rates. Thus, error resiliency tools for broadcast applications should be designed and optimized differently than those for wireless multiaccess systems. Nevertheless, it is instructive to review the error resiliency features included in MPEG-2. In Section 6.4.3, we illustrate a much different approach.

MPEG-2 is a service layer standard and today is used with a range of bitways: direct broadcast satellite, digital switched line, cable television, ATM, and more. As a suite of service layer standards, MPEG-2 does not provide bitway QoS-enhancing functionality such as data interleaving, selective packet discard, or FEC. Still, MPEG-2 does provide a range of functionality to help resynchronize and recover quickly from bitway errors and to configure to trade off efficiently between bandwidth, delay, loss, and service quality.

6.4.2.1 Features

MPEG-2 contains three subparts that define bitstream formats: The audio specification defines a compressed representation of a multichannel audio signal. The video specification defines a compressed representation of a moving picture sequence. The systems specification defines, among other things, how to multiplex multiple audio, video, and data streams into a single packetized bit stream.

The audio and video compression methods defined by MPEG-2 contain many predictive coding steps. For example, the video specification includes interframe motion-compensated DPCM, predictive coding of motion vectors within a frame, predictive coding of DCT brightness coefficients within a frame. Predictive coding, which represents a signal's difference from a predicted value rather than representing the signal value directly, removes signal redundancy very effectively but suffers from error propagation. Errors cause predictive decoders to incorrectly render data which is used in future predictions. These subsequent predictions with errors lead to more incorrectly decoded data; this propagates errors throughout spatio-temporally nearby audio or video. Fortunately, MPEG-2 allows an encoder to define "resynchronization points" almost as often or infrequently as the designer desires, to trade between bandwidth efficiency and rapid error recovery. At a resynchronization point, signal values are coded directly rather than via a predictor.

Both the audio and video compression algorithms use Huffman coding, which uses short bitstrings to represent frequently occurring values and long bitstrings to represent infrequent values. A property of Huffman codes is that they cannot be decoded without some context—knowledge of the bit position of the start of some bitstring. Huffman codes also suffer error propagation: one bit error destroys the decoder's context, and the decoder may incorrectly decode a long sequence of values. To alleviate this problem, the audio and video specifications both define "startcodes," which are patterns in the bit streams that decoders can find easily and at which Huffman codes are known to be at the start of a bitstring. These startcodes do consume a small but non-zero percentage of audio and video stream bandwidth, but they enable decoder recovery after a transmission error. The video specification allows bitstream encoders to insert "slice" startcodes at either a default minimum rate or at a higher rate to enable faster-than-default error recovery.

The systems specification defines a bitstream format ("transport streams") that consists of short (compared to TCP/IP) fixed-length packets. Short packets ensure that if a receiver identifies a packet as corrupted, comparatively little data is suspect. Fixed-length packets facilitate rapid identification of packet delineators after errors.

MPEG-2 transport streams can contain "duplicate packets." A duplicate packet is a copy of the previous packet with the same source identifier. Duplication is a simple (but not efficient) flavor of FEC. Combined with bitway interleaving, duplication greatly reduces the occurrence of uncorrectable burst errors, however. A sensible strategy is to duplicate all packets that contain the highest-level start-codes.

Example

Suppose we transmit an MPEG-2 transport stream with 30 packets per second that contain critical video headers (one such packet per picture). Suppose the bitway layer uses interleaving to ensure that the probability of bit error is uniform and identically distributed and uses FEC to ensure a probability of bit error of 10^{-8}. The probability that any packet contains an error is $1 - (1 - 10^{-8})^{188 \times 8} = 1.5 \times 10^{-5}$ since there are 188 bytes in an MPEG-2 transport stream packet. This means that without duplicate packets, we can expect a packet with a critical header to be corrupted about once every 2,216 seconds or every 37 minutes. If we send duplicate packets for each packet with a critical header, the probability that both an original critical packet and its duplicate are corrupted is $(1 - (1 - 10^{-8})^{188 \times 8})^2 = 2.3 \times 10^{-10}$. Both an original critical packet and its duplicate are lost about once every 1.5×10^8 seconds, or once per 4.7 years.

The MPEG-2 systems specification defines a timing recovery and synchronization method that utilizes two types of timestamps. "Program clock references" (PCRs) allow receivers to implement accurate phase-locked loop clock recovery. Stringent limitations on the encoder clock frequency accuracy and drift rate allow decoders to identify and discard corrupted PCRs. Video frames and audio segments are identified by decode/presentation timestamps (DTS/PTS), which tell the decoder the proper time to decode and present the associated video and audio data. Since each frame has a fixed (and known to the decoder) duration, there is a lot of redundancy in the DTS/PTS values. However, the small amount of bandwidth spent on PCRs and DTS/PTSs helps decoders properly prefill their input buffers and properly synchronize their video and audio outputs after errors.

The MPEG-2 systems, video, and audio subparts specify only bitstream formats. Another part of MPEG-2, the RTI, defines constraints on real-time delivery of systems bit streams to actual decoders. The RTI defines a method for measuring the delay jitter present when a bit stream is delivered to a decoder; this approach aids in ensuring interoperability between bitstream providers and decoders. The RTI does not mandate a specific delay jitter value; the designer chooses a value suitable for the system, e.g., 50 μs for a low-jitter connection between a digital VCR and a decoder, or more than 1 ms for an international ATM connection. The RTI defines decoder memory requirements and bitstream delivery constraints based on the chosen jitter value. The specification of decoder memory requirements as a function of delay jitter is very important for many MPEG-2 applications, where decoder cost, largely driven by memory, is the biggest determinant of commercial viability.

6.4.2.2 Scalability Tools

Hierarchical or layered coders are good candidates for use with QoS-impaired bitways. As shown in Figure 6.15, the base layer of a hierarchical coder represents the input signal at some coarse fidelity. Higher layers code the residual between the base layer decoded output and the original input; the output of the base layer combined with higher-layer decoded output is more accurate than the base layer output alone. For a given service quality level, the aggregate bit rate of a good hierarchical coder is close to that achievable by the best nonhierarchical coders.

A hierarchical coder's outputs have different QoS requirements; often it is acceptable for only the base layer to be decoded for short periods of time. Thus, only the base substream requires high QoS in order to achieve acceptable application quality; if data from other substreams is lost, the decoded signal is corrupted, but not catastrophically.

A simple example of a hierarchical video coder simply transmits the most significant bits of a picture's pixels on one substream and the least significant bits on another. If some of the least significant bits are lost, affected picture regions appear coarsely quantized but certainly recognizable.

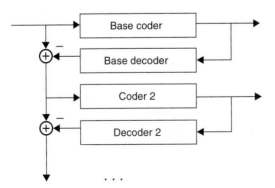

Figure 6.15 Structure of a hierarchical encoder.

The MPEG-2 video specification defines several much more sophisticated ways by which an encoder can produce a two- or three-layer hierarchically encoded bit stream. MPEG-2 video allows an encoder to generate hierarchical bit streams that decompose a signal into low frame-rate vs. high frame-rate components, small picture size vs. large picture size, or low image fidelity vs. high image fidelity (called "SNR scalability") components. Several of these decompositions can be used in tandem as well.

MPEG-2 video defines another scalability tool, called *data partitioning*. Data partitioning defines how to split a non-hierarchically-encoded bit stream into two substreams. The high-priority substream contains video headers and other important syntax elements such as motion vectors. The low-priority bit stream contains lower-priority syntax elements such as high-frequency discrete cosine transform (DCT) coefficients. The encoder can choose its definition of high-priority and low-priority syntax elements to achieve its best trade-off between high-priority bandwidth, high-priority QoS, low-priority bandwidth, and low-priority QoS.

6.4.3 JSCC for Delay: Delay-Cognizant Video Compression

Having described MPEG, an established standard, let us now illustrate a dramatically different approach motivated by the need for efficient use of wireless channels. The design of today's CM services are a holdover from the circuit switched era, when bitways did not introduce significant delay jitter. Existing compression standards for both audio and video thus assume a fixed-delay transport model, imposing on the bitway the need to emulate a fixed-delay circuit. This emulation requires the artificial delay of packets arriving early. In the context of delay-critical interactive services, it seems intuitively unattractive to artificially add delay, and one wonders if it is not possible to take advantage of these early-arriving packets.

Since substreams have different delay characteristics, it is inherent that they are asynchronous at the receiving terminal. They can be resynchronized by an appropriate medley transport protocol, but *not* resynchronizing them allows the delay characteristics of the bitway to be exploited. To this end, the medley gateway abstractions offer several key benefits:

- The service is allowed to specify different delay characteristics for different substreams. This is a direct way for the service to control *which* packets arrive earlier and which arrive later, which in turn makes the differential delays more useful. The model also encourages the medley bitway to deliver certain packets earlier, whereas conventional approaches do not.

- For a fixed traffic capacity, bitways generally trade higher reliability for increased delay (through techniques like FEC, interleaving, retransmission, etc.). The medley gateway model allows the service to explicitly control as well as exploit this trade-off and to force this trade-off to be quantitatively different for different packets.

- Since the overall delay is no longer determined by the *worst-case* delay, the bitway worst-case delay can be relaxed, which can in turn be traded for increased traffic capacity through traffic smoothing.

What we have just described is JSCC in the delay dimension. By making the source coding delay-cognizant, that is, segmenting its information into delay classes, we hope to achieve a more desirable combination of perceptual delay and traffic capacity.

An early example of delay-cognizant video coding is *asynchronous video* [68], a coding technique that exploits variations in the temporal dimension of video to segment information into distinct delay classes. We leave the details to other references [68] but illustrate the basic idea in Figure 6.16. The frame is block-segmented into different delay and reliability classes in accordance with motion estimation (three classes are shown). These different classes are allowed to be offset at the receiver by one or more frames in the reconstruction process. The hope is that low-motion blocks are less susceptible to multiple-frame delay jitter at the receiver than are high-motion blocks and that the user perception of delay will be dominated by the high-motion blocks. If this is the case, low-motion blocks can be assigned to a medley bitway substream with a relaxed delay objective, and the bitway can

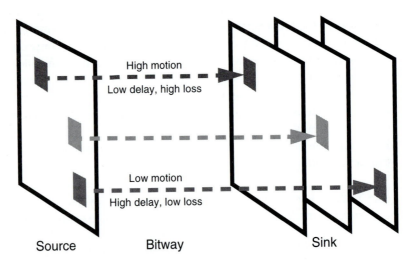

Figure 6.16 Asynchronous video as an example of delay-cognizant video coding. Blocks of video are reconstructed in different frames at the sink, based on motion segmentation.

exploit this relaxed delay jitter objective to achieve higher traffic capacity. In addition, high-motion blocks are assigned to substreams with a relaxed reliability objective since the motion tends to subjectively mask losses or corruption. Fortuitously, the bitway naturally provides precisely the needed exchange of higher reliability for higher delay.

6.4.4 Multiple-Delivery Transport Protocol

The importance of the transport protocol as a way to change the characteristics of the bitway to the benefit of the application was discussed in Section 6.2.5. An example of a transport protocol tailored to the needs of CM services is a multiple delivery service [70, 71]. Interference-limited wireless access links typically have two undesirable characteristics: restricted bandwidth and low reliability. Error control techniques to compensate for the latter increase the rate (for redundancy or retransmissions), and this rate increase trades unfavorably against delay because of the restricted bandwidth. Thus, reliable delivery mechanisms increase delay substantially. This will be problematic for interactive applications, for example refreshing a graphics window in a WWW browser. If graphics are treated as a pixel map (as in the InfoPad™ system [72]), it is advantageous to display corrupted information early, but it is also important that corruption artifacts do not stay on the screen indefinitely (asymptotic reliability). This corruption control can be accomplished *without a traffic capacity penalty* by exploiting the redundancy needed anyway to deliver two or more copies of a packet to the receiver, each with increasing reliability, as illustrated in Figure 6.17. The application delivers a single copy of each packet to the transport protocol. The transport delivers, in general, more than

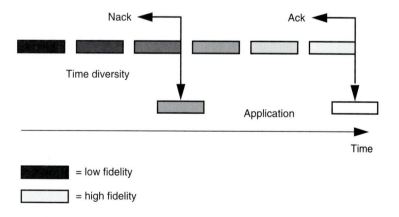

Figure 6.17   A multiple-delivery transport protocol.

one copy of the packet to the receiver, where it is agreed that each copy has statistically greater fidelity (fewer bit errors) than the previous copy. Internally, the transport protocol can utilize packet combining techniques, where it transmits the packet as many times as required and caches all received renditions of the packet, delivering to the application its best estimate of the packet based on all the cached information. Acknowledgments built into the protocol allow the number of transmissions to be adjusted dynamically to channel conditions. This protocol has proven useful for video [32].

Protocols such as the multiple-delivery transport protocol should also have a mechanism to purge stale packets; that is, packets that will not be used by the receiver if they are delivered.

6.5 CONCLUDING REMARKS

The most important point of this chapter is that in an integrated-services multimedia network, it is advantageous to take an overall systems perspective, rather than designing wireless access networks in isolation. We have seen how, by coordinating the design of the backbone network, terminals, and servers with the wireless access network, greater traffic capacity can be achieved subject to subjective quality objectives. At the same time, it is important to adhere to good principles of complexity management and ensure that the different parts of the multimedia network are made modular and as independent as possible, with appropriate levels of scalability and configurability. Achieving modularity requires a carefully crafted architecture for the network. We have proposed the medley gateway model based on substreams or flows (supported by existing or emerging protocols in both IP and ATM networks) as a basic unifying principle of the architecture. Once an architectural approach is chosen, many opportunities for research in the various modules open up. We have illustrated the design of a video source coder, a variable QoS wireless CDMA media access layer, and a transport protocol within the context of this architecture.

The considerations covered in this chapter suggest many opportunities for research, which include:

- The design of medley services that take advantage of the medley bitway (such as delay/loss trade-offs and segmentation) and that have the needed level of scalability and configurability.
- The design of medley bitways that maintain the structural integrity of the substreams, which have the ability to configure to different impairment profiles for different substreams and which exploit the substream structure to achieve higher traffic capacity.

- An understanding of JSCC, as constrained by the structure of the medley gateway model. Similarly, an understanding of the design of hierarchical compression algorithms for multicast heterogeneous terminals, as constrained by the same substream structure.

- An understanding of issues inherent in the aggregation of concatenated bitway links for CM services.

- Development of negotiation strategies for resolving the trade-off between subjective quality vs. bitway QoS and cost.

- The upgrade of signaling systems to provide the needed capabilities in support of the edge architecture, including aggregation of bitway links and negotiation between the endpoint terminals and the aggregated links.

REFERENCES

[1] G. M. Parulkar and J. S. Turner, "Towards a framework for high-speed communication in a heterogeneous networking environment," *IEEE Network*, vol. 4, no. 2, March 1990, pp. 19–27.

[2] L. Zhang, S. Deering, D. Estrin, S. Shenker, and D. Zappala, "RSVP: a new resource ReSerVation Protocol," *IEEE Network*, vol. 7, no. 5, Sept. 1993, pp. 8–18.

[3] National Research Council, Computer Science and Telecommunications Board, *Realizing the Information Future; The Internet and Beyond.* Washington, DC: National Academies Press, 1994.

[4] D. G. Messerschmitt, "Complexity management: A major issue for telecommunications," *International Conference on Communications, Computing, Control, and Signal Processing in Honor of Prof. Thomas Kailath*, A. Paulraj, V. Roychowdhury, C. Schaper, editors, Boston: Kluwer Academic Press, 1996.

[5] *IEEE Commun. Mag.*, Issue on Intelligent Networks, Feb. 1992, vol. 30:2.

[6] M. Lengdell, J. Pavon, M. Wakano, M. Chapman, and others, "The TINA network resource model," *IEEE Communications Magazine*, March 1996, vol. 34:3, pp. 74–79.

[7] F. Dupuy, C. Nilsson, and Y. Inoue, "The TINA consortium: toward networking telecommunications information services," *IEEE Commun. Mag.*, Nov. 1995, vol. 33:11, pp. 78–83.

[8] D. G. Messerschmitt, "The future of computer telecommunications integration," *IEEE Commun. Mag.*, Special Issue on "Computer-Telephony Integration," vol. 34, no. 4, April 1996, pp. 66–69.

[9] D. G. Messerschmitt, "The convergence of telecommunications and computing: What are the implications today?," *Proc. IEEE*, vol. 84, no. 8, August 1996, pp. 1167–1186.

[10] D. G. Messerschmitt, "Convergence of telecommunications with computing," Special Issue on Impact of Information Technology, *Technology in Society*, Elsevier Science Ltd., to appear.

[11] A. G. MacInnis, "The MPEG systems coding specification," *Signal Processing: Image Communication*, April 1992, vol. 4:2, pp. 153–159.

[12] C. Holborow, "MPEG-2 Systems: a standard packet multiplex format for cable digital services," *Proc. 1994 Conference on Emerging Technologies, Society of Cable Television Engineers*, Phoenix, AZ., Jan. 1994.

[13] J. Massey, "An introduction to contemporary cryptology," *Proc. IEEE*, Special Section on Cryptology, vol. 76, no. 5, May 1988, pp. 533–549.

[14] D. J. Le Gall, "The MPEG video compression algorithm," *Signal Processing: Image Communication*, April 1992, vol. 4:2, pp. 129–140.

[15] D. J. Le Gall, "MPEG: a video compression standard for multimedia applications," *Commun. ACM*, April 1991, vol. 34:4, pp. 46–58.

[16] ISO/IEC Standard 11172, "Coding of Moving Pictures and Associated Audio at up to about 1.5 Mbits/s." (MPEG-1).

[17] ISO/IEC Standard 13818, "Generic Coding of Moving Pictures and Associated Audio." (MPEG-2).

[18] J. E. Natvig, S. Hansen, and J. de Brito, "Speech processing in the pan-European digital mobile radio system," *Proc. GLOBECOM*, Dallas, TX, vol. 2, Nov. 1989, pp. 1060–1064.

[19] T. H. Meng, B. M. Gordon, E. K. Tsern, and A. C. Hung, "Portable video-on-demand in wireless communication," *Proc. IEEE*, April 1995, vol. 83:4, pp. 659–680.

[20] D. Ferrari, "Real-time communication in an internetwork," *J. High Speed Networks*, 1992, vol. 1:1, pp. 79–103.

[21] D. Ferrari, "Delay jitter control scheme for packet-switching internetworks," *Computer Communications*, July-Aug. 1992, vol. 15:6, pp. 367–373.

[22] S. Vembu, S. Verdu, and Y. Steinberg, "The source-channel separation theorem revisited," *IEEE Trans. Inform. Theory*, Jan. 1995, vol. IT-41:1, pp. 44–54.

[23] A. Goldsmith, "Joint source/channel coding for wireless channels," *1995 IEEE 45th Vehicular Technology Conference. Countdown to the Wireless Twenty-First Century*, Chicago, IL, July 1995.

[24] S. McCanne, and M. Vetterli, "Joint source/channel coding for multicast packet video," *Proceedings International Conference on Image Processing*, Washington, DC, 23–26 Oct. 1995.

[25] M. Khansari, and M. Vetterli, "Layered transmission of signals over power-constrained wireless channels," *Proceedings International Conference on Image Processing*, Washington, DC, vol. 3, Oct. 1995, pp. 380–383.

[26] M. W. Garrett, and M. Vetterli, "Joint source/channel coding of statistically multiplexed real-time services on packet networks," *IEEE/ACM Trans. Networking*, Feb. 1993, vol. 1:1, pp. 71–80.

[27] K. Ramchandran, A. Ortega, K. M. Uz, and M. Vetterli, "Multiresolution broadcast for digital HDTV using joint source/channel coding," *IEEE J. Select. Areas Commun.*, Jan. 1993, vol. 11:1, pp. 6–23.

[28] L. Yun, and D. G. Messerschmitt, "Power control and coding for variable QoS on a CDMA channel," *Proc. IEEE Military Communications Conference*, vol. 1, Oct. 1994, pp. 178–182.

[29] N. Chaddha, and T. H. Meng, "A low-power video decoder with power, memory, bandwidth and quality scalability," *VLSI Signal Processing, VIII*, Sakai, Japan, Sept. 1995, pp. 451–460.

[30] T. H. Meng, E. K. Tsern, A. C. Hung, S. S. Hemami, and others, "Video compression for wireless communications," *Virginia Tech's Third Symposium on Wireless Personal Communications Proceedings*, Blacksburg, VA, June 1993.

[31] R. Han, L. C. Yun, and D. G. Messerschmitt, "Digital video in a fading interference wireless environment," *IEEE Int. Conf on Acoustics, Speech, and Signal Processing*, Atlanta, GA, May 1996.

[32] J. M. Reason, L. C. Yun, A.Y Lao, and D. G. Messerschmitt, "Asynchronous video: coordinated video coding and transport for heterogeneous networks with wireless access," *Mobile Computing*, H. F. Korth and T. Imielinski, editors, Boston: Kluwer Academic Press, 1995.

[33] D. Anastassiou, "Digital television," *Proceedings of the IEEE*, April 1994, vol. 82:4, pp. 510–519.

[34] S.-M. Lei, "Forward error correction codes for MPEG2 over ATM," *IEEE Transactions on Circuits and Systems for Video Technology*, vol. 4, no. 2, April 1994, pp. 200–203.

[35] L. Montreuil, "Performance of coded QPSK modulation for the delivery of MPEG-2 stream compared to analog FM modulation," *Proc. National Telesystems Conference*, 1993.

[36] R. J. Siracusa, K. Joseph, J. Zdepski, and D. Raychaudhuri, "Flexible and robust packet transport for digital HDTV," *IEEE J. Select. Areas Commun.*, Jan. 1993, vol. 11:1, pp. 88–98.

[37] A. Albanese, J. Blomer, J. Edmonds, and M. Luby. "Priority encoding transmission," International Computer Science Institute Technical Report TR-94–039, Berkeley, CA, Aug. 1994.

[38] J. Hagenauer, N. Seshadri, and C. E. Sundberg, "The performance of rate-compatible punctured convolutional codes for digital mobile radio," *IEEE Trans. Commun.*, July 1990, vol. 38:7, pp. 966–980.

[39] J. Proakis, *Digital Communications*, 2nd edition., New York: McGraw Hill, 1989.

[40] G. Ungerboeck, "Channel coding with multilevel/phase signals," *IEEE T. on Information Theory*, Jan. 1982, vol. IT-28:1, pp. 55–67.

[41] E. Zuk, "GSM security features," *Telecommunication Journal of Australia*, 1993, vol. 43:2, pp. 26–31.

[42] D. Gollmann, and DW. G. Chambers, "Clock-controlled shift registers: a review," *IEEE J. Select. Areas Commun.*, May 1989, vol. 7:4, pp. 525–533.

[43] M. Smid, and D. Branstad, "The Data Encryption Standard: past and future," *Proc. IEEE*, Special Section on Cryptology, vol. 76, no. 5, May 1988, pp. 550–559.

[44] A. F. Webster and S. E. Tavares, "On the design of S-boxes," in *Advances in Cryptology - Proc. of CRYPTO '85*, H. C. Williams, editor, New York: Springer-Verlag, 1986, pp. 523–534.

[45] P. Haskell, "Flexibility in the Interaction Between High-Speed Networks and Communication Applications," Electronics Research Laboratory Memorandum UCB/ERL M93/83, University of California at Berkeley, Dec. 2, 1993.

[46] E. A. Lee and D. G. Messerschmitt, *Digital Communication*, 2nd Edition, Boston: Kluwer Academic Press, 1993.

[47] C. Topolcic, "Experimental Internet Stream Protocol, Version 2 (ST-II)," Internet RFC 1190, October 1990.

[48] C. Bradner and A. Mankin, "The Recommendation for the IP Next Generation Protocol," Internet Draft, NRL, October 1994.

[49] P. Pancha and M. El Zarki, "MPEG coding for variable bit rate video transmission," *IEEE Commun. Mag.*, May 1994, vol. 32:5, pp. 54–66.

[50] P. Pancha and M. El Zarki, "Prioritized transmission of variable bit rate MPEG video," *Proc. IEEE GLOBECOM*, vol. 2, 1992, pp. 1135–1139.

[51] Q.-F. Zhu, Y. Wang, and L. Shaw, "Coding and cell-loss recovery in DCT-based packet video," *IEEE Transactions on Circuits and Systems for Video Technology*, June 1993, vol. 3:3, pp. 248–258.

[52] H. Eriksson, "MBone: the multicast backbone," *Commun. ACM*, Aug. 1994, vol. 37:8, pp. 54–60.

[53] "RTP: A transport protocol for real-time applications," Internet Engineering Task Force Draft Document, July 18, 1994.

[54] B. Schneier, *Applied Cryptography, Protocols, Algorithms, and Source Code in C*. New York: John Wiley & Sons, 1994.

[55] H. R. Liu, "A layered architecture for a programmable data network," *Proc. Symposium on Communications Architectures & Protocols*, Austin, TX, March 1983.

[56] D. A. Keller and F. P. Young, "DIMENSION AIS/System 85-the next generation meeting business communications needs," *Proc. IEEE International Conference on Communications*, Boston, MA, vol. 2, June 1983, pp. 826–830.

[57] H. O. Burton and T. G. Lewis, "DIMENSION AIS/System 85 system architecture and design," *Proc. IEEE International Conference on Communications*, Boston, MA, vol. 2, June 1983, pp. 831–836.

[58] J. M. Cortese, "Advanced Information Systems/NET 1000 service," *Proc. IEEE International Conference on Communications*, Boston, MA, vol. 2, June 1983, pp. 1070–1074.

[59] S. A. Abraham, H. A. Bodner, C. G. Harrington, and R. C. White, Jr., "Advanced Information Systems (AIS)/Net 1000 service: technical overview," *Proc. IEEE INFOCOM*, San Diego, CA, April 1983, pp. 87–90.

[60] L. C. Yun, *Transport of Multimedia on Wireless Networks*, Ph.D. dissertation, University of California at Berkeley, 1995.

[61] L.-F. Wei, "Coded modulation with unequal error protection," *IEEE Trans. Commun.*, Oct. 1993, vol. 41:10, pp. 1439–1449.

[62] R. V. Cox, J. Hagenauer, N. Seshadri, and C.-E. W. Sundberg, "Subband speech coding and matched convolutional channel coding for mobile radio channels," *IEEE Trans. Signal Processing*, Aug. 1991, vol. 39:8, pp. 1717–1731.

[63] A. J. Viterbi, "Very low rate convolutional codes for maximum theoretical performance of spread-spectrum multiple-access channels," *IEEE J. Select. Areas Commun.*, May 1990, vol. 8:4, pp. 641–649.

[64] D. Divsalar and M. K. Simon, "The design of trellis coded MPSK for fading channels: performance criteria," *IEEE Trans. Communications*, Sept. 1988, vol. 36:9, pp. 1004–1012.

[65] M. B. Pursley, "Performance evaluation for phase coded spread-spectrum multiple access communication - Part I: System Analysis," *IEEE Trans. Commun.*, August 1977, vol. COM-25, pp. 795–799.

[66] L. C. Yun and D. G. Messerschmitt, "Variable quality of service in CDMA systems by statistical power control," *Proc. International Conf. on Communications*, Seattle, WA, vol. 2, June 1995, pp. 713–719.

[67] ISO/IEC standard 14496, "Coding of Audio-Visual Objects." (MPEG-4).

[68] A. Lao, J. Reason, and D. G. Messerschmitt, "Layered asynchronous video for wireless services," *IEEE Workshop on Mobile Computing Systems and Applications*, Santa Cruz, CA., Dec. 1994.

[69] M. Kawashima, C.-T. Chen, F.-C. Jen, and S. Singhal, "Adaptation of the MPEG video-coding algorithm to network applications," *IEEE Transactions on Circuits and Systems for Video Technology*, Aug. 1993, vol. 3:4, pp. 261–269.

[70] R. Han and D. G. Messerschmitt, "Asymptotically reliable transport of text/graphics over wireless channels," *Proc. Multimedia Computing and Networking*, San Jose, CA, January 1996.

[71] R. Han and D. G. Messerschmitt, "Asymptotically reliable transport of text/graphics over wireless channels," *ACM/Springer Verlag Multimedia Systems Journal*, to appear 1998.

[72] S. Sheng, and others, "A portable multimedia terminal," *IEEE Commun. Mag.*, Dec. 1992, vol. 30:12, pp. 64–75.

Acknowledgments

The authors appreciate the contributions of their colleagues Jonathan Reason, Richard Han, and Yuan-Chi Chang to the insights reported in this chapter. This research is supported by Bell Communications Research, Pacific Bell, Tektronix, MICRO, and the Defense Advanced Research Projects Agency.

7

Multiresolution Joint Source-Channel Coding

Kannan Ramchandran
Martin Vetterli

With the rapid growth of wireless communications systems, there is an increasing demand for wireless multimedia services. Wireless image and video transmission, an essential component of wireless multimedia, poses a particularly important challenge that deserves attention for several reasons. In particular, image and video transmission is the main system bottleneck because it requires far more bandwidth than the transmission of other information sources such as speech or data. Moreover, it is a more difficult problem due to the inherent complexity of the coding methods.

While Chapter 6 discussed multimedia networks broadly, in this chapter we specifically focus on the image/video source. For such sources it is important to consider the end-to-end image communication problem very carefully. Current communication link designs are typically mismatched for wireless video because they fail to take into account important considerations such as (i) highly time-varying source and channel characteristics, (ii) high source tolerance to channel loss, and (iii) unequal importance of transmitted bits. This mismatch comes from a long history of data communications, where loss of bits is disastrous (e.g., data files), and where every bit is equally important. Some salient attributes of the image/video source are summarized below.

- The performance metric is the delivered visual quality (e.g., mean-square error, or more correctly, the perceptual distortion) of the source due to both source quantization *and* channel distortion under constraints of fixed system resources like bandwidth and transmission energy. This metric contrasts with commonly used performance criteria like bit error rates, which are appropriate for data communications.

- The existence of unequal error sensitivities in a typical video bit stream (e.g., bits representing motion vectors or synchronization/header information versus bits representing high-frequency, motion-compensated error residue or detail in textured image areas) emphasizes the desirability of a layered approach to both source and channel coding, and calls for a rehauling of conventional "single resolution" digital transmission frameworks with their multiresolution counterparts.

- Due to the stringent delay requirements of synchronous video applications, there is a need to include finite buffer constraints (efficient rate control strategies). These requirements will influence the choice of error control coding strategies like forward error correction (FEC) versus automatic repeat request (ARQ) techniques, as well as more powerful hybrid FEC/ARQ choices [3].

Wireless image and video transmission comes in various application-driven varieties. Of particular interest are point-to-point transmission and broadcast/multicast scenarios. A key challenge of wireless transmission is the time-varying nature of the mobile point-to-point transmission on the one hand and the diversity of channels in the broadcast case on the other hand. There are fundamental differences between the time-varying point-to-point case and the broadcast case. In the broadcast channel, the same information is sent and read by two or more different users, each seeing another corrupted version of the data that was sent at the same time. In the point-to-point time-varying channel, the data sent is seen by the same user but is corrupted differently at different times. As we will see, despite these differences, similar techniques, based on multiresolution ideas, can be used to achieve good performance at reasonable cost.

The problem of transmitting image and video signals naturally involves both source coding and channel coding. The image or video source has an associated rate-distortion characteristic that quantifies the optimal trade-off between compression efficiency and the resulting distortion. The classical goal of source coding is to operate as closely as possible to this rate-distortion bound. Then comes the task of reliably transmitting this source-coded bit stream over a noisy channel, which is characterized by a channel capacity that quantifies the maximum rate at which information can be reliably transmitted over the channel. The classical goal of channel coding is to deliver information at a rate that is as close to the channel capacity as possible. For point-to-point communications with no delay constraints, one can theoretically separate the source and channel coding tasks with no loss in performance. This was shown by Shannon [2] and is discussed shortly in more depth. In the presence of delay constraints, however, or for broadcast or multicast scenarios, Shannon's results do not apply and there is a need for closer interaction between the source and channel coding functions. A key issue in the design of efficient image and video transmission systems for these cases involves the investigation of joint design of these source and channel coding components. This issue is the theme of this chapter. Our perspective is primarily from a lossy source coding viewpoint but with an eye on channel coding issues. We emphasize the elegance of multiresolution-based techniques for both source coding and channel coding and show how one can efficiently "match" these resolutions to improve the overall system performance, measured as the delivered image quality.

The framework for both source coding and channel coding is, of course, Shannon's groundbreaking work on information theory [2]. An important information-theoretic result, which goes back to Shannon's work, is the *separation principle*, alluded to earlier, which allows the separate design of a source compression/decompression scheme and a channel coding/decoding scheme, as long as the source code produces a bit rate that can be carried by the channel code. The separation principle is illustrated in Figure 7.1.

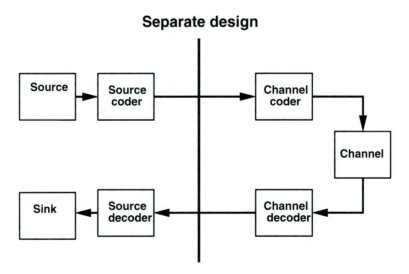

Figure 7.1 Separation principle. Optimality is achieved by separate design of the source and channel codecs.

It is nevertheless important to recall that Shannon's result relies on two important assumptions, namely, (i) the use of arbitrarily long block length for both source and channel codes and (ii) the availability of arbitrarily large computational resources (and associated delays). It is obvious that such conditions are not met in practice, both because of delay constraints, and practical limits on computational resources.

As an example, speech transmission for interactive communication is delay sensitive, and thus block sizes need to be bounded for source coding. Complexity is a major cost factor in compression and channel coding systems, while the strategies implied by usual random coding arguments used in information-theoretic proofs generally have exponential complexity. Thus, many practical source compression algorithms have evolved quite differently from what pure information theory would suggest. The random coding argument of source coding leads to a situation where all bits from the source code are of equal importance (they represent indices into coding tables), but, as mentioned, practical image and video coders typically produce some bits that are more important than others for the reconstruction quality.

Channel coding practice is also quite different from the information-theoretic idealization. While large block sizes allow vanishingly small block error probabilities, practical transmission schemes operate at some finite block error or bit error probabilities.

Because both source and channel coding differ in practice from their information-theoretic idealizations, it is unclear whether the separation principle still holds under these conditions. What is certain is that examples abound of practical systems where a coupling between source and channel coding has led to substantial gains in performance or reduction in complexity. We provide examples of this coupling in this chapter. In the context of time-varying point-to-point transmission as applicable for wireless communications, we consider two important cases of interest related to the presence or absence of a feedback channel from the receiver to the sender: (i) where the receiver alone is informed of the channel state information (CSI) and (ii) where both the transmitter and receiver are informed of the CSI.

The outline of the chapter is as follows. Section 7.1 reviews multiresolution source compression methods. These are source coding schemes that are naturally suited for rate adaptation and unequal error protection. Representative examples are subband coders and wavelet coders, which are popular in image compression. Section 7.2 concentrates on multiresolution channel codes, which allow varying degrees of noise immunity, or decoding qualities. Section 7.3 ties the source and channel coding mechanisms together by matching their respective multiresolutions. The result is a natural and efficient paradigm for joint source-channel coding.

7.1 MULTIRESOLUTION SOURCE CODING FOR IMAGES AND VIDEO

This section gives a brief overview of multiresolution-based techniques in source coding for images and video. Multiresolution-based methods in source coding (such as those based on wavelets, pyramids, etc.) are theoretically elegant. At least as importantly, a key motivation for considering multiresolution environments comes from the fact that due to exploding demands on interconnectivity and heterogeneity (e.g., driven by the Internet and wireless communications), traditional single-resolution coding and transmission are being replaced by flexible scalable source-coding architectures and multirate transmission capabilities.

7.1.1 Successive Approximation of Information

The problem of lossy source compression is one of a trade-off between representation complexity and quality, that is, a trade-off between bit rate and distortion.[1] Given a source with certain statistical characteristics, rate-distortion theory [3, 4] states that there exists a distortion-rate function $D(R)$ such that, at a given bit rate per sample R_0, it is possible to represent the source with a distortion $D_0 = D(R_0) - \varepsilon$ with $\varepsilon > 0$ arbitrarily small.

The important point to note is that in general the distortion-rate curve cannot be travelled in the sense of retaining the successive approximation property. More precisely, each point of the curve is constructed independently of the others. That is, if we pick two rates R_1 and R_2, $R_2 > R_1$, and $\delta_{12} = R_2 - R_1$, then knowledge of the coded version at rate R_1 plus an additional information of δ_{12} bits is generally not enough to construct the coded version at rate R_2 that meets the rate-distortion bound [5]. This property follows because being able to construct the rate R_2 version incrementally from a rate R_1 version is an added constraint, which in general will decrease the quality. As a simple example, consider scalar quantizers. It is clear that an optimal Lloyd-Max quantizer for $2N$ bits does not necessarily have a subset of levels corresponding to an optimal quantizers for N bits. (Of course, exceptions are possible, for example, the uniform distribution when N is even.)

More generally, a successive approximation code is optimal if and only if the successive versions form a Markov chain. More precisely, assume X is the original signal, X_1 a fine approximation, and X_2 a coarse approximation. Then, if $X \rightarrow X_1 \rightarrow X_2$ form a Markov chain, successive approximation coding is optimal, since knowing X_1 to code X_2 is as good as knowing X.

---

[1]For the sake of this discussion, we assume the distortion to be the squared norm of the error between the original and the approximation.

Despite the fact that most sources do not strictly meet this Markovian property, it turns out that many practical compression schemes achieve successive approximation with a negligible loss of quality. That is, many source compression algorithms can produce a bit stream that can be decoded successively, up to some very fine granularity. Such algorithms thus allow the possibility of traveling a practical distortion-rate curve that is not much worse than that achievable with the best coders that do not have the successive approximation property (e.g., about 1–2 dB worst case and typically less than 1 dB loss in performance on typical test images). We therefore briefly consider a number of such successive approximation source codes.

7.1.2 Practical Successive Approximation Source Coders

7.1.2.1 Pyramid Coding

Pyramid coding is the first instance of hierarchical source coding that had an impact on compression practice. First formally proposed by Burt and Adelson [6], pyramid coding codes two versions of the signal. First, a coarse version of the original signal or image is derived using, for example, lowpass filtering and subsampling. Based on this coarse approximation, the original signal is predicted (using upsampling and interpolation, for example) and a prediction error or difference signal is calculated. This prediction error represents the detail features that are missing in the coarse approximation. An example is shown in Figure 7.2.

The scheme can be iterated on the coarse image, leading to a sequence of lower and lower resolution images of geometrically decreasing sizes, hence the name pyramid. Each time a lower-resolution image is derived, a difference image is created. The compression problem is now reduced to finding methods to efficiently represent the various difference images, as well as the last low-resolution version. The difference images are usually treated, in a first approximation, as independent identically distributed samples, for example, having a Laplacian distribution.

A simple counting argument shows that the pyramid representation is over-complete. If the original is a picture of size $N \times N$, the difference image is of the same size, while the coarse image is of size $N/2$ by $N/2$. This is an increase of 25% in the number of pixels. If iterated, it can lead to an increase of up to 33% in the number of pixels. While theoretically this extra redundancy can be removed through compression, it creates an additional burden.

The idea of pyramid coding can be applied to three-dimensional data as well, leading to video pyramids. This multiresolution representation of video is useful for compatibility and for robustness with unequal error protection [7].

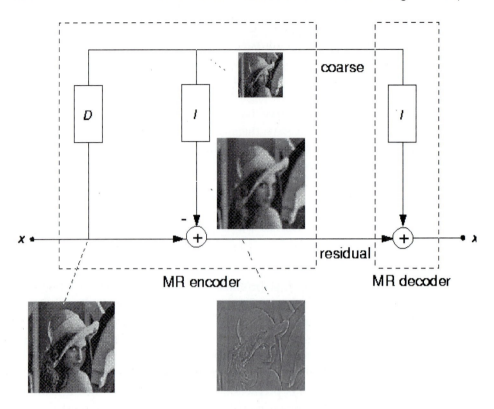

Figure 7.2 Pyramid coding scheme showing coarse-to-fine hierarchy.

7.1.2.2 Subband Coding

The problem of overcompleteness—i.e., the redundancy in the pyramid representation—is solved by subband coding. In that scheme, an orthonormal or biorthogonal expansion is calculated using filter banks [8], and the expansion coefficients are quantized appropriately. Subband coding is a generalization of transform coding, a popular method in image compression [9]. Unlike transform coding, however, subband coding is not plagued by blocking artifacts. Subband coding has been used in speech compession [10, 11] and was proposed for images in [12]. Practical coders followed [13, 14, 15], and the results spurred substantial work in subband image coding.

Several questions arise around subband image coding:

(i) What subband decomposition is best suited to images: how many subbands should be used, what frequency partitioning, what time resolution?

(ii) What filters are best used: orthogonal or biorthogonal filters, linear phase filters, what extension schemes should be used at boundaries?

(iii) What subband quantization scheme should be applied: scalar quantization, vector quantization, uniform or nonuniform (Lloyd-Max) quantization [16]?

(iv) What are efficient lossless compression schemes to handle the quantized subband coefficients: static entropy codes, adaptive entropy codes?

These issues have been explored extensively in the literature and results are summarized in, for example, [8]. The idea of subband coding has been extended to video in [17] and has led to interesting compression results recently [18, 19].

7.1.2.3 Wavelet Coding

An important variant of subband coding is what is now popularly called wavelet coding [20, 21, 22]. This variant uses an octave-band tree structure and filters that possess a regularity property (related to a notion of smoothness of an iterated function). Such schemes have the ability to compactly capture the space-frequency characterization of natural images. Early wavelet-based image coding algorithms were patterned after standard transform-coding principles in that they were designed to exploit (only) the wavelet transform's ability to do *frequency* compaction, i.e., to efficiently pack most of the image energy into a few, low-frequency coefficients. Coding gain was achieved by optimizing the matching of quantizers (scalar and vector) to the statistics of the frequency subbands. These techniques reported modest gains over standard block transform-coded algorithms, with the primary source of the gains coming from the improved frequency energy compaction property of the wavelet over the block transform.

A new class of algorithms developed recently has achieved significantly improved performance over the previous class [21, 23, 24, 25, 26] by exploiting the wavelet's space-frequency compaction properties. The wavelet is able *both* to "frequency compact" energy into a small set of low-frequency coefficients and to "spatially compact" energy into a small set of localized high-frequency coefficients, with the exact extent of the localization depending on the spatial support of the wavelet filters. The incorporation of a mechanism to exploit this spatial characterization is vital to improving efficiency: the most popular data structure for this is the zerotree structure developed by Shapiro [23].

A wavelet image representation can be thought of as a tree-structured spatial set of coefficients, providing a hierarchical data structure for representing images, with each wavelet transform coefficient corresponding to a spatial area in the image. Figure 7.3 illustrates the parent-children dependencies in the tree-structured representation of a typical wavelet image decomposition.

As shown, each parent node has 4 children nodes, the coefficients of the 2×2 region that corresponds to the same spatial location and orientation but in the immediately finer scale of the decomposition. A zerotree symbol refers to a spatial

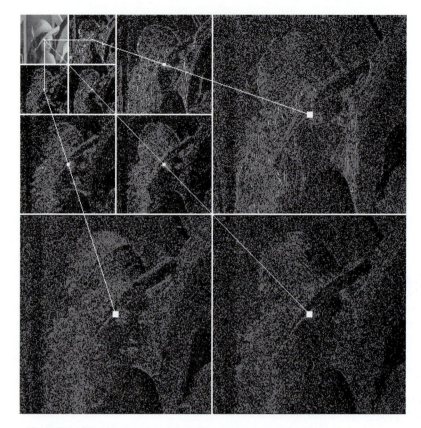

Figure 7.3   Wavelet decomposition of Lena image: spatial tree structure.

tree of zeros: it is designed to capture the high conditional probability of low-energy coefficients at a particular scale, given the low-energy state of parent coefficents at the same spatial orientation at a coarser scale. This approach is obviously well suited to capturing the typical decaying spectral energy profile exhibited by natural images (barring localized singularities like edges and textures). Shapiro combined the use of this zerotree symbol (as a sort of spatial pointing mechanism to indicate low-variance spatial regions) with the concept of bitplane coding (which dyadically refines the threshold with respect to which the zerotree "zeros" are defined) to devise a remarkably efficient embedded wavelet coding algorithm that could refine its resolution by the bit, and yet which thoroughly outperformed the existing (nonembedded) JPEG standard (see Figure 7.4).

Also shown in Figure 7.4 is a rate-distortion optimized version of Shapiro's zerotree coder, dubbed the space-frequency quantization (SFQ) based zerotree coder in [25, 27], that gives up the embedding feature for attaining rate-distortion optimality (within the zerotree framework). The SFQ coder is so named because it optimizes the application of two simple quantization modes—a zerotree spatial

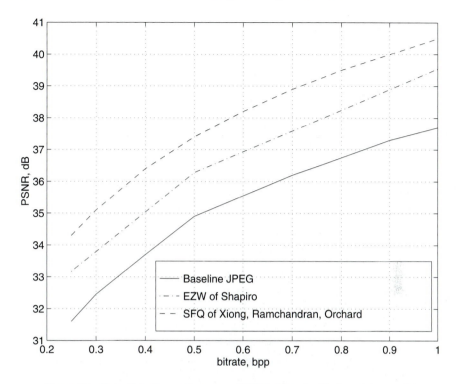

Figure 7.4   Rate-distortion performance of JPEG, the embedded zerotree wavelet coder of Shapiro [23], and the space-frequency-quantization wavelet coder of Xiong, Ramchandran, and Orchard [27], which is a rate-distortion optimized version of Shapiro's zerotree coder. The image is again "Lena."

quantizer and a uniform scalar frequency quantizer—designed to capture the space-frequency characterization of the wavelet image decomposition. The basic idea arises from the desirability, in a rate-distortion sense, of "creating" zerotrees where there are none in reality: i.e., where certain tree-structured sets of coefficients are not *all* zeros, as needed to qualify for the privileged zerotree status. The SFQ coder can realize sizable performance gains, on the order of 1–2 dB in peak signal-to-noise ratio (PSNR), over Shapiro's generic zerotree-based framework and ranks among the best coders in the image coding literature. These gains are achieved by (i) optimally weighing the savings in bit-rate cost from the extra zerotrees against the distortion from the "killing" of coefficients that enable these new zerotrees; and by (ii) balancing this zerotree quantizer with a simple scalar frequency quantizer applied to coefficients that survive the zerotree pruning operation.

Although the zerotree structure is the most celebrated way of exploiting the space-frequency characterization of the wavelet decomposition, it is by no means the only one. Efficient alternative structures are complementary to, and often outperform, the zerotree structure. Such structures include spatial classification-based methods [28, 29, 30] and a novel framework based on morphological dilation [31]. The latter has the added advantage over the zerotree structure of being "object oriented," and is capable of a flexible embedded/nonembedded mode of operation that may be suitable for many evolving multimedia-driven applications, such as supporting object-indexing, etc., without giving up compression performance [32].

7.1.3 Successive Approximation Source Coding in the Context of Standards

Successive approximation source coding is found in several image and video coding standards. These are reviewed briefly in the following paragraphs.

7.1.3.1 Scalability in JPEG

The standard image compression method JPEG [33] has a hierarchical mode. A multiresolution representation is obtained by scaling the size of the encoded image. The base layer carries the data for the lowest spatial resolution. Each subsequent layer is constructed from the previous layer using a predictor (known to the encoder and decoder), and the prediction error is encoded as the data corresponding to this layer. Obviously, all the preceding layers have to be decoded correctly in this case.

Other examples are the spectral selection (SS) and successive approximation (SA) modes in JPEG [33].[2] Spectral selection allows the separate encoding of transform coefficients for each block of data (this separation is useful because the human visual system has different sensitivity to different spatial frequencies). The SA option allows one to separately encode transform coefficient bits. Both modes can be used together, thus providing a possibility for very flexible system design (which has not yet been fully explored, even by researchers).

7.1.3.2 Scalable Modes in Video Compression Standards

The dominant video compression standards are H.261/H.263 and MPEG. The H.261 video compression algorithm is a motion predictive coding method that is akin to the adaptive differential pulse coded modulation (ADPCM) loop over time. It is therefore difficult to make the algorithm hierarchical without loss of performance, and thus, the standard does not currently include any scalability feature. The reason is the following: in an ADPCM prediction loop, both encoder and decoder need to compute the same prediction, that is, they need to be in the same state. To facilitate low-resolution (only) operation, low-resolution reconstruction has to be possible on its own, and therefore this is the only state usable in the prediction loop (see also Chapter 6). This restriction leads in general to a subopti-

---

[2]This mode of progressive JPEG is popular in Internet applications. For example, the Netscape® image browser uses progressive JPEG.

mal prediction for the full-resolution mode of operation and, therefore to a loss in full-resolution quality.

The video coding standard MPEG [34] comes in two varieties, namely, MPEG-1 for 1 Mbits/s, and MPEG-2 for the 5–20 Mbits/s range. Only the MPEG-2 version includes scalable or hierarchical modes, which we briefly describe below. There are two main purposes for an embedded bit stream in MPEG-2 coding:

- Layering for prioritizing of video data (e.g., unequal error protection).
- Scalability for complexity division (e.g., a standard TV set that can decode only a part of the HDTV bit stream).

A brief summary of the MPEG-2 video scalability modes follows.

(i) **Spatial Scalability** is essentially a pyramid coding mode, in which a low-resolution (subsampled) version is coded first, and its decoded version is interpolated and used as a prediction for the full resolution.

(ii) **Data Partitioning** is similar to JPEG's frequency progressive mode. This frequency domain method breaks motion information and transform coefficients into two bit streams. The first, higher-priority bit stream contains the more critical, lower-frequency coefficients and side information (such as DC values, motion vectors). The second, lower-priority bit stream carries higher-frequency data.

(iii) **SNR Scalability** is a spatial domain method where channels are coded at identical sampling rates but different picture qualities (through control of quantization step sizes). The higher-priority bit stream contains base layer data to which a lower-priority refinement layer can be added to construct a higher-quality picture.

(iv) **Temporal Scalability** allows one to play with frame rates and is thus based on multiresolution in time. A first, higher-priority bit stream codes video at a lower frame rate, and the intermediate frames can be coded in a second bit stream by use of the first bit stream reconstruction as prediction.

7.2 MULTIRESOLUTION CHANNEL CODING

When dealing with an image or video source, we have seen that having a multiresolution architecture enables efficient adaptation to changing bit rates and bandwidths. Just as it is important to have an adaptive source coding strategy to efficiently adapt to changing bit rate requirements, it is equally important to match this strategy with an adaptive channel coding framework that can offer a hierarchy of "resolutions" of noise immunity to adapt to varying channel conditions. We refer to these codes as *multiresolution channel codes*. Our goal here is not to survey the state-of-the-art in channel coding theory, which is a very active area of research that has resulted in significant technological advancements, e.g., in modem technology. Rather, our motivation is to establish the conceptual dual in channel coding to the multiresolution source coding framework covered in Section 7.1. Multiresolution channel codes allow efficient adaptation to different levels of channel impairment *and* to different levels of "importance" of the source (e.g., motion vectors and low-frequency information versus high-frequency residue detail). This strategy leads naturally and efficiently to joint source and channel coding in a multiresolution framework, where the multiresolution source and channel codes are efficiently matched to each other—more on this in Section 7.3.

Multiresolution channel codes offer unequal error protection. Unequal error protection channel codes have been studied extensively in the forward error-correction coding literature over the past 10–20 years [1], and there exist systematic ways of designing efficient unequal error protection codes for several families of channel codes, including block codes and convolutional codes. We are additionally interested in unequal error protection codes that are endowed with the "embedding" property. Families of embedded unequal error protection codes have the desirable property that high-rate codes in the family (offering lower noise immunity) can be embedded in their low-rate members (offering increased noise immunity). This property is very useful for applications requiring compatibility of a single transmitted information stream with several channel capacities or receiver resolutions, as in broadcast or multicast or even point-to-point communication over time-varying channels where there is no feedback, i.e., where the transmitter is uninformed of the instantaneous channel "capacity." For example, in a two-resolution case, both receiver resolutions have access to the "coarse" information layer, while the stronger receiver can additionally extract the embedded "detail" information layer. This has been shown to be superior to naive multiplexing using non-embedded unequal error protection codes [35, 36].

We shall see that the idea of multiresolution channel coding can be extended to cover modulation and demodulation systems, for example, in defining embedded modulation schemes like embedded quadrature amplitude modulation (QAM) [36]. Indeed, embedded modem schemes can coexist with unequal error protection, forward-error correction codes to form a powerful multiresolution channel coding infrastructure.

7.2.1 Error Control Coding

In the following paragraphs, we describe briefly the two types of error-control codes mentioned above—namely, unequal error-protection codes and embedded codes.

7.2.1.1 Unequal Error Protection Codes

As described above, unequal error protection codes are families of channel codes that can change their coding strengths flexibly and efficiently. These codes are obviously useful for protecting information sources such as images and video that are typically characterized by unequal importance of bits (due to the properties of practical coding algorithms). They are also useful for dealing with time-varying channel conditions, where the degree of needed protection varies. Unequal error protection codes have been studied for both block codes (e.g., BCH codes [37]) and convolutional codes (e.g., rate-compatible punctured convolutional (RCPC) codes [38]). RCPC codes have become very popular due to their combination of flexibility and efficiency. The basic idea is that one can define an array of codes of differing strengths by simply "puncturing" parity bits appropriately to control the rate of the code. This method is implementationally attractive because the codes come from a single family defined by an underlying finite state machine (FSM), with the various decoding resolutions being based on the same Viterbi decoding trellis structure [38]. As an example, consider an RCPC code family of rates ranging from 8/9 to 8/24. In this family, all codes are nested versions of the 8/24 code and are derived from the same FSM. Puncturing tables specifying the parity bits to be suppressed are used to control the rate of the code. RCPC codes have become popular of late for protecting image and video sources, where different components of the bit stream have different error sensitivities and are therefore deserving of different degrees of protection [39].

7.2.1.2 Embedded Codes

While the unequal error protection codes such as the RCPC codes are powerful, they are not optimally suited to applications where the transmitter is unable to alter its configuration dynamically, and one needs to have a code that is simultaneously decodable at multiple resolutions. Examples of such scenarios are broadcast and multicast (a single transmitter to several users, each having different channel capacities) or a point-to-point link without channel feedback information (uninformed transmitter). For example, if one were to use an RCPC family of codes as described above in a broadcast scenario, the worst channel would become the bottleneck for the system, and parity bits needed for the worst channel would waste the higher bandwidth of the better channels. In such scenarios, embedded codes are an attractive alternative. These codes would allow each receiver to extract the amount of information commensurate with the available capacity.

    For example, a two-level embedded unequal error protection code can be described as an (n, k_1, k_2, t_1, t_2) code (where t_i represents the number of channel errors the code can withstand for the k_i information bits). Note that one could use multiplexing schemes to achieve unequal error protection, e.g., to combine two separate

(n_1, k_1, t_1) and (n_2, k_2, t_2) codes into a composite $(n_1 + n_2, k_1, k_2, t_1, t_2)$ code. However, multiplexing is inferior to embedding [35, 36]. In other words, the combined $(n_1 + n_2, k_1, k_2, t_1, t_2)$ code above can be potentially outperformed by an (n, k_1, k_2, t_1, t_2) embedded code. As an example, consider a $(63, 12, 24, 5, 3)$ binary cyclic unequal error protection embedded code listed in [40]. Alternatively, one can consider two smaller BCH codes with characteristics $(31, 11, 5)$ and $(31, 12, 3)$. Combining these codes to yield a $(62, 11, 12, 5, 3)$ code is clearly inferior to the embedded option.

While embedded unequal error protection codes are more efficient than those derived from separate foward error-correction codes, unequal error protection codes are hard to find, and no structured method has been described to design them. Lin et al. [40] tabulate all possible embedded codes of odd lengths up to 65, using exhaustive computer search. The list is fairly sparse, meaning that only a limited discrete set of rates and correction capabilities exist with embedding in the error correction coding domain. We will see that if one turns to doing unequal error protection in the modulation domain, more flexible and efficient possibilities exist.

7.2.2 ARQ in Universal Channel Coding

In our discussion of channel coding, we stressed the need to adapt the transport mechanism to changing channel conditions. It is worthwhile pointing out that for channels with feedback, one can use automatic repeat request (ARQ), which perfectly adapts to channel conditions. From an information theoretic standpoint (i.e., where infinite complexity and delay are permissible), there is surprisingly no advantage to having feedback when communicating over memoryless channels [4], i.e., the capacity is unchanged.[3] In fact, the ARQ scheme reaches channel capacity over a binary memoryless erasure channel,[4] irrespective of what the erasure probability actually is [4], i.e., it is a universal code for this channel. This feature is very powerful since without feedback, universality would be hard to achieve. Other coding schemes in the literature, e.g., [41], are universal in the sense that they can be applied to channels with unknown parameters. In [41], chaotic sequences for encoding of analog sources are shown to be better (in the end-to-end mean-square error (MSE) sense) than any digital (finite-alphabet) codes for additive white Gaussian noise (AWGN) channels in some power-bandwidth regimes.

In the case of channels with memory, feedback is known to provide theoretical advantages even in the Shannon sense, although feedback increases the capacity for a nonwhite (i.e., correlated) Gaussian additive noise channel by at

---

[3]However, the complexity needed to attain this capacity may well be significantly lower if feedback is present [85].

[4]Recall that in an erasure channel, the receiver knows which symbols are erased: this is typical of packet losses in networks, for example.

most half a bit [4]. It is useful to note that if delay constraints are stringent, then feedback may be useless. Similarly, in the case of a single transmitter and multiple receivers (e.g., broadcast or multicast), the issue of ARQ is rendered moot by the presence of several unequal capacity receivers but only one transmitter.

7.2.3 Embedded Modulation

Although embedded channel codes are difficult to design, one need not restrict oneself to the forward error correction domain in order to achieve embedding. The justification for embedded transmission has its roots in Cover's classic work on broadcast channels [35]. Consider a typical broadcast scenario, where a source wishes to convey information $\{r, s_1\}$ to a stronger receiver and $\{r, s_2\}$ to a weaker one. Note that r represents the common message to be conveyed to both receivers. Cover established in an information-theoretic setting that the optimal strategy involves superimposing the detail information intended for the stronger receiver in the coarse information intended for the noisier receiver. That is, the superior receiver 1, in an optimal scenario, necessarily has access to the information $\{r, s_2\}$ meant for the weaker receiver 2. While this is a theoretical result (in the Shannon sense), a practical way of realizing this embedding gain was described in [36].

Cover's concept of embedding is generic in scope and places no restrictions on the domain in which this embedding should be performed. The benefits of embedding in the modulation domain, versus in the forward error protection channel coding domain, were laid out in [36] in the form of a novel, multiresolution, embedded modulation structure that offers the embedded unequal error protection property *directly in the modulation domain.* The idea is simple: see Figure 7.5 for examples of "generic" two-level multiresolution constellations that have "clouds" of "satellites," characterized by an intracloud-to-intercloud distance ratio μ, as shown. Unequal levels of noise immunity are offered by these constellations, as represented by the satellites and the clouds in which they are embedded.[5] The goal is to match these optimally to different levels of bit error rate requirements; e.g., for the appropriate choice of μ, the clouds may offer a bit error rate of 10^{-7}, whereas the satellites may offer only 10^{-3}.

A modulation-domain-based unequal error protection scheme, similar to that outlined above, has been considered for European digital audio and video broadcast [42]. Each layer of different error protection corresponds to the specific type of the receiving monitor (typically, there are three layers or resolutions) and each has

---

[5]For example, in the multiresolution-64QAM example shown, for every two important cloud bits/symbol having increased protection, four less important satellite bits/symbol get decreased protection.

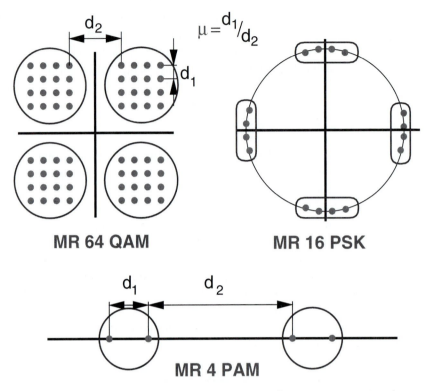

MR 64 QAM   **MR 16 PSK**

MR 4 PAM

Figure 7.5 Some multiresolution modulation constellations. μ parametrizes the intracloud-to-intercloud ratio of the constellation.

different bit error rate requirements. Thus, the quality of the received video varies gracefully with the receiver type as well as with its distance from the transmitter.

The added attraction of the embedded modulation scheme comes from being able to combine it naturally with standard error correction techniques, as was proposed in [36] and later in [43], as well as with other techniques described in the following section.

7.2.4 Hybrid Embedded Options

One can combine the concept of embedded modulation with other tools and strategies to obtain powerful hybrids. Examples include the combination of embedded modulation with unequal error protection channel codes (e.g., to increase the number of resolution layers), with Ungerboeck's trellis coded modulation (TCM) [44] to achieve embedded TCM [36], with multicarrier systems to attain embedded multicarrier modulation, etc. To illustrate the power of these hybrids, we pick multicarrier modulation as a representative example [45].

We begin with a very brief review of multicarrier modulation, referring the reader to [46, 47, 48] for details. Multicarrier modulation has become a topic of great interest recently due to the demand for high-speed data transmission over twisted-pair copper wiring, an environment where severe intersymbol interference (ISI) can occur. Instead of employing single-carrier modulation with a very complex adaptive equalizer, the channel is divided into many subchannels that are essentially ISI-free. One multicarrier method in particular, discrete multitone modulation (DMT), has become extremely popular due to its efficient implementation, which uses fast Fourier transforms (FFTs) to modulate and demodulate data. DMT has been adopted as the standard for asymmetric digital subscriber loop (ADSL) telecommunications transmission technology [49, 50].

The fundamental idea behind multicarrier modulation is to divide a single communications channel into a large number of QAM subchannels that could be treated as independent additive Gaussian noise (AGN) channels—but with white, rather than colored, noise. Since each of these channels is memoryless, an equalizer is not needed. Data is divided among the subchannels, which are then modulated and summed to form a composite channel signal. At the receiver, each subcarrier is demodulated, and the data from each subchannel is combined to reconstruct the original. The "loading" problem of optimizing the power and deliverable bit rate per symbol for each QAM subchannel for a given total power budget can be done elegantly in theory with the "inverse water-pouring" principle [51]. The idea is to invert the multichannel SNR spectral profile and pour water into it. The optimal energy allocation in each subchannel is simply the amount of water it contains, with the water-level depending on the total energy budget.

Adaptivity to changing channel conditions is relatively simple in multicarrier modulation and involves periodic reloading. Thus, if a particular subchannel should become extremely noisy, it can be shut down very easily by allocating no power or bits to it, as the waterpouring algorithm would dictate. Current DMT systems [49] are channel-adaptive but are single-resolution based, i.e., there is a notion of a single, fixed quality of service (QoS), which is typically measured by a single fixed bit error rate for the entire bit stream (e.g., typically 10^{-7} bit error rate for xDSL [48]). In keeping with the multiresolution theme of this chapter, it is both useful and interesting to consider extensions of the single-bit error-rate regime to a multiple-bit error-rate regime, equivalent to having multiple qualities of service. This extension would result in the formulation of a multiresolution-DMT scheme that is useful whenever the source representation, due to practical constraints, has layers of unequal importance that are deserving of unequal QoS, e.g., MPEG-compressed video bit streams that have critical components, like motion vectors, and supplemental components, like high-frequency DCT coefficients. As the video-over-DSL application is one of the main catalysts for this technology, a multiresolution-DMT framework is all the more relevant and useful to consider. In such a case, a hierarchy of appropriate bit error rates are of interest. Overlaying the concept of embedded constellations of Figure 7.5 on the multicarrier modulation framework gives an embedded, multicarrier modulation system that is characterized by *multiple embedded constellations,* one for each multicarrier subchannel. Figure 7.6 gives a simple example showing a three-carrier embedded modulation system. An optimal way of designing the embedded modulation system based on a fast, table-lookup power allocation (or loading) method has been described in [45], with performance gains of the order of 25% in actual delivered throughput (or equivalently 1–2 dB in delivered image quality for image transmission applications) for an embedded two-resolution multicarrier system over "naive" time-division multiplexing of the two priorities.

7.2.5 Channel Models Used in Multiresolution Channel Coding

An obviously important question is the choice of realistic yet analytically tractable channel models in our communication system analysis. Consider the general setup of the digital communication system shown in Figure 7.7. The data from the information source enters the source encoder, where it is compressed to achieve a more compact digital representation. The channel coder usually adds some controlled redundancy to the source-encoded data to mitigate the effect of channel errors. The modulator converts the discrete-alphabet sequence into a waveform suitable for transmission over the channel by means of digital modulation. The demodulator and source/channel decoders perform the inverse operations and deliver the

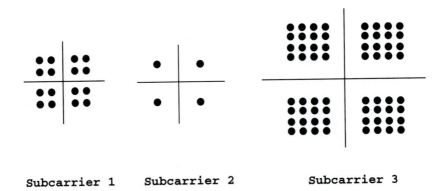

Subcarrier 1   Subcarrier 2   Subcarrier 3

Figure 7.6 Multiresolution QAM constellations for a typical multicarrier system. Each subchannel supports embedded modulation that can carry more important and less important bits, that can be optimally allocated across the subchannels, using multiresolution power loading algorithms [45].

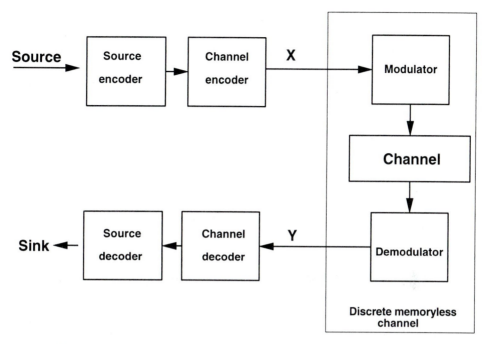

Figure 7.7   Block diagram of a standard communication system.

reconstructed data to the destination. The modulator/demodulator (modem), to-gether with the continuous waveform channel, may be considered to be a discrete-alphabet noisy channel with input X and output Y taken from finite sets $\mathcal{X}$ and $\mathcal{Y}$, respectively. The discrete channel is characterized by the collection of transitional probability mass functions $p_{Y|X}(y \mid x)$, one for each $x \in \mathcal{X}$ [4]. This channel is called a *discrete memoryless channel* if the following condition is satisfied:

$$\text{Prob}(Y_k = y_k \mid X_k = x_k, Y_{k-1} = y_{k-1}) = \text{Prob}(Y_k = y_k \mid X_k = x_k).$$

This model is almost exclusively used for designing error-correction codes. It also includes the binary symmetric channel as a special case.

Although the discrete memoryless channel model is certainly a useful high-level channel model, it has the drawback of not explicitly including critical low-level communications channel parameters, such as power and bandwidth, which are hidden, and only indirectly captured in the symbol transitional probabilities. A family of somewhat more informative models is obtained if the analog waveform channel is considered. The simplest example is an AWGN channel where a signal waveform is assumed to be corrupted by AWGN independently from the signal. The performance of the communication system in AWGN is completely character-ized by the signal-to-noise power ratio. The AWGN channel model and its gener-alization—AGN, where the noise is not necessarily white—are well suited for stationary communication channels, when parameters of the channel are constant or changing very slowly (e.g., telephone lines).

If the channel parameters are not constant in time or frequency (such as in typical AWGN channels with fading, a popular model for wireless communications), it is reasonable to approximate the channel as a mixture of several AWGN channels indexed by states. Transition between states may be modeled by a Markov process [52]. Consider a typical fading channel for wireless communications, as in Figure 7.8. The distribution of the fading parameter a depends on the actual communication channel. For example, as discussed in Chapters 1–3, in cellular communications, a good model for a is that it has a Rayleigh distribution, while for line-of-sight communications, a Rice distribution is often used [53].

Each particular channel state is modeled as a discrete-time, continuous amplitude, AWGN channel with a distinct noise variance, and it is assumed that all channels are statistically independent. To justify this model, interleaving/deinterleaving is assumed to be present and to have sufficient depth to enable independent treatment of the received signals. Interleaving shuffles the sequence of data before transmission to ensure that, at the receiver after deinterleaving, two consecutive signals will have noise components that are almost independent. The reason for using interleaving is that the channel conditions in real-life fading channels have high temporal correlation and may cause bursts of errors. Since these bursts of errors are very challenging for error-correction codes, interleaving is a popular technique used in practice to randomize errors, which are easier to handle. No loss of performance occurs with this method if perfect CSI is available at the receiver. This multistate model allows approximation of a wide variety of realistic channels, such as slow fading channels or multicarrier channels, while maintaining only a moderate complexity of the analysis. To understand the multistate model, consider the following example. Suppose it is desired to approximate a memoryless Rayleigh channel by a multistate AWGN channel. By allowing $h^2 = \mathcal{E}/N_0$ to be the received carrier-to-noise power ratio, the instantaneous h parameter has Rayleigh probability density function:

$$f(h) = 2h/h_0^2 \exp(-h^2/h_0^2),$$

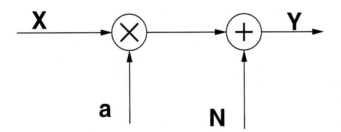

Figure 7.8 Fading communication channel. The transmitted signal X is multiplied by a fading factor a. Then, an AWGN sample N is added to the result to yield the received signal Y.

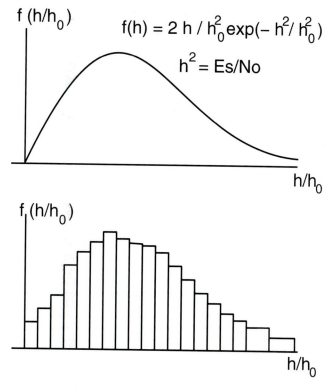

Figure 7.9   Example of approximation of memoryless Rayleigh channel by a multistate AWGN channel model.

where $h_0 = \frac{2}{\sqrt{\pi}}E\{h\}$. By approximating this distribution by a piecewise constant probability density having N regions, we achieve a multistate AWGN representation that becomes exact as N becomes arbitrarily large (see Figure 7.9). This approximation can be made for other carrier-to-noise ratio distributions as well, as long as the channel is memoryless.

7.3 MULTIRESOLUTION JOINT SOURCE-CHANNEL CODING

As mentioned above, Shannon's pioneering result showing the separation without loss of optimality of source and channel coding [2] does not apply in practice in the face of finite delay and complexity constraints. The understanding of the superiority of a joint approach to source and channel coding in such cases has recently initiated numerous research activities in this area, a partial list of which can be found among [54, 55, 56].

In keeping with the theme of this chapter, we focus on multiresolution-based frameworks for joint source-channel coding. A natural and efficient starting point is to combine the multiresolution-based source and channel coding frameworks of Sections 7.1 and 7.2, respectively.

7.3.1 Brief History of Joint Source-Channel Coding of Images/Video

A joint approach to source and channel coding has been motivated primarily by emerging communications applications involving speech, image, and video transport. These applications typically involve strict delay and complexity requirements that do not meet the assumptions of Shannon's separation principle. Source-channel coding schemes have been relatively recent in the literature since they have been primarily driven by these emerging applications. At a high level, two approaches present themselves naturally when joint source-channel coding is considered historically. Very roughly, these approaches may be classified as being inspired by digital versus analog transmission methods.

The digital class of techniques is based on optimally allocating bits between digital source and channel codes. Source-coding bits correspond to a digitally compressed and entropy-coded stream. Channel-coding bits correspond to the parity information of a digital error-correction code. It is of historical importance to note that the idea of unequal error protection-based joint source-channel coding has been around for quite a while. For example, the idea of unequal error protection channel codes (such as unequal-strength convolutional codes) that are appropriately designed to unequally protect different source components having different error sensitivities (e.g., quantized source bit-planes) goes back at least to [54]. More efficient frameworks fashioned after similar principles have become popular recently [57]. It should be noted that this "digital" approach, while allowing higher source compression because of entropy coding, can also lead to increased risk of error propagation. The popular solution is to insert periodic resynchronization capabilities by means of packetization. The resulting synchronization and packetization overheads that are needed to increase error resilience obviously reduce the compression efficiency. The problem becomes one of optimizing this balance.

The other approach has been inspired essentially by the "graceful degradation" philosophy reminiscent of analog transmission. Thus, while the single-resolution digital philosophy adopts an all-or-nothing approach (within the packetization operation) resulting in the well-known "cliff effect," the analog-inspired approach carries a "bend but do not break" philosophy. The idea is to do intelligent mappings of source codewords into channel constellation points, so as to have a similarity mapping between "distances" in the source coding domain and "distances" in the channel modulation domain [55, 58, 59, 60, 61]. Thus, large source distortions are effectively mapped to high noise immunity, i.e., to low probability error events, and vice versa, with intelligently chosen index assignments. Among the advantages of such an approach are increased robustness and graceful degradation. The disadvantage is the lack of a guaranteed QoS (there is no notion of "perfect" noise immunity).

It is interesting to consider hybrid versions of these two philosophies that are aimed at exploiting the best of both worlds advocated above. We dedicate Section 7.3.5 to this idea, but we provide a preview here to underline the historical perspective. A practical hybrid analog/digital transmission scheme was first proposed by Schreiber [62], using a novel analog-under-digital scheme. An all-digital solution in the same spirit was proposed in [36], where the goal was to merge the benefits of both approaches without giving up an all-digital representation. These methods enjoy the important embedding property covered in Section 7.2, making them ideally suited to applications like broadcast and multicast (see Section 7.3.3).

Joint source-channel coding schemes in the literature have been based on a variety of channel models. Historically, discrete memoryless channels and especially binary symmetric channels came first. The typical assumption was to fix the channel symbol transition probability matrix (i.e., specifying the channel symbol error probabilities of receiving symbol j given that i was sent: $\text{Prob}(j \mid i)$), which fully characterize these channel models. Based on this, the optimization of source coders or mappings from source to channel alphabets was performed. This approach was advocated for example, in [63], where an algorithm similar in spirit to the celebrated Lloyd-Max design algorithm for an optimal scalar quantizer [16] was proposed to design an optimal scalar quantizer for a discrete memoryless channel. Recall that the Lloyd-Max quantization algorithm consists of iterative optimizations of the encoder and decoder, respectively, while holding the other fixed. The encoder function deals with finding cell (or Voronoi region) partitions. The decoder function deals with finding representative codewords (or centroids) for each partition. The algorithm iteratively updates the cell partitions (for fixed representative levels), using a weighted nearest neighbor condition, and the representative levels (for fixed cell partitions), using a weighted centroid condition. The "channel-optimized" extension to the Lloyd-Max clean-channel source quantization algorithm basically involves the incorporation of the symbol error probabilities appropriately in defining the weights for the nearest neighbor and centroid rules of the algorithm.

The joint design of channel-optimized vector quantizers (VQ) and the mapping of these VQ indices to channel alphabets were presented in [58] and later in [64], where the value of "intelligent" mappings between source and channel domains was demonstrated. Due to the use of a high-dimensional VQ framework, finding these intelligent mappings is, however, generally a computationally intensive process that usually results in only locally optimal solutions.

Joint source-channel coding approaches based on the discrete memoryless channel model are useful but somewhat limiting, as this model is basically a black-box approach that does not directly and fully address the physics of the communication system, i.e., parameters like bandwidth, average/peak transmission power, delay, etc. are not explicitly included in this approach. Several joint source-channel coding schemes that include more basic communication channel parameters like transmission power and bandwidth have been studied. AWGN channels have been studied in [65, 66, 67], and extensions to Rayleigh channels were considered in [56] with a view to optimizing joint performance subject to an average transmission power constraint. The framework in [56] was based on maximum a posteriori estimation of the input source (based on tractable statistical signal models like autoregressive processes, etc.). By considering appropriate extensions of the generalized Lloyd algorithm [16] (which is a VQ extension of the scalar Lloyd-Max algorithm described earlier), an optimized signal constellation design was included in the optimization loop, leading to performance improvement over traditional separately designed systems. These schemes use fixed-length codewords, i.e., they sacrifice the increased performance of variable-length codewords (through entropy coding) for robustness. A relevant attribute of these schemes, e.g., the scalar quantizer design for binary symmetric channels in [63] and the VQ design for AWGN channels in [56], has to do with the notion of reducing the number of codewords when the channel gets noisier, i.e., to correctly match the source and channel resolutions. This matching can be very efficiently performed with the multiresolution representations, as we describe below.

The trend in the evolution of joint source-channel coding systems in recent times has been to use more accurate source and channel models, as well as more sophisticated state-of-the-art source- and channel-coding techniques. Further, as more system components are jointly designed, better performance has resulted, but at the expense of higher system complexity. As an example, inclusion of the modem in the optimization loop results in increased performance (see Section 7.3.3.3) but with increased complexity.

Let us now take a brief look at a few recent examples that consider source-channel coding in more sophisticated frameworks. An interesting combination involves the joint design of trellis coded quantization (TCQ) and TCM in [61] by exploiting the fact that the two frameworks are duals of each other, one in the modulation (channel) domain and the other in the quantization (source) domain. The idea is to maintain a similarity relationship between distances in the source codeword space and the channel codeword space, implementationally facilitated through the use of the same finite state machine (or trellis structure). This is conceptually akin to the analog transmission philosphy introduced earlier.

Binary symmetric channel transmission of still images with RCPC codes [68] was addressed in [57] and more recently in [69], where a popular embedded wavelet-based zerotree coder [24] was combined with RCPC codes and a series-concatenated channel coder based on the "list-Viterbi decoding" principle [70]. The list-Viterbi decoding paradigm is a conceptually simple but powerful extension of the conventional Viterbi decoding paradigm in that it keeps, at each state of the trellis, a list of the N best-metric paths, with the traditional Viterbi decoder having $N = 1$. The work in [70] reported substantial performance improvement over the conventional Viterbi decoder by combining a first-stage list-Viterbi convolutional coder with a second-stage error-detection coder, using a block cyclic-redundancy-check (CRC) code. This combination was achieved by declaring the correct Viterbi path as the highest-ranked candidate Viterbi path in the trellis (from the ranked top-N list) that additionally passes the CRC test. A more sophisticated extension of the list-Viterbi framework has been recently considered in [71] through the use of a continuous-error-detection second stage based on arithmetic coding that dynamically checks the validity of subtrellis paths and maintains a legal top-N list at each state.

By simply cascading this sophisticated channel coder with a sophisticated embedded wavelet zerotree image source coder (such as [24]) as is done in [69], one can obtain robust image transmission systems. With more powerful channel coders (such as that of [71]) and more powerful source coders (such as that of [29]), even better systems can result, as pointed out in [71]. Note that these are examples of separately designed systems that have state-of-the-art source and channel coding components. However, a subtle point is that the use of an embedded source representation, such as the wavelet coders of [23, 24], results naturally in a gracefully degraded digital system because each succeeding bit effectively represents a resolution layer. Decoding can be stopped when an uncorrectable error is detected, with the resulting quality depending on how far along the bit stream the error has occurred. This capability points to the advantages of having an embedded source representation (see Section 7.1.1). In [72], a joint design that used the distortion-rate characteristics of scalar wavelet-based source coder and RCPC channel coder was suggested. Joint source-channel coding schemes for images based on many-to-many mappings between source domain and modulation domain have also been studied, for example, in [67].

7.3.2 Basic Infrastructure of Multiresolution Joint Source-Channel Coding

We now explore a basic, high-level philosophy for multiresolution joint source channel coding, which is useful in addressing a number of relevant scenarios of interest.

Recall our terminology from previous sections of source resolution as the source quality level, and of channel resolution as the noise immunity level. A useful conceptualization involves looking at these in terms of trees having various depths or resolutions. In this framework, the set of resolutions in both channel and source coders in a very general setup can be associated with this tree structure. This multiresolution infrastructure can be illustrated compactly, as in Figure 7.10. Two source resolutions are mapped into two channel resolutions. If a feedback

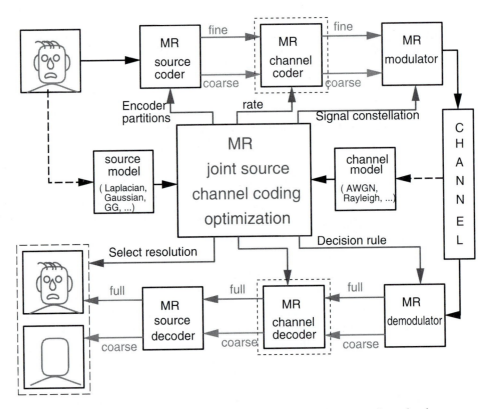

Figure 7.10   General infrastructure for multiresolution joint source-channel coding (MR-JSCC). Channel and source models are used to jointly optimize the system parameters: source encoder/decoder, channel modem, and encoder/decoder. A multiresolution representation is used at all stages in matching different system components.

channel is present, the transmitter chooses whether to send both resolutions, only the coarse one, or perhaps neither. If there is no feedback, then all resolutions are sent simultaneously.

To summarize, we assume, as is often realistic, that the receiver is always informed of the channel state information,[6] and propose the following strategy:

(i) If no feedback channel is present (we call this the uninformed transmitter case), then at the transmitter, optimally design a multiresolution source and channel encoder whose layers are optimized to the average channel conditions (i.e., statistically optimized for good and bad channel states). At the receiver, match the decoder resolution optimally to the instantaneous channel state information. See Figure 7.11(a).

(ii) If a feedback channel is present (we call this the informed transmitter case), then both transmitter and receiver match the resolution of the encoder and decoder, respectively, to the instantaneous channel state information in a synchronized fashion. See Figure 7.11(b). Note that in this case, the use of a multiresolution framework is not theoretically optimal (except if the source obeys a certain Markov property [5]—see Section 7.1.1), but a multiresolution design is more flexible and attractive from an engineering perspective (due to its architectural simplicity). There is little loss in performance typically (see, for example, [73]), with the encoder and decoder both selecting the resolutions they want, based on the channel state information.

At a high level, the proposed infrastructure of Figure 7.10 supports the key idea of efficiently matching source and channel resolution trees to maximize system performance and transcends the details of any algorithmic or implementation details that are scenario- and application-specific.

7.3.3 Uninformed Transmitter Scenarios

We consider the following representative scenarios under this category. Recall that this category refers to the case where the receiver is informed of the channel state information, but the transmitter is not.

7.3.3.1 Broadcast: Embedded Modulation

As stated earlier, the idea of embedded modulation derives its roots from Cover's information-theoretic results on broadcast channels [35]. The merging of the compression advantages of digital transmission systems with the natural robustness of

---

[6]The receiver in general has access to relevant parameters that allow it to infer the channel quality.

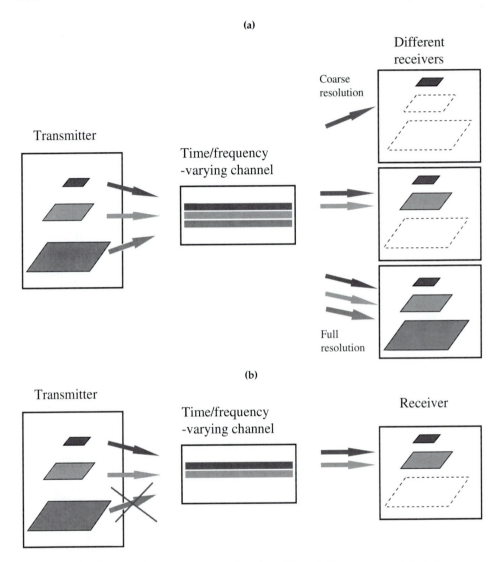

Figure 7.11 (a) Uninformed transmitter case: All resolutions are communicated to the receiver, which decides at which resolution to operate. (b) Informed transmitter case: Both transmitter and receiver select the resolution that matches the channel state.

analog systems was proposed by Schreiber [62] in a hybrid analog-under-digital transmission scheme for the broadcast of HDTV. The idea was to have analog information "ride on top" of digital modulation information (to form analog "clouds") with the essential information being sent digitally (to ensure lossless delivery) and the detail information riding in the analog clouds. The partitioning between analog and digital modes of transmission was, however, done in a somewhat ad hoc manner. In the same spirit, [36] described an all-digital solution, which was aimed at more efficiently retaining the compression advantages of digital systems and optimally matching the source and channel resolutions by a stepwise, graceful degradation philosophy. Embedding was done directly in the modulation domain, using the idea of clouds and satellites depicted in Figure 7.5, offering unequal levels of protection in an efficient and continuously controllable way (through the μ parameter for a two-resolution system). Using a multiresolution embedded approach rather than naive multiplexing of the resolutions (such as by time-division-multiplexing or frequency-division-multiplexing methods), significant gains were obtained in HDTV broadcast applications. This approach can be viewed as a practical way of realizing the theoretical results of Cover.

A modulation-domain-based, unequal error protection scheme similar to that of [36] has been considered for European digital audio and video broadcast [42]. Each layer of different error protection corresponds to the specific type of the receiving monitor (typically, there are three layers or resolutions) and has different bit error rate requirements. Thus, the quality of the received video varies gracefully with the receiver type and with distance from the transmitter.

7.3.3.2 Multicast: Layered Coding

An area where joint source-channel coding ideas have had an impact is in communicating over heterogeneous networks. In particular, the case of multicast in a heterogeneous environment is well suited for multiresolution source and channel coding.

The idea is very simple: give each user the best possible quality by deploying a flexible networking infrastructure that will reach each user at its target bit rate. More precisely, a multicast transmission can be conceptualized as existing on a multiresolution tree. Each user then reaches as many levels of the multiresolution tree as is possible, given its access capabilities. Such a scheme was proposed in [74] for a heterogeneous packet environment, such as the Internet. Figure 7.12 succinctly captures the basic idea.

While currently mostly wired links are involved in multicast applications, it is clear that mobile components are also becoming more and more important. Such a scheme would be suitable for such an environment as well, possibly with bridges between wired and wireless components.

7.3.3.3 Point-to-Point Image Transmission with No Feedback

Consider the example of point-to-point communication over slowly time-varying channels, efficiently modeled as an AWGN mixture multistate channel with no feedback. This case is conceptually similar to the broadcast case discussed in Section 7.3.3.1 in that it is a multichannel communications problem where the

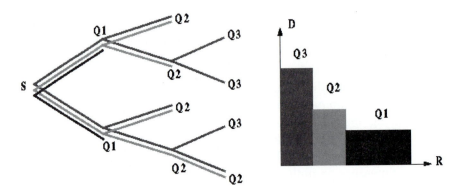

Figure 7.12 Illustration of joint source-channel coding for multicast scenario.

transmitter cannot adapt individually to different receiver capacities. An efficient (and, in practice, near-optimal) solution can be attained by using a multiresolution design of source and channel coders, and by an optimal matching of the source-channel coding resolutions to the instantaenous channel state. At a high level, we propose the following strategy: at the transmitter, design an optimal multiresolution source and channel encoder whose layers are matched to a discrete set of target channel states having known, long-term statistical weights corresponding to the channel model. At the receiver , optimally match the decoder resolution to the channel state information.

Example We will now use a simple but illustrative example (from [60]) to show the usefulness of having a multiresolution approach to joint source-channel coding that incorporates the techniques advocated in the previous sections, i.e., multiresolution embedded signal constellations, a modified Lloyd-Max design for source scalar quantization, and a multistate channel model. This example is based on jointly designing source and channel coders in a tree-structured way and optimally matching the resolutions of these trees.

Consider the example of point-to-point communication over a flat fading Rayleigh channel, which can be modeled as an N-state AWGN channel characterized by different noise variances, with the approximation improving as N increases.

Consider the source-channel coder illustrated in Figure 7.13. Suppose the channel can be in one of only two different states. In each state s (with p_s denoting the probability of occurrence of the state s), the channel is AWGN with a given noise variance σ_s^2, with the states being labeled good and bad. Suppose that the receiver knows the actual channel state whereas the transmitter has knowledge only of long-term channel statistics, i.e., the state probabilities. Suppose we want to transmit an i.i.d. (scalar) source X with probability density function $f(x)$ quantized to four levels through this channel, using a 4-PAM modulation constellation, assuming the optimal one-to-one mapping between the encoder-partitioning $\{\gamma_i\}_{i=0}^{3}$ and the constellation points $\{m_i\}_{i=0}^{3}$, as also shown in Figure 7.13. The joint encoder/modulator operation is therefore to partition the source x into intervals $\{\gamma_i\}_{i=0}^{3}$ and map each γ_i to the corresponding constellation point m_i. Suppose that at the receiver a hard-decision demodulator is used to declare, based on optimized demodulator thresholds $\{t_i\}$, which m_i was transmitted. Finally, the decoder performs a one-to-one mapping between the m_i's and the source reconstruction codewords c_i.

Given this setting, the question is then how to choose *all* the system design parameters to maximize the delivered image quality for a fixed, average transmission energy (per channel symbol). The system parameters include:

- Source encoder partitions $\Gamma = \{\gamma_i\}$
- Channel modulation constellation $\mathcal{M} = \{m_i\}$

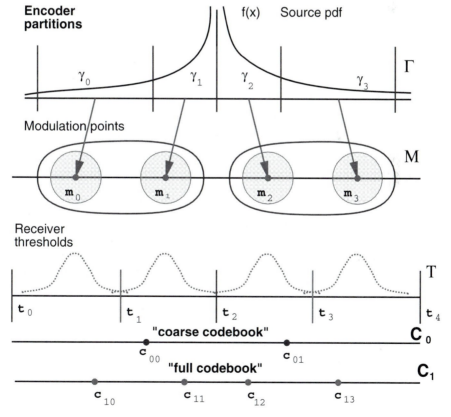

Figure 7.13   Two-resolution, joint source-channel coding scheme for multiresolution 4-PAM and source codebooks having 2/4 levels.

- Receiver decision thresholds $\mathcal{T}_s = \{t_{s,i}\}$ $s = 0, 1$
- Source decoder codebooks $\mathcal{C}_s$, $s = 0, 1$, containing the reconstruction code-words $\{c_{s,i}\}$

Our goal is to minimize the expected distortion

$$E\{D(X, \hat{X})\} = \sum_{s=0}^{1} p_s \sum_{i=0}^{3} \int_{\gamma_i} f(x) \sum_{j=0}^{3} (x - c_{s,j})^2 \text{Prob}_s(j \mid i) \, dx, \qquad (7.1)$$

subject to the fixed energy constraint

$$\mathcal{E}_{av} = \sum_i m_i^2 \int_{\gamma_i} f(x) \, dx, \qquad (7.2)$$

where the transitional probabilities of decoding $c_{s,j}$, given that $x \in \gamma_i$ in the channel state s, are in the form

$$\text{Prob}_s(j \mid i) = Q((t_{s,j} - m_i)/\sigma_s) - Q((t_{s,j+1} - m_i)/\sigma_s), \qquad (7.3)$$

with $Q(x) = \dfrac{1}{\sqrt{2\pi}} \int_x^\infty e^{-t^2/2} \, dt$.

The (i, s)th integral in (7.1) is equal to the individual expected distortion if index i is sent (i.e., given that $x \in \gamma_i$). These individual contributions are summed among all source symbols and averaged over two possible channel states, yielding the total expected distortion.

To solve the constrained problem of (7.1, 7.2) we introduce a Lagrange multiplier λ [75] and solve the unconstrained problem of the form:

$$\min_{\mathcal{M}, \Gamma, \mathcal{T}_s, \mathcal{C}_s} [J(\mathcal{M}, \Gamma, \mathcal{T}_s, \mathcal{C}_s) = E\{D(X, \hat{X})\} + \lambda \mathcal{E}_{av}], \qquad (7.4)$$

where $\lambda > 0$ is chosen to satisfy the energy constraint (7.2) and can be interpreted as a coefficient that trades off energy for distortion in the optimization process. If a solution for the unconstrained problem (7.4) exists, it is also a solution to the constrained optimization problem (7.1), (7.2). Given the expression for the cost function (7.4) the following encoding and decoding rules can be derived. Assign x to γ_i where

$$i = \arg \min_l [\lambda(m_l)^2 + \sum_{s=0}^{1} p_s \sum_{j=0}^{3} \text{Prob}_s(j \mid l)(x - c_{s,j})^2], \qquad (7.5)$$

i.e., where i is a value of l for which the expression in square brackets is minimum. Then reconstruct the jth codeword with

$$c_{s,j} = \frac{\sum_{i=0}^{3} \text{Prob}_s(j \mid i) \int_{\gamma_i} x f(x) \, dx}{\sum_{i=0}^{3} \text{Prob}_s(j \mid i) \int_{\gamma_i} f(x) \, dx}. \qquad (7.6)$$

It is insightful to note that the cost of assigning x to γ_i in (7.5) is a weighted sum of distortion and energy associated with this decision, which is a variation of the entropy-constrained vector quantization (ECVQ) [76] problem in which an equivalent entropy term in ECVQ is replaced by an energy term. At this step, we find the optimal position for the boundary between regions γ_i and γ_{i+1}, which minimizes the cost function (7.4) with all other parameters being fixed. The decoding rule (7.6) is a variation of the weighted centroid condition in the channel-optimized Lloyd-Max quantizer. The decoding rule (7.6) simply assigns the expected value of the transmitted signal X to the reconstruction level $c_{s,j}$, thereby minimizing the mean-squared error. (Recall that the minimum MSE estimator for a random variable X from the observation Y is the one that estimates X as $E\{X \mid Y\}$.) It is possible to formulate the optimal encoding (values of m_i) and decoding ($\mathcal{T}_s$) rules given the cost-function in (7.4), but it appears that an analytical solution is not feasible because of nonlinear relations (through the Q-function) between transitional probabilities $\text{Prob}_s(j \mid i)$ in the optimal receiver and channel coder parameters. On the other hand, if a suboptimal receiver (for example, a maximum-likelihood receiver) is used, then the performance of the system degrades dramatically because the constellation points are not equiprobable.

For our simple example, it is clear that the optimal solution would be the one that allows two *different* distortion-optimized decoders, one for each channel state. The receiver should use the channel state information to switch between these two decoders. Such a design, while manageable for a two-state AWGN channel model, is clearly impractical when the number of states gets large (as needed to approximate the desired channel arbitrarily closely) since this would require a separate design for each channel state and hence would significantly increase the complexity. This problem can be addressed by allowing only a few multiresolution codebooks and devising an optimal decoding strategy for deciding which codebook should be used in each channel state (see Figure 7.11(a)). Embedded modulation is used to provide different levels of noise immunity. In our example, the lowest resolution codebook $\mathcal{C}_0$ has only two reconstruction levels and is used during the bad channel state, while the full-resolution codebook $\mathcal{C}_1$ is used in the good state. Thus, in this simplified example, we can guarantee that the all-important sign information will be delivered in both channel states, while refinement of this information will be possible only during good channel states. The important message is that this two-mode approach is better than trying to decode at full-resolution at all times.

The degradation in performance due to constraining the number of codebooks to a multiresolution codebook, rather than having separate codebooks for each channel state, has been shown in [73] to be insignificant—the key observation is that in bad channel conditions, *the optimal design naturally chooses a lower source codebook resolution,* validating the simpler and more elegant multiresolution approach. The results of this simplified example are easily extended to more practical real-world scenarios by a similar concept of optimally matching the source resolution and signal constellation resolution "trees" in accordance with the time-varying channel. This matching leads to improved performance over separately designed source and channel coders with the real-time operation needing only table lookups. The results, when applied to wavelet image transmission, reveal substantial gains, of the order of 2–3 dB in PSNR, over conventional systems in fading channels.

We note that for the special case of a Gaussian memoryless source and an AWGN channel, the mapping from the source domain into the signal constellation will tend to become linear as the number of levels in the source-channel coder increases (see Section 7.3.5.1 below). Actually, linear one-to-one mapping (when the channel signal amplitude is determined by scaling the source sample amplitude) will achieve the optimal performance theoretically attainable (OPTA) for this source and AWGN channel within an energy constraint [3]. Unfortunately, for other source distributions, linear mappings are no longer optimal, and optimal mappings are not known in most cases; in such cases, numerical techniques are necessary [56, 60].

7.3.4 Informed Transmitter Case

We now turn to the situation in which a feedback channel is available so that both the transmitter and the receiver have knowledge of the channel state.

7.3.4.1 Point-to-Point Image/Video Transmission with Feedback

If the current channel state or feedback from the receiver is available at the transmitter, this information can often be used to improve the system performance for channels with memory. In the case of perfect channel state information availability to both transmitter and receiver, they can both tune to the instantaneous channel state information (see Figure 7.11(b)). Clearly, with other conditions being equal, the performance for this case will be superior to that of the uninformed transmitter case, where the receiver alone can tune the resolution, while the transmitter has to transmit all resolutions. As noted earlier, the use of a multiresolution framework is not theoretically optimal, but a multiresolution design is far more flexible and convenient, and in practice, there is little performance loss [73, 77]. The use of channel-matched hierarchical vector quantization for image transmission over noisy channels with feedback has been described in [77], building on the hierarchical VQ source coding framework of [78].

A significant advantage of having feedback channels is that for applications where delay requirements are not overly stringent, they can be used to signal errors in the received digital stream if additional redundancy is spent on error detection (ARQ schemes). Indeed, ARQ can be used as a universal channel coding scheme, as pointed out in Section 7.2.2. One of the most effective ARQ schemes is called the Class 2 hybrid ARQ [40], which basically uses an ARQ-based scheme to alternately send data and parity. (A half-rate invertible code is used from which, given the parity information, the data can be obtained by simple inversion.)

Note that improved versions of this hybrid ARQ scheme can be obtained according to a multiresolution concept by using the idea of incremental parity. For example, this idea can be implemented very efficiently by using systematic versions of a family of UEP channel codes like RCPC, as was shown in [79]. The idea is simple. The data is sent first with a few parity bits (highest rate code). If this data is acknowledged without error (as detected with a block CRC check), the next data packet is transmitted; otherwise, only incremental parity bits (corresponding to the next level of channel code) are sent. Thus, the strength of the channel code is gradually adapting to the channel conditions, permitting the channel bandwidth to be used efficiently. Note that there is a price to be paid in terms of increased redundancy (i.e., bits spent on error detection) and, more importantly for real-time applications, delays due to retransmissions. There exists an interesting trade-off between delay and redundancy that can be explored in a continuous way that uses a framework based on error-detection via arithmetic coding [80]. Typically, a 16-bit CRC code performs error detection in ARQ protocols [40]. Though efficient, the CRC can detect errors only after an entire block of data is received. In [80], a method of error detection is used that provides a continuous trade-off between the amount of redundancy added and the amount of time before an error is detected. The method of detection, achieved with an arithmetic codec, has the attractive feature that it can be combined very easily with an arithmetic-coding based source coder, as is popular in state-of-the-art image coders [23, 24]. When this method of error detection is applied to ARQ protocols, significant gains in throughput performance (or equivalently, delivered image quality) are obtained over conventional ARQ schemes [80].

Recent results [71] also demonstrate the gains of continuous error detection applied to serial concatenated coding schemes with convolutional codes. As mentioned in Section 7.3.1, continuous error detection can be integrated into list-Viterbi decoding to improve system performance in the face of limited memory/complexity constraints. By combining both ARQ and serial, concatenated forward error correction with continuous error detection, powerful hybrid ARQ schemes can be devised.

7.3.5 Hybrid Techniques in Image Transmission

We have hinted at the potential of hybrid analog/digital techniques (or, in other words, compressed/uncompressed methods) for image transmission. Before we explore this topic in more detail, let us summarize the facts. The compressed mode of operation has a much more compact representation but is maximally vulnerable to channel errors. Therefore, this mode needs controlled redundancy via channel coding to ensure that these errors are not catastrophic, but this insurance may come at a steeper price than is necessary. On the other hand, the uncompressed mode has increased error-resilience due to the lack of error propagation, but compression efficiency is potentially compromised. This situation leads to the obvious question of whether it is possible to get the best of both worlds.

As a prelude to addressing this interesting question, we consider a fundamental but little-referenced result from Berger's book on rate-distortion theory [3] that is highly relevant to the approach one might adopt in devising hybrid solutions; see also [41].

7.3.5.1 Transmission of Gaussian Signals over Gaussian Noise Channels

Shannon taught us that a theoertically optimal way of transmitting information from point to point is via the separation theorem. This approach involves the source coder doing the best it can (in the rate-distortion sense) in tandem and in separation from the channel coder doing the best it can (in the sense of transmitting at a rate that is within an arbitrary positive ε of the channel capacity). Thus, the smallest distortion attainable by any source-channel system where the channel capacity is C is given by $D(C)$, i.e., the source distortion-rate $D(R)$ function evaluated at $R = C$. As with all Shannon-like results, this result comes with the usual package of potentially infinitely complex codes for both source and channel coding.

There is an illuminating special case, however, that dispenses with this infinite complexity argument without losing theoretical optimality. This case serves as a great motivator for considering hybrid joint source-channel coding methods in image transmission, as we describe shortly. Suppose we wish to communicate over an AWGN channel with X being a discrete-time, continuous-amplitude source, and N the AWGN noise that is uncorrelated with X. Suppose $Y = f(X)$ is the encoder function such that Y is the transmitted signal. The received signal is $\hat{Y} = Y + N$. The decoder function $g(\cdot)$ performs an optimal estimation of X, i.e., it reconstructs $\hat{X} = g(\hat{Y})$.[7] The problem is to determine $f(\cdot)$ and $g(\cdot)$ to minimize the expected distortion $E[(X - \hat{X})^2]$ subject to the variance of Y being fixed (the last constrains the transmitted energy). In [3] (p. 162), Berger showed that:

"The optimum PAM system for transmitting a memoryless $N(0, \sigma_s^2)$ source at the Nyquist rate over an ideal bandlimited channel with additive, zero mean, white Gaussian noise and an average input power constraint achieves the least MSE theoretically attainable with any communication system whatsoever."

---

[7]Note that X, Y, $\hat{Y}$, and $\hat{X}$ are vectors of possibly different dimensions. In our discussion, we suppress the issue of different dimensionalities for clarity of presentation and assume that they are all scalar random variables.

The version of this result applied to discrete memoryless AWGN channels rather than continuous Gaussian noise channels can easily be shown by observing that the distortion-rate function for the Gaussian source evaluated at the capacity of a power-constrained AWGN channel is equal to

$$E[(X - \hat{X})^2] = \sigma_s^2(1 + P/N_0)^{-1},$$

where $N_0 = E[N^2]$ is the variance of the noise and $P = E[Y^2]$ is the energy of the transmitted signal per source symbol. The smallest possible theoretical MSE distortion (above) is, interestingly, at the same time that of the optimal PAM system for the same power constraint, where the encoder function $f(\cdot)$ is a simple linear scaling of the signal by the factor $\sqrt{P/\sigma_s^2}$, and the function $g(\cdot)$ is the linear minimum mean square error estimator of X:

$$\hat{X} = \sqrt{\sigma_s^2/P}(1 + N_0/P)^{-1}\hat{Y}.$$

These results clearly show that *no digital compression and channel coding scheme can outperform the simple linear mappings* for a Gaussian source and an ideal, band-limited AWGN channel. Also observe that source and channel coding are not performed separately but in a joint linear mapping operation. This is a simple but powerful example of the potential of joint source-channel coding. When the source is not Gaussian, however (e.g., even generalized Gaussian [81]), linear mappings are no longer optimal. For this case, optimal analytical mappings are not known, and one has to resort to numerical methods [60], as illustrated in Section 7.3.3.3.

Further, the optimality of linear mappings does not hold for the case of composite or mixture sources, even in the Gaussian case [82]. This case is of interest because Gaussian mixture distributions have been shown to be very accurate for modeling wavelet image coefficients [29], where a coding algorithm based on this model attains performance that ranks among the very best in the cited literature. For clarity of the presentation, we consider a slightly simpler version of this problem here. Suppose the source (typically, the wavelet image decomposition) consists of independently distributed Gaussian random variables having different standard deviations σ_i, $i = 1,2,...N$. This source is to be transmitted over N identical parallel Gaussian channels with a total power constraint. Let P and D_{OPTA} be the average energy and distortion per source sample. Then, the optimal performance theoretically attainable is obtained by "inverse waterfilling" for source coding and "waterfilling" for channel capacity [4], and, in a low target distortion case, can be shown to be

$$D_{\mathrm{OPTA}} = \left(\prod_{i=1}^{N}\sigma_i\right)^{2/N}\frac{1}{(1 + P/N_0)}.$$

Unfortunately, the optimal PAM for each channel will not achieve this performance, and the distortion will increase by a factor K given by

$$K = \frac{D_{\text{PAM}}}{D_{\text{OPTA}}} = \frac{\left(\frac{1}{N}\sum_{i=1}^{N}\sigma_i\right)^2}{\left(\prod_{i=1}^{N}\sigma_i\right)^{2/N}}.$$

Nevertheless, K typically is small (empirical results reveal that it is typically about 0.5 dB for wavelet coefficients of typical natural images [82]); then, considering the complexity needed to approach the theoretical upper bound by use of conventional Shannon-type arguments, the simplicity of a joint source-channel mapping approach becomes very attractive.

7.3.5.2 Applications to Image Transmission

For recovery from channel errors, some amount of the communication link load has to be reserved for redundant information. This fact, however, does not necessarily imply that this redundancy should be inserted *in toto* by channel-coding techniques into a maximally compressed data stream. In [2], Shannon mentioned the possibility of using the redundancy left in the source to combat channel errors. For many practical image and video applications, it may well be better to optimally split the redundant information between source data and channel code parity. Indeed, the task of assigning all redundancy to the channel coder is a corollary of the separation principle and may well be suboptimal for finite complexity systems.

We illustrate by a specific example the advantage of this approach [83]. The basic idea is that the source redundancy, which is still present in practical state-of-the-art image and video source coders (and which enables postprocessing, error concealment, etc.) can be used to improve the performance of the channel decoder or to recover from channel errors in the decoding process *as early as possible.* One promising approach to using source redundancy is based on the philosophy that if there is any a priori information about the source, obtained by observing the previously received channel signals, this information can be used to aid the channel decoding. Instead of making errors in the channel decoder and then trying to correct for them with postprocessing techniques, it may be smarter to put forth the best effort up-front to avoid having to do error concealment in the first place. This approach was advocated in [84], where the a priori information about the source was used to help the Viterbi decoder in estimating the correct path for RCPC-encoded data. This idea was developed further in [79], where the usefulness of digital compression via entropy coding was questioned. It was demonstrated that, for certain situations, it pays to leave the redundancy in the source instead of trying to get rid of it by compression and than reinserting it with the channel coding. However, the question of optimal partitioning of the parity information between source and channel coders was not addressed. Also, lossless transmission was enforced with a hybrid ARQ scheme, which is typically suboptimal for video communications applications due to the high loss tolerance of the source and strict delay requirements.

The question of efficient allocation of redundancy between source and channel coders for delivery of visual data over low-power wireless channels has been recently addressed in [83] as the question of how to efficiently represent an image or video source into compressed and uncompressed subsets. The motivation is based on integrating the two classes of joint source-channel coding techniques (based on analog and digital transmission techniques, respectively) mentioned in Section 7.3.1.

To summarize, the first class consists of a quantized, entropy-coded, compressed layered source representation illustrated with the conventional digital system (upper path) in Figure 7.14. The goal here is to minimize the total distortion

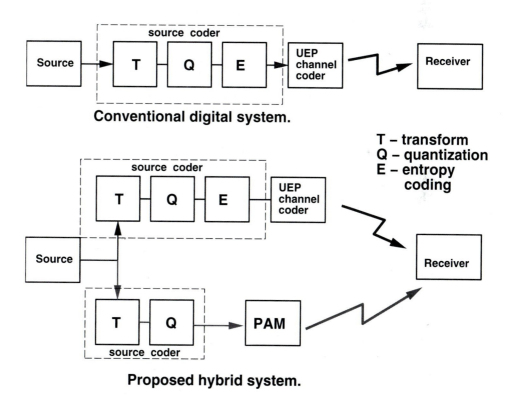

Figure 7.14   Conventional digital and hybrid system diagrams. The upper path of the hybrid system corresponds to the conventional entropy-coded digital system (first approach). The lower path of the hybrid system corresponds to a non-entropy-coded system (second approach).

due both to source quantization *and* to channel noise, subject to a total bit rate constraint on source and channel codes. This minimization could be done with a state-of-the-art image compression platform based on entropy coding (variable code lengths), such as the zerotree wavelet-based SFQ coder of Section 7.1.2.3, followed by a power-efficient modulation scheme such as BPSK [53]. The second class of methods involves the combined design of source quantizers *without entropy-coding* (fixed code lengths) jointly with a PAM modulation scheme (recall Section 7.3.5.1), with a view to minimizing the total distortion subject to a total transmission power constraint (lower path in Figure 7.14). The joint source-channel coding algorithm of [60] (see the example of Section 7.3.3.3) based on energy-optimized, multiresolution, codebook design is a representative of this class.

The intuition behind trying to integrate these classes comes from understanding the fundamental trade-offs associated with a mixed mode of operation. When the channel is clean, high compression ratios are desirable because bit errors are rare, and the compressed mode is preferred. When the channel quality degrades, however, the unequal importance of the source bits that is typical of compressed imagery requires unequal error protection techniques. When the channel degrades to a certain point, lowering the source resolution and transmitting it more reliably is more important (see the example of Section 7.3.3.3). A drawback of the uncompressed mode is that there is no notion of guaranteed QoS (i.e., bit error rate) as is typical of conventional digital communication systems. This property limits its utility in multistage transmission systems, where, for example, perfect error recovery is possible with the use of line repeaters. However, for a large class of single-hop end-user links, e.g., mobile terminals and "backpack" mobile units, this lack of guaranteed QoS is not an issue. Another drawback of the uncompressed mode of operation is that the number of channel uses is likely to be needlessly large. For example, an $N \times N$ image sent using scalar quantization and a 1 to 1 mapping between quantization and modulation points as in [60]—see lower path of Figure 7.14—would require N^2 channel uses, which could exceed available resources for even moderate values of N. This problem could be tackled with vector mappings from source domain to modulation domain as in [67]. An alternate and simpler method based on scalar mappings has been described in [83], where the key idea is to dispense with the need to transmit N^2 coefficients through the use of the wavelet zerotree structure, and combine this structure with simple scalar quantization of the significant image wavelet coefficients. This approach results in a high-performance image coder for slow-fading, energy-constrained channels.

The idea in [83] is to induce a source decomposition into two components, each suited to the appropriate mode of transmission outlined. Thus, while a subset of the image decomposition is treated much more effectively with the compressed mode above, its complementary subset can fairly accurately approximate the above assumptions for which PAM is theoretically optimal (see Section 7.3.5.1). Taking as a platform the zerotree-based, wavelet SFQ image coder of [27] (see Section 7.1.2.3), the coded data, consisting of the digital zerotree significance map information and the residue subband coefficients, is partitioned into two subsets, one of which is compressed, using entropy-coding, and unequally error-protected, using appropriate RCPC channel coding. The other component is transmitted uncompressed, using a one-to-one PAM-mapping. The key idea is that the number of analog channel uses can be reduced with the aid of the digital zerotree data structure, which requires the transmission of only significant wavelet coefficients. By efficiently controlling the trade-off between these two modes in a simple way, it was shown that significant gains (2–3 dB in PSNR) can be attained by this hybrid system over fully compressed UEP digital systems. An interesting observation is that the lower the delay requirement, i.e., the lower the rate of channel usage per source symbol, the higher the fraction of the uncompressed mode that is used by the optimal hybrid system [83]. See the reconstructed images of Figure 7.15 and

Figure 7.15 Reconstructed image in the conventional reference system in the Rayleigh channel 31.9 dB PSNR. $E_s/No = 17$dB. BPSK with soft decisions and perfect channel state information is assumed. The total rate is 0.2 channel usages per source symbol.

Figure 7.16 Reconstructed image in the hybrid system in Rayleigh channel 34.7 dB PSNR. $\mathcal{E}_s/No = 17$ dB. Perfect channel state information is assumed. The total rate is 0.2 channel usages per source symbol.

Figure 7.16 for a subjective comparison of the conventional and hybrid systems, respectively.

7.4 CONCLUDING REMARKS

In this chapter, we have outlined some of the key research issues related to the design of practical, efficient, robust image communications systems for applications like wireless communications and video transport over heterogeneous networks. We have stressed the use of flexible and elegant multiresolution frameworks for designing joint source-channel coding algorithms for such systems. The field of joint source-channel coding, as pertaining to the research, design, and deployment of fundamentally sound and practical end-to-end efficient image communications systems and algorithms, is in its relative infancy. There are many interesting open questions, and in this multimedia age, where one is in constant danger of being unable to separate hype from substance, this field stands out as offering an attractive arena for the fusion of sound theory with important applications and promises to be a research area with high impact.

REFERENCES

[1] S. Lin and D. J. Costello, *An Introduction to Error-Correcting Codes.* Englewood Cliffs, NJ: Prentice-Hall, 1984.

[2] C. E. Shannon, "A mathematical theory of communication," *Bell Syst. Tech. Journal,* vol. 27, pp. 379–423, 1948.

[3] T. Berger, *Rate Distortion Theory.* Englewood Cliffs, NJ: Prentice Hall, 1971.

[4] T. M. Cover and J. A. Thomas, *Elements of Information Theory.* New York: Wiley Inter-science, 1991.

[5] W. H. Equitz and T. M. Cover, "Successive refinement of information," *IEEE Trans. Inform. Theory,* vol. 37, pp. 269–275, March 1991.

[6] P. J. Burt and E. H. Adelson, "The Laplacian pyramid as a compact image code," *IEEE Trans. Commun.,* vol. 31, pp. 532–540, April 1983.

[7] K. M. Uz, M. Vetterli, and D. LeGall, "Interpolative multiresolution coding of advanced television with compatible subchannels," *IEEE Trans. Circuits Syst. Video Technol.,* Special Issue on Signal Processing for Advanced Television, vol. 1, pp. 86–99, March 1991.

[8] M. Vetterli and J. Kovačević, *Wavelets and Subband Coding.* Englewood Cliffs, NJ: Prentice Hall, 1995.

[9] A. K. Jain, *Fundamentals of Digital Image Processing.* Englewood Cliffs, NJ: Prentice Hall, 1989.

[10] R. E. Crochiere, S. A. Webber, and J. L. Flanagan, "Digital coding of speech in sub-bands," *Bell Syst. Tech. J.,* vol. 55, pp. 1069–1085, Oct. 1976.

[11] D. Esteban and C. Galand, "Application of quadrature mirror filters to split band voice coding schemes," in *Proc. Int. Conf. Acoust., Speech, Signal Processing,* pp. 191–195, May 1977.

[12] M. Vetterli, "Multidimensional subband coding: Some theory and algorithms," *Signal Processing,* vol. 6, pp. 97–112, February 1984.

[13] J. W. Woods and S. D. O'Neil, "Sub-band coding of images," *IEEE Trans. Acoust., Speech, Signal Processing,* vol. 34, pp. 1278–1288, May 1986.

[14] P. H. Westerink, J. Biemond, D. E. Boekee, and J. W. Woods, "Subband coding of images using vector quantization," *IEEE Trans. Commun.,* vol. 36, pp. 713–719, June 1988.

[15] J. W. Woods, *Subband Image Coding.* Boston: Kluwer Academic Press, 1991.

[16] A. Gersho and R. M. Gray, *Vector Quantization and Signal Compression.* Boston: Kluwer Academic Press, 1992.

[17] G. Karlsson and M. Vetterli, "Three-dimensional subband coding of video," in *Proc. IEEE Int. Conf. Acoust., Speech, Signal Processing,* (New York), pp. 1100–1103, April 1988.

[18] J.-R. Ohm, "Three-dimensional subband coding with motion compensation," *IEEE Trans. Image Processing,* Special issue on Image Sequence Compression, vol. 3, pp. 559–571, September 1994.

[19] D. Taubman and A. Zakhor, "Multi-rate 3-D subband coding of video," *IEEE Trans. Image Processing,* Special Issue on Image Sequence Compression, vol. 3, pp. 572–588, September 1994.

[20] J. M. Shapiro, "An embedded wavelet hierarchical image coder," in *Proc. Int. Conf. Acoust., Speech, Signal Processing,* (San Francisco), pp. 657–660, March 1992.

[21] A. S. Lewis and G. Knowles, "Image compression using the 2-D wavelet transform," *IEEE Trans. Image Processing,* vol. 1, pp. 244–250, April 1992.

[22] M. Antonini, M. Barlaud, P. Mathieu, and I. Daubechies, "Image coding using wavelet transform," *IEEE Trans. Image Processing,* vol. 1, pp. 205–220, April 1992.

[23] J. M. Shapiro, "Embedded image coding using zerotrees of wavelet coefficients," *IEEE Trans. Signal Processing,* Special Issue on Wavelets and Signal Processing, vol. 41, pp. 3445–3462, December 1993.

[24] A. Said and W. A. Pearlman, "A new fast and efficient image coder based on set partitioning in hierarchical trees," *IEEE Trans. Circuits Syst. Video Technol.,* pp. 243–250, June 1996.

[25] Z. Xiong, K. Ramchandran, and M. T. Orchard, "Joint optimization of scalar and tree-structured quantization of wavelet image decomposition," in *Proc. Asilomar Conf. Signals, Syst., Computers,* vol. 2, (Pacific Grove, CA), pp. 891–895, November 1993.

[26] R. L. Joshi, V. J. Crump, and T. R. Fisher, "Image subband coding using arithmetic and trellis coded quantization," *IEEE Trans. Circuits Syst. Video Technol.,* vol. 5, no. 6, pp. 513–523, December 1995.

[27] Z. Xiong, K. Ramchandran, and M. T. Orchard, "Space-frequency quantization for wavelet image coding," *IEEE Trans. Image Processing,* vol. 6, pp. 677–693, May 1997.

[28] R. L. Joshi, T. R. Fisher, and R. H. Bamberger, "Optimum classification in subband coding of images," *Proc. Int. Conf. Image Processing,* (Austin, TX), pp. 883–887, November 1994.

[29] S. LoPresto, K. Ramchandran, and M. T. Orchard, "Image coding based on mixture modeling of wavelet coefficients and a fast estimation-quantization framework," *Proc. Data Compression Conf.,* (Snowbird, Utah), pp. 221–230, 1997.

[30] Y. Yoo, A. Ortega, and B. Yu, "Adaptive quantization of image subbands with efficient overhead rate distortion," *Int. Conf. Image Processing,* vol. 2, (Lausanne, Switzerland), pp. 361–364, September 1996.

[31] S. Servetto, K. Ramchandran, and M. T. Orchard, "Morphological representation of wavelet data for image coding," in *Proc. Int. Conf. Acoust., Speech, Signal Processing,* (Detroit, MI), pp. 2229–2232, May 1995.

[32] S. Servetto, K. Ramchandran, and T. S. Huang, "Image and video coding with object indexing support," *J. VLSI Signal Processing,* 1998. To appear, Special Issue on Multimedia Signal Processing.

[33] "JPEG technical specification: Revision (DRAFT), Joint Photographic Experts Group, ISO/IEC JTC1/SC2/WG8, CCITT SGVIII," August 1990.

[34] "MPEG video simulation model three, ISO, coded representation of picture and audio information," 1990.

[35] T. Cover, "Broadcast channels," *IEEE Trans. Inform. Theory,* vol. IT-18, pp. 2–14, January 1972.

[36] K. Ramchandran, A. Ortega, K. M. Uz, and M. Vetterli, "Multiresolution broadcast for digital HDTV using joint source-channel coding," *IEEE J. Select. Areas Commun.*, vol. 11, pp. 6–23, January 1993.

[37] A. R. Calderbank, "Multilevel codes and multistage decoding," *IEEE Trans. Commun.*, vol. 37, pp. 222–229, March 1989.

[38] R. V. Cox, J. Hagenauer, N. Seshadri, and C. Sundberg, "Variable rate sub-band speech coding and matched convolutional channel coding for mobile radio channels," *IEEE Trans. Signal Processing*, vol. 39, pp. 1717–1731, 1991.

[39] Y. H. Kim and J. Modestino, "Adaptive entropy coded subband coding of images," *IEEE Trans. Image Processing*, vol. 1, pp. 31–48, January 1992.

[40] M.-C. Lin, C.-C. Lin, and S. Lin, "Computer search for binary cyclic VEP codes of odd length up to 65," *IEEE Trans. Inform. Theory*, vol. 36, no. 4, pp. 924–935, July 1990.

[41] B. Chen and G. W. Wornell, "Efficient channel coding for analog sources using chaotic systems," in *Proc. IEEE GLOBECOM*, pp. 131–135, 1996. [Also, "Analog error-correcting codes based on chaotic dynamical systems," to appear in *IEEE Trans. Commun.*]

[42] R. Schafer, "Terrestrial transmission of DTVB signals—the European specification," in *Int. Broadcasting Convention*, no. 413, September 1995.

[43] K. Fazel and M. Ruf, "Combined multilevel coding and multiresolution modulation," in *Proc. ICC*, (Geneva), vol. 2, pp. 1081–1085, May 1993.

[44] G. Ungerboeck, "Channel coding with multilevel/phase signals," *IEEE Trans. Inform. Theory*, vol. IT-28, pp. 55–67, January 1982.

[45] S. S. Pradhan and K. Ramchandran, "Efficient layered video delivery over multicarrier systems using optimized embedded modulations," in *Proc. Int. Conf. Image Processing*, vol. 3, pp. 452–455, 1997.

[46] W. L. Zou and Y. Wu, "COFDM : An overview," *IEEE Trans. Broadcasting*, vol. 41, pp. 1–8, March 1995.

[47] J. A. C. Bingham, "Multicarrier modulation for data transmission: An idea whose time has come," *IEEE Commun. Mag.*, pp. 5–14, May 1990.

[48] P. S. Chow, J. C. Tu, and J. M. Cioffi, "A discrete multitone receiver system for HDSL applications," *IEEE J. Select. Areas Commun.*, vol. 9, pp. 895–908, August 1991.

[49] J. M. Cioffi, "A multicarrier primer," in *ANSI T1E1.4 Committee Contribution*, pp. 91–157, November 1991.

[50] I. Kalet, "The multitone channel," *IEEE Trans. Commun.*, vol. 37, no. 2, pp. 119–124, 1989.

[51] R. G. Gallager, *Information Theory and Reliable Communication*. New York: John Wiley & Sons, 1968.

[52] H. S. Wang and N. Moayeri, "Finite-state Markov channel–a useful model for radio communication channels," *IEEE Trans. Veh. Technol.*, vol. 44, pp. 163–171, February 1995.

[53] J. G. Proakis, *Digital Communications*. New York: McGraw-Hill, 1989.

[54] J. Modestino, D. G. Daut, and A. Vickers, "Combined source-channel coding of images using the block cosine transform," *IEEE Trans. Commun.*, vol. COM-29, pp. 1261–1274, September 1981.

[55] N. Farvardin and V. Vaishampayan, "On the performance and complexity of channel-optimized vector quantizers," *IEEE Trans. Inform. Theory*, vol. IT-37, pp. 155–160, Jan. 1991.

[56] F. H. Liu, P. Ho, and V. Cuperman, "Joint source and channel coding using a non-linear receiver," in *Proc. ICC*, vol. 3, pp. 1502–1507, June 1993.

[57] N. Tanabe and N. Farvardin, "Subband image coding using entropy-coded quantization over noisy channels," *IEEE J. Select. Areas Commun.*, vol. 10, pp. 926–943, June 1992.

[58] H. Kumazawa, M. Kasahara, and T. Namekawa, "A construction of vector quantizers for noisy channels," *Electron. Eng. Japan*, vol. 67-B, pp. 39–47, 1984.

[59] K. A. Zeger and A. Gersho, "Zero redundancy channel coding in vector quantization," *Electron. Lett.*, pp. 654–655, June 1987.

[60] I. Kozintsev and K. Ramchandran, "Multiresolution joint source-channel coding using embedded constellations for power-constrained time-varying channels.," in *Proc. Int. Conf. Acoust., Speech, and Signal Processing*, pp. 2345–2348, May 1996.

[61] T. R. Fischer and M. W. Marcellin, "Joint trellis coded quantization/modulation," *IEEE Trans. Commun.*, vol. 39, pp. 172–176, February 1991.

[62] W. F. Schreiber, "All-digital HDTV terrestrial broadcasting in the U.S. : Some problems and possible solutions," *Workshop on Advanced Television, ENST, Paris*, May 1991.

[63] N. Farvardin and V. Vaishampayan, "Optimal quantizer design for noisy channels: an approach to combined source-channel coding," *IEEE Trans. Inform. Theory*, vol. IT-33, pp. 827–838, November 1987.

[64] K. Zeger and A. Gersho, "Pseudo-gray coding," *IEEE Trans. Commun.*, vol. 38, pp. 2147–2158, December 1990.

[65] P. G. M. de Bot, "Multiresolution transmission over the AWGN channel. Technical report TN 181/92," Tech. report, Philips Research Laboratories, Eindhoven, The Netherlands, June 1992.

[66] M. Polley, S. Wee, and W. Schreiber, "Hybrid channel coding for multiresolution HDTV terrestrial broadcasting," in *Proc. Int. Conf. Image Processing*, vol. 1, pp. 243–247, November 1994.

[67] T. Ramstad, "Efficient and robust communication based on signal decomposition and approximative multidimensional mappings between source and channel spaces," in *NORSIG'96 tutorial*, 1996.

[68] J. Hagenauer, N. Seshadri, and C.-E. W. Sundberg, "The performance of rate-compatible punctured convolutional codes for digital mobile radio," *IEEE Trans. Commun.*, vol. 38, pp. 966–980, July 1990.

[69] P. G. Sherwood and K. Zeger, "Progressive image coding on noisy channels," *IEEE Signal Processing Lett.*, vol. 4, pp. 654–655, July 1997.

[70] N. Seshadri and C.-E. W. Sundberg, "List-Viterbi decoding algorithms with applications," *IEEE Trans. Commun.*, vol. 42, pp. 313–323, February/March/April 1994.

[71] I. Kozintsev, J. Chou, and K. Ramchandran, "Image transmission using arithmetic coding based continuous error detection.," in *Data Compression Conf.*, (Snowbird, Utah), April 1998. To appear.

[72] M. Ruf and J. Modestino, "Rate-distortion performance for joint source and channel coding of images," in *Proc. Int. Conf. Image Processing*, pp. 77–80, 1995.

[73] I. Kozintsev and K. Ramchandran, "Robust image transmission over energy-constrained time-varying channels using multiresolution joint source–channel coding," *IEEE Trans. Signal Processing*, vol. 46, April 1998. To appear.

[74] S. McCanne, M. Vetterli, and V. Jacobson, "Low-complexity video coding for receiver-driven layered multicast," *IEEE J. Select. Areas Commun.*, vol. 15, pp. 983–1001, August 1997.

[75] D. Bertsekas, *Nonlinear Programming*. Belmont, MA: Athena Scientific, 1995.

[76] P. Chou, T. Lookabaugh, and R. Gray, "Entropy-constrained vector quantization," *IEEE Trans. Acoust. Speech Signal Processing*, vol. 37, pp. 31–42, January 1989.

[77] H. Jafarkhani and N. Farvardin, "Channel-matched hierarchical table-lookup vector quantization for transmission of video over wireless channels," *Proc. Int. Conf. Image Processing*, vol. 3, (Lausanne, Switzerland), pp. 755–758, September 1996.

[78] M. Vishwanath and P. Chou, "An efficient algorithm for hierarchical compression of video," *Proc. Int. Conf. Image Processing*, vol. 3, (Austin, TX), pp. 275–279, November 1994.

[79] G. Buch, F. Burket, J. Hagenauer, and B. Kukla, "To compress or not to compress?," in *Proc. IEEE GLOBECOM*, pp. 198–203, 1996.

[80] J. Chou and K. Ramchandran, "Arithmetic-coding based continuous error detection for efficient ARQ-based image transmission," in *Proc. Asilomar Conf. Signals, Syst. Computers*, (Pacific Grove, CA), November 1997.

[81] K. A. Birney and T. R. Fischer, "On the modeling of DCT and subband image data for compression," *IEEE Trans. Image Processing*, vol. 4, pp. 186–193, February 1995.

[82] I. Kozintsev and K. Ramchandran, "Robust image transmission using multiresolution joint source-channel coding," in *Proc. IEEE Inform. Theory Workshop*, (San Diego, CA), pp. 58–59, February 1998.

[83] I. Kozintsev and K. Ramchandran, "A hybrid compressed-uncompressed framework for wireless image transmission," in *Proc. Int. Conf. Image Processing*, (Santa Barbara, California), vol. 2, pp. 77–80, October 1997.

[84] J. Hagenauer, "Source-controlled channel decoding," *IEEE Trans. Commun.*, vol. 43, pp. 2449–2457, September 1995.

[85] J. Ooi and G. W. Wornell, "Fast iterative coding techniques for feedback channels," to appear in *IEEE Trans. Inform. Theory*.

ACKNOWLEDGMENT

The authors would like to acknowledge Igor Kozintsev for his insightful comments and discussions, as well as for his help in improving the quality of this document.

8

Underwater Acoustic Communications

David Brady

James C. Preisig

Much of the development of the preceding chapters is applicable to a broad range of wireless networks, although there is a strong emphasis on radio-frequency (RF) systems and their applications. And while underwater acoustic channels and systems share many features with their RF counterparts, there are also important differences. In this chapter, we describe the special characteristics of underwater acoustic channels and show how they impact system design.

The underwater acoustic channel (UAC) is quite possibly nature's most unforgiving wireless communication medium. Multipath delay spreads exceed 60 ms for horizontal medium-range channels, and frequency responses may exhibit deep nulls. Fading processes may be fast or slow, frequency selective or frequency nonselective, depending on the direction of propagation and conditions of the water column. Battery and mission lifetimes restrict the transmitter powers of practical modems to 30W. High-frequency absorption losses and low-frequency ship noise confine the transmission bandwidth to less then 40 kHz for medium-range shallow-water applications. The UAC is a broadcast channel, and the aggregate bandwidth must be shared by an asynchronous group of noncooperating users. The low propagation velocity of sound in water (1500 m/s) translates wave or modest platform motion to significant Doppler compression and permits the use of a slowly time-varying channel model only if phase and synchronization tracking are given special attention.

Underwater communication has been used since the beginning of the 20th century to permit a data link between the surface and the water column [20].[1] Today underwater acoustic telemetry is used to communicate between untethered platforms, such as underwater vehicles (UV) and data logging stations. Manned or unmanned oceanographic exploration, ocean monitoring, and the offshore oil industry all rely on underwater acoustic telemetry.

This chapter presents an overview of the state of underwater acoustic communications and the role of signal processing in this field. We begin with the development of an appropriate channel model. We describe the relationship between environmental conditions and the characteristics of the channel model, then we analyze the impact of these characteristics on the communications problem. We then turn our attention to digital modulation and signal processing techniques that facilitate demodulation: adaptive equalization, estimation, and detection in sparse systems, multiuser detection, and multisensor detection. We present adaptive techniques that are especially suited for the UAC to reduce convergence time, tracking error, or computational complexity. The chapter also highlights current problems in this active research area and reviews approaches to their solutions.

8.1 THE UNDERWATER ACOUSTIC CHANNEL

From the communications perspective, the underwater acoustic channel poses many challenges to the realization of reliable, high-rate communications. In this section, we summarize many of the salient physical characteristics of the channel and their effects on the communications problem. The section begins with a discussion of the ray model for acoustic propagation, its dependence on the speed of propagation of sound in the ocean, and the dominant mechanisms for sound speed variability in the ocean. Then, we address the bandwidth constraints imposed by the absorption and spreading of sound in the ocean, the sources of noise in the ocean, and their effects on achievable signal to noise ratios. Finally, we discuss the dominating influence of both time-invariant and time-varying multipath propagation and relate these effects to the input/output response and statistical characterization

---

[1]However, the possibilities of underwater communication were envisioned by Leonardo da Vinci as early as 1490: "If you cause your ship to stop, and place the head of a long tube in the water and place the outer extremity to your ear, you will hear ships at a great distance from you." (See [1].)

of the communications channel. Excellent material covering the general topic of sound propagation in the ocean is contained in [2, 3].

8.1.1 The Ray Propagation Model

In most conditions, sound waves at middle and high frequencies (> 5 kHz) can be reasonably modeled as propagating along paths or rays through the ocean. A good rule of thumb is that this model is valid when the spatial scale of the inhomogeneities in the ocean is larger than the wavelength of the sound. Since the speed of sound is approximately 1500 m/s in the ocean, the wavelength in this region is less than one-third of a meter. In a homogeneous environment, the rays' paths would follow straight lines radiating from the source. However, the sound speed structure of the ocean is highly variable, both spatially and temporally. In accordance with Snell's Law, the spatial variability induces a bending of the rays referred to as *refraction*. Consider a simplified, two-dimensional model for the ocean, shown in Figure 8.1, in which the sound speed c is a function of depth z.

In this case, the path of a refracted ray in this environment is described by the pair of differential equations

$$\frac{d\theta}{dx} = c(z)^{-1}\frac{dc(z)}{dz} \tag{8.1}$$

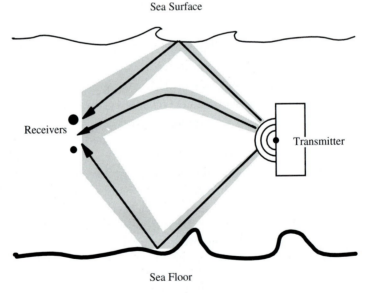

Figure 8.1   The ray model for acoustic propagation.

and

$$\frac{dz}{dx} = -\tan(\theta(x)). \tag{8.2}$$

Here, x denotes the horizontal distance from the source, $z(x)$ denotes the depth of the ray at that range, and $\theta(x)$ denotes the angle of the path with respect to the horizontal ($\theta(x) < 0$ indicates a ray pointed downward). In an environment in which the speed of sounds varies in all three physical dimensions (x, y, and z), the ray paths refract horizontally as well as vertically. However, the spatial gradient of the speed of sound tends to be much smaller in the horizontal than in the vertical, so the horizontal refraction is usually much smaller than the vertical refraction.

    In addition to refraction within the water column, rays experience reflection from the sea surface and sea floor. The nature and strength of the reflections or scattering depend primarily on the amplitude and spatial scale of the roughness of the water/air or water/bottom interface and the density, sound speed, and absorption properties of the bottom material. Accounting for both the refraction of rays as described by Snell's Law and the reflection of rays at the sea surface and sea floor, the sound propagates along many paths from a source to a receiver, as shown by the solid lines in Figure 8.1. The nature and effects of this multipath propagation are discussed more fully in Section 8.1.4.

8.1.2 Sources of Sound Speed Variability

The spatial and temporal variability of the speed of sound in water is a result of the inhomogeneity of the physical properties of the water. A reasonable approximation of the sound speed in sea water is given by [2]

$$c(T, S, z) = 1449.2 + 4.6T - 0.055T^2 + 0.00029T^3 \tag{8.3}$$
$$+ (1.34 - 0.01T)(S - 35) + 0.016z.$$

Here, c is the sound speed in meters/second, T is the water temperature in degrees Celsius, S is the salinity in parts per thousand, and z is the depth in meters of the point at which the sound speed is evaluated. The dependence of sound speed on depth is due to the dependence of sound speed on the hydrostatic pressure. The functional dependence of the speed of sound on environmental conditions yields different sound speed characteristics in different environments.

    In the deep oceans at mid-latitudes, the approximately isosaline water combined with solar heating of the upper portion of the water column yields the characteristic sound velocity profile shown in Figure 8.2. When this type of sound velocity profile is encountered, the refraction of rays towards the region of minimal

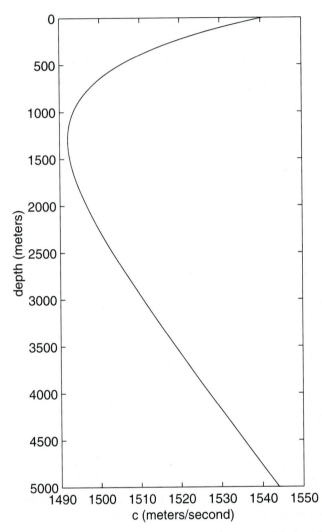

Figure 8.2 Typical deep water sound speed profile.

sound speed creates a natural waveguide in which rays are trapped. This region is referred to as the *sound channel*, or *SOFAR channel*.

In shallow waters, a greater variety of sound speed conditions is often encountered. Following the passage of a storm when the entire water column has been well mixed or in the winter months when solar heating is minimal, the sound speed is nearly constant with depth. At other times, solar heating, the presence of less saline water flowing from rivers and bays, and the tidally driven mixing of this water with oceanic water masses yield a characteristic sound speed profile, shown in Figure 8.3. With no mid-water sound channel present, sound propagating in shallow water experiences a greater extent of interaction with the sea surface and

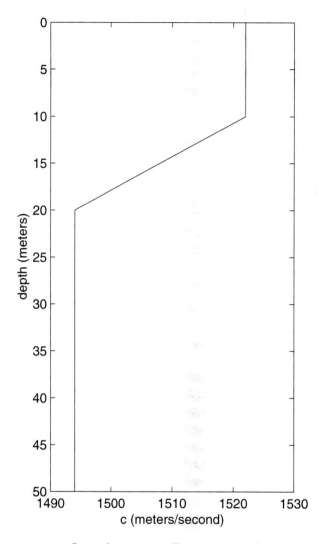

Figure 8.3 Typical shallow water sound speed profile.

sea floor than generally occurs in the deep oceans. In general, this interaction results in higher signal losses and temporal variability.

8.1.3 Signal Losses and Ambient Noise

The signal losses encountered by propagating sound and the ambient noise present in the ocean significantly influence the received signal-to-noise ratio (SNR). The primary mechanisms of signal loss are *spreading loss, absorption loss* in both the water and the bottom, and *scattering loss* at the sea surface and sea floor. The spreading loss associated with the propagation of sound occurs in primarily two types. In regions close to the sound source where the wavefront radiates spherically, the conservation of energy results in an attenuation of the signal energy by a factor of r^{-2}, where r is the range from the source. Further from the source, the vertical propagation of the sound energy reaches the limits imposed by the sea bottom, sea surface, or sound channel. At this point, the wavefront begins to radiate in a cylindrical fashion from the source, and the resulting attenuation of the signal energy is by a factor of r^{-1}.

While the speed of sound and spreading losses are independent of the frequency of the sound, the absorption of sound by the water is highly dependent on frequency. The absorption of sound by the water is the result of the conversion of acoustic energy into heat. A number of physical mechanisms govern this conversion (see [3] for a complete description). Formulas such as Eq. (3.3.6) in [3] are available to compute the attenuation rate of sound due to absorption in sea water. Figure 8.4 shows the attenuation coefficient in dB/kilometer as a function of frequency for sound in sea water at the sea surface with a temperature of 14 degrees Celsius and salinity of 35 parts per thousand. As can be seen, the attenuation coefficient rises rapidly with frequency, effectively limiting the channel bandwidth at all but very short ranges.

Acoustic signals are attenuated by interaction with the sea surface and sea floor. The losses are caused by rough surface scattering at both interfaces and absorption losses within the bottom. When the sea surface is rough, the reflection of the acoustic energy from the surface is not specular but is scattered in a multitude of directions. Most of the energy that is scattered in a direction other than the direction of the receiver is effectively lost. While the sea floor does not perfectly reflect acoustic signals, the reflected signals also contain a mixture of specularly reflected and scattered signals. Once again, the energy in the scattered signal is lost. In addition, since the reflection of energy at the sea floor is incomplete, sound will penetrate into the bottom. The absorption of sound in the bottom is significantly higher than that in the water, resulting in further signal losses.

In the deep oceans, the presence of the sound channel limits the surface and bottom interaction. Sound propagating in the SOFAR channel interacts with neither the sea surface nor sea floor and can travel for long distances with no attenuation other than spreading losses and absorption by the water. Any sound that leaves the sound channel and interacts with the sea surface or sea floor is attenuated quickly and is often ignored at ranges past a few times the water depth. However, in shallow water the lack of a mid-water sound channel makes surface and bottom reflected signals a significant portion of the propagating sound. On all but the calmest of days, the rough characteristic of the sea surface yields nonspecular scattering of the sound. The scattering loss associated with sound interaction with the sea surface

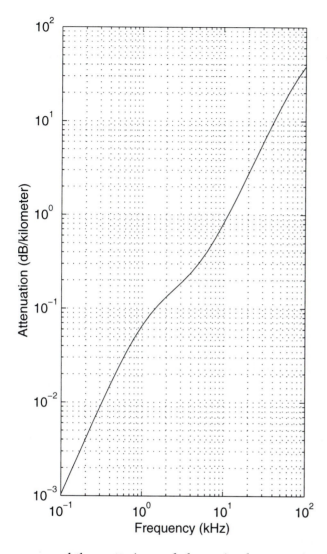

Figure 8.4 Attenuation of sound in sea water.

and the scattering and absorption loss associated with sound interaction with the sea floor conspire to reduce the effective range over which sound will travel in shallow water. Bottom interaction losses are most prevalent in the presence of downward refracting sound speed profiles, such as that shown in Figure 8.3. See [2] for a more complete treatment of these effects.

A second major determinant of the received signal-to-noise ratio is the *ambient noise* present in the ocean. There are many sources of ambient noise in the ocean, including breaking waves, marine life, and passing ships. The nature of the noise depends strongly on its source. Significant noise can be generated by marine organisms such as whales, fish, and shrimp. In the case of shrimp, the noise is highly impulsive and can severely disrupt the operation of a communications system [4, 5]. Another significant natural source of ambient noise is that generated near the surface of the ocean by breaking waves and rain. Since the generation of breaking waves is primarily driven by the winds, noise generated by these sources is highly dependent on the weather conditions [38]. Man-made sources of noise in the ocean contribute to the ambient noise as well. Predominant among these sources are ships' propulsion machinery. With a large number of possible sources in the ocean, the level of the total ambient noise field can vary widely. In one location over a several-week period, measured noise levels ranging from 90 dB to 120 dB were reported [6].

Ignoring the losses due to surface and bottom interaction, signal fading due to multipath effects, and the effects of shadow zones (regions where rays of sound will not reach), the average signal power at a range r from a narrowband source can be estimated by

$$SL \approx 169 \text{ dB} + 10 \log_{10}(P) - \alpha_s r - 20 \log_{10}(d/2) - 10 \log_{10}(r - d/2). \quad (8.4)$$

Here, SL is the signal level in dB, P is the radiated signal power in watts, α_s is the absorption coefficient in dB/meter, r is the range in meters, and d is the ocean depth in meters. It has been assumed here that the radiating transducer is omnidirectional and that $r > d$. The first two terms account for the radiated power, the third term accounts for absorption loss, the fourth term approximates spherical spreading loss, and the final term approximates cylindrical spreading loss. Assuming a center frequency of 15 kHz, the water conditions used to generate Figure 8.4, a water depth of 50 m, and a radiated power of 20 W, the average signal power as a function or range (in kilometers) is shown in Figure 8.5. Assuming that ambient noise conditions range between 90 dB and 120 dB, as reported above, signal-to-noise levels can be seen to range from 32 dB to 2 dB at 1-km range and 8 dB to -22 dB at 10-km range.

8.1.4 The Multipath Propagation Model

As in the case of RF systems discussed in earlier chapters, multipath propagation is one of the dominant environmental influences on the performance of acoustic communications systems in the ocean. This multipath is usually time-varying, and there are many sources of the temporal fluctuations. These include internal waves (i.e., vertical movement of the inhomogeneous layers of the water mass), internal turbulence, tidal flows, surface waves, and platform motion. Researchers have long attempted to develop a stochastic framework for characterizing and analyzing the effect these fluctuations have on acoustic signals. The seminal work in this

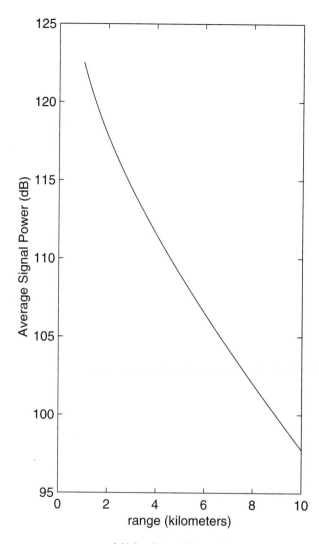

Figure 8.5 Example of signal level vs. range.

area was published in 1979 and reviewed in 1983 [7, 8]. These works concentrate on analyzing the effect of internal waves on acoustic propagation in deep water. However, the paradigm developed therein of decomposing the path followed by sound into micro- and macro-multipath structure serves as a useful starting point for analyzing more general environmental fluctuations. Analyzing these fluctuations in the shallow-water environment continues to be an active area of research. Chapter 10 of [2] also contains a good treatment of the subject. The analysis in [7] and the many derivative publications consider the propagation fluctuations for a propagating monochromatic signal. In Section 8.1.4.1, this method of analysis is detailed. In Section 8.1.4.2, the results from Section 8.1.4.1 are extended to include wideband signals and yield a description of the fluctuations in the input/output response of the channel.

8.1.4.1 Micro- and Macro-Multipath

The micro- and macro-multipath classification decomposes the acoustic channel into that portion whose characteristics depend on slowly varying and quasi-deterministic properties of the ocean and that portion whose characteristics depend on rapidly varying stochastic properties of the ocean. We begin by representing the spatially and temporally varying sound speed structure of the ocean, using

$$c(\mathbf{z}, t) = c_0(1 + U_0(\mathbf{z}, t) + \mu(\mathbf{z}, t)). \tag{8.5}$$

Here, c_0 represents the nominal sound speed, $\mathbf{z}$ is the spatial position vector, and $U_0(\mathbf{z}, t)$ represents the spatially and temporally variant changes to the index of refraction caused by static or slowly changing environmental factors. Such factors include pressure changes with depth and the variations in water temperature and salinity caused by tidal cycles, daily heating and cooling, seasonal changes, and large-scale geographic variations. The term $\mu(\mathbf{z}, t)$ represents the rapid changes to the index of refraction caused by such sources as internal waves and turbulence.

Suppose a source at a particular location transmits a signal $e^{j\omega t}$, and suppose that signal is received by a sensor at another location. Assume that there are no rapid stochastic fluctuations in the environment (i.e., $\mu(\mathbf{z},t) = 0$), that the sea surface is flat, and that the sea floor has only large-scale features. Here, the terms small and large scale refer to spatial scales that are small or large with respect to the nominal acoustic wavelength, $2\pi c_0/\omega$. Let $\tilde{X}_l(t, \omega)$ be the portion of the received signal that would propagate along the lth ray connecting the source to the receiver. We can express $\tilde{X}_l(t, \omega)$ as

$$\tilde{X}_l(t, \omega) = H_l(t, \omega)e^{j\omega t}, \tag{8.6}$$

where $H_l(t, \omega)$ is a complex-valued function that accounts for the slowly varying phase delay and attenuation of the signal propagating along the lth ray. $H_l(t, \omega)$ depends not only on c_0 and $U_0(\mathbf{z}, t)$ but also on the source and receiver positions and the large-scale bathymetric features of the sea floor. The rays defined under the above assumptions (e.g., the solid lines in Figure 8.1) are referred to as the *macro-multipath structure* of the channel.

If we remove the three aforementioned assumptions, the sound would not necessarily follow the rays defined by the macro-multipath structure. Instead, the sound would be refracted and scattered by the small-scale features represented by $\mu(\mathbf{z}, t)$ and the roughness of the sea surface and sea floor. In this case, the actual path followed by the sound is modeled as staying within a ray tube surrounding the nominal ray. In Figure 8.1, the gray regions represent the ray tubes surrounding each nominal ray. For small-amplitude or large-scale environmental fluctuations, the sound may follow a single perturbed path within a ray tube. As the amplitude of the fluctuations increases and their scale decreases, the single path followed by the sound may split into many micropaths within the ray tube. Thus, the portion of the received signal that propagates within the *l*th ray tube will be the sum of signals that propagate along the one or more micropaths within the tube. These micropaths constitute the *micro-multipath structure* of the channel. Accounting for the combination of signals propagating along the micropaths within the ray tube, we can express the portion of the received signal that propagates through the *l*th ray tube as

$$X_l(t, \omega) = \Psi_l(t, \omega)\tilde{X}_l(t, \omega) = \Psi_l(t, \omega)H_l(t, \omega)e^{j\omega t}. \tag{8.7}$$

Here, $\Psi_l(t, \omega)$ accounts for the fluctuations in the received signal due to the micro-multipath structure of the *l*th ray tube.

Accounting for all of the rays between the source and receiver, the received signal can be expressed as

$$X(t, \omega) = \sum_{l=1}^{L} X_l(t, \omega), \tag{8.8}$$

where it is assumed that there are L rays. For different rays between the source and receiver, the ray tubes in the channel generally have little overlap. Thus, unless the scale of the environmental fluctuations is as large as the separation between rays, the micro-multipath-induced signal fluctuations in different ray tubes will show little correlation. This behavior is discussed more fully in Section 8.1.4.3.

The amplitude and spatial scale of the stochastic component of the environmental fluctuations determine the nature of the fluctuations in the micro-multipath structure of the channel. When these environmental fluctuations have a very small amplitude or the scale of the environmental fluctuations is greater than or equal to the radius of a ray tube, the signal will follow a single perturbed path within each ray tube. The channel in this case is said to be *unsaturated*. For slightly stronger fluctuations and with the same scale of fluctuations, the signal will simultaneously propagate along several perturbed paths within each tube and the perturbations in the paths will be coherent. A channel showing these characteristics is said to be *partially saturated*. If the strength of the fluctuations increases beyond this level or if the spatial scale of the environmental fluctuations becomes smaller than the radius of a ray tube while the strength of the fluctuations remains modest, the channel will be *fully saturated*. In this case, the signal will follow many perturbed paths within the ray tube, and the path perturbations will be independent.

The likelihood of encountering a fully saturated channel will increase if either the range from source to receiver is increased or the frequency of the signal is increased. Results in [25] indicate fully saturated channel conditions at a frequency of

50 kHz and range of 1 km. Reference [38] indicates the onset of full channel saturation at a range of 60 km for a frequency of 1 kHz. Analysis in [9, 10] indicates this onset at a range of around 5 km for a frequency of 20 kHz. Reference [40] reports experimental results indicating either an unsaturated or partially saturated channel at ranges of 2 to 5 km and frequency of 10 kHz. While these results provide rough guides as to what channel conditions yield saturation, the actual degree of saturation will be highly dependent on the dominant local sources of micro-multipath fluctuations in the region of propagation.

The theoretical analysis in [9, 10] does not consider the fluctuations caused by surface scattering. References [11, 12, 13, 14] as well as Chapter 9 of [2] contain explicit treatments of surface scattering effects. Experimental evidence [15] also indicates that bottom scattering can have a significant effect on the micro-multipath fluctuations in the shallow-water acoustic communications channel.

8.1.4.2 Wideband Channel Characterization

The monochromatic signal results from the preceding section can be extended to yield a wideband characterization of the channel between source and receiver [16]. The particular characterization we use is the *input delay-spread function* [17] denoted by $g(t, \tau)$ and discussed in Chapters 1 and 3. The channel input/output relationship using the input delay-spread function is

$$x(t) = \int_{-\infty}^{\infty} g(t, \tau)s(t - \tau)\, d\tau. \tag{8.9}$$

Here, $s(t)$ is the channel input and $x(t)$ is the channel output. We can see that if the input delay-spread function is independent of the time t, $g(t, \tau)$ reduces to the time-invariant impulse response. Denote the Fourier transform of $s(t)$ by $S(\omega)$ and define the *time-variant transfer function*, $G(t, \omega)$, using the Fourier transform relation

$$G(t, \omega) = \int_{-\infty}^{\infty} g(t, \tau)e^{j\omega\tau}\, d\tau. \tag{8.10}$$

It is then straightforward to show that

$$x(t) = \frac{1}{2\pi} \int_{-\infty}^{\infty} G(t, \omega)S(\omega)e^{j\omega t}\, d\omega. \tag{8.11}$$

With the Fourier transform of the transmitted signal denoted by $S(\omega)$, we can exploit the linearity of the channel to express the portion of the received signal that propagated through the lth ray tube as

$$x_l(t) = \frac{1}{2\pi} \int_{-\infty}^{\infty} S(\omega)X_l(t, \omega)\, d\omega. \tag{8.12}$$

Substituting (8.7) into (8.12) and defining

$$G_l(t, \omega) = \Psi_l(t, \omega)H_l(t, \omega) \tag{8.13}$$

yields

$$x_l(t) = \frac{1}{2\pi} \int_{-\infty}^{\infty} G_l(t, \omega)S(\omega)e^{j\omega t}\, d\omega. \tag{8.14}$$

Comparing (8.11) and (8.14), we see that $G_l(t, \omega)$ is the time-variant transfer function of the lth ray tube.

To calculate the input delay-spread function of the lth ray tube, $g_l(t, \tau)$, we begin with

$$g_l(t, \tau) = \frac{1}{2\pi} \int_{-\infty}^{\infty} G_l(t, \omega)e^{j\omega\tau}\, d\omega = \frac{1}{2\pi} \int_{-\infty}^{\infty} \Psi_l(t, \omega)H_l(t, \omega)e^{j\omega\tau}\, d\omega. \tag{8.15}$$

Denote the *micro-multipath input delay-spread function* of the lth ray tube as $\psi_l(t, \lambda)$. Then,

$$\Psi_l(t, \omega) = \int_{-\infty}^{\infty} \psi_l(t, \lambda)e^{-j\omega\lambda}\, d\lambda. \tag{8.16}$$

Substituting (8.16) into (8.15) and rearranging yields

$$g_l(t, \tau) = \int_{-\infty}^{\infty} \psi_l(t, \lambda)\left[\frac{1}{2\pi} \int_{-\infty}^{\infty} H_l(t, \omega)e^{j\omega(\tau-\lambda)}\, d\omega \right] d\lambda. \tag{8.17}$$

Note that the term in the brackets is the *macro-multipath input delay-spread function of the lth ray*, which we denote as $h_l(t, \tau - \lambda)$. We can then rewrite (8.17) as

$$g_l(t, \tau) = \int_{-\infty}^{\infty} \psi_l(t, \lambda)h_l(t, \tau - \lambda)\, d\lambda. \tag{8.18}$$

The input delay-spread function of the lth ray tube is therefore equal to the convolution in the delay variable of the macro- and micro-multipath input delay-spread functions for the lth ray and ray tube, respectively.

Combining the input delay-spread functions for all ray tubes, the channel input delay-spread function is given by

$$g(t, \tau) = \sum_{l=1}^{L} g_l(t, \tau). \tag{8.19}$$

Figure 8.6 shows a snapshot in time of the input delay-spread functions that could correspond to the propagation paths shown in Figure 8.1. The solid black impulses represent the macro-multipath input delay-spread function for each of the rays, and the gray shaded regions represent the scaled and translated micro-multipath input delay-spread functions for each of the ray tubes. For each ray tube,

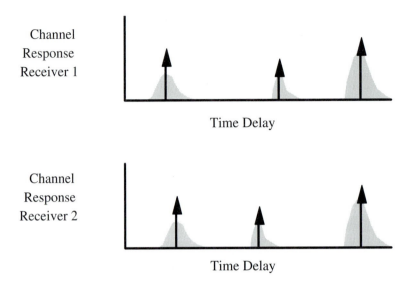

Figure 8.6   Snapshots of acoustic channel input delay-spread functions.

the total input delay-spread function is the convolution of the micro and macro-multipath functions. Note that the slowly varying macro-multipath structure influences primarily the delay and amplitude of the cluster of arrivals for each ray tube. Thus, the relative delays and amplitudes of the clusters of arrivals at one or more sensors will change on a time scale commensurate with that of the macro-multipath structure. The rapidly varying micro-multipath structure for each ray tube influences the detailed shape of the arrival for that ray tube and induces a temporal spreading of the arrivals for each tube. Thus, the particular structure of the arrivals within each cluster will change on a time-scale commensurate with the micro-multipath structure. Note that the temporal spreading function is different for each of the tubes.

Figure 8.7 shows a snapshot of the amplitude of the estimate of the complex baseband input delay-spread function for a channel in shallow water. For this example, the water depth was approximately 20 m, the distance from transmitter to receiver was approximately 200 m, and the signal covered a frequency range of 11.5 to 17.5 kHz.

Figure 8.8 shows the temporal evolution of the amplitude of the same estimated input delay-spread function over a 1.75-s interval. The estimates were made with a deterministic least squares algorithm with an averaging window of 0.02 s. The horizontal axis represents delay, the vertical axis represents absolute time, and the amplitude is represented in the gray scale. Here, there are two major clusters of arrivals. The first shows small temporal fluctuations while the second path fades in and out over intervals of 0.4 s.

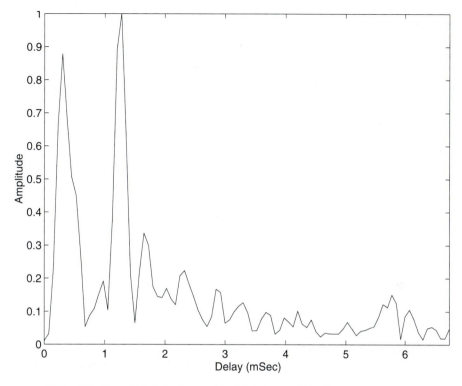

Figure 8.7   Snapshot of short range input delay-spread function.

Finally, Figure 8.9 shows the temporal evolution of the phase of the same estimated input delay-spread function.

Notice that the taps corresponding to the first significant cluster of arrivals as identified in Figure 8.8 show a slow but continuous phase drift. However, during the intervals when the taps corresponding to the second cluster of arrivals have significant amplitude, those taps show no such phase drift. The phase drift in the taps for the first cluster is the result of a Doppler shift along the corresponding ray tube and is discussed in Section 8.1.5. While the delay spread for this very short-range channel is only 6 ms, spreads exceeding 80 ms are common in longer-range channels (e.g., see Figure 8.10).

8.1.4.3 Spatial/Temporal Channel Statistics

The micro- and macro-multipath decomposition of the channel allows us to predict the spatial and temporal channel statistics as a function of the level of saturation of the channel. We can represent the input delay-spread function for a channel by clusters of taps in a tapped delay line, each tap having a time-varying and complex-valued weight applied to its output. Each cluster corresponds to the input

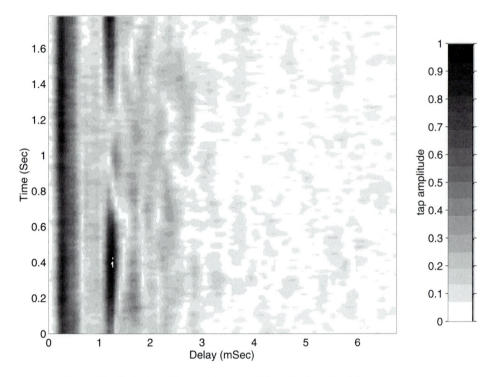

Figure 8.8   Evolution of input delay-spread function (amplitude).

delay-spread function for a particular ray tube. When the channel is either unsaturated or partially saturated, the fluctuations of the tap weights within a cluster will be coherent. However, when the channel is fully saturated, the incoherent fluctuations on each micropath within the ray tube will induce independent fluctuations in the tap weights within a cluster [25]. Whenever the spatial scale of environmental fluctuations is smaller than the separation between ray tubes, the fluctuations of the tap weights in different clusters will be independent of one another.

When examining the coherence among tap weights corresponding to the channels from a transmitter to two spatially separated receivers, we must first establish a pairing of the clusters of arrivals in each of the channels. Assume that we start with the two receivers located at the same point and establish a pairing between the individual clusters of arrivals at each receiver. Then, as we move one of the receivers and adjust the ray tubes to follow that receiver, we maintain the same pairings between the clusters of arrivals at that receiver and the clusters of arrivals at the stationary receiver. As long as the adjusted ray tube in a pair can be treated as a continuous perturbation of the original ray tube, we may continue to treat the two clusters of arrivals corresponding to the two ray tubes as belonging

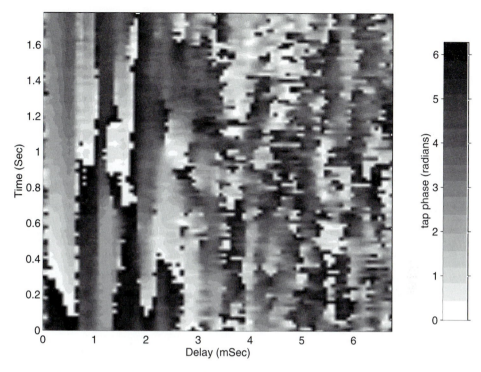

Figure 8.9   Evolution of input delay-spread function (phase).

to the same basic ray. When the channel is either unsaturated or partially saturated, there will exist some coherence between the tap weight fluctuations for the clusters of arrivals at each receiver belonging to the same basic ray. This coherence will decrease as the spatial separation of the receivers increases. At some point, when the separation between the paired ray tubes exceeds the spatial scale of the environmental fluctuations, this coherence will be lost. When the channel is fully saturated, coherence between the corresponding taps of paired clusters will be lost whenever the receivers are separated.

For the case of unsaturated or partially saturated channels, it is interesting to compare the interpath coherence at a single sensor (i.e., the coherence between taps in different clusters at one receiver) with the intersensor coherence for a single path (i.e., the coherence between taps in paired clusters at different receivers). Our analysis above tells us that for channels that are not fully saturated, the intersensor coherence for a single path will be greater than the interpath coherence at a single sensor. This coherence structure can be exploited in the development of multichannel demodulation algorithms, as discussed in Section 8.4.4.

Correlation analysis of the estimates of the input delay-spread function, using the same experimental data used to generate Figures 8.7 through 8.9, gives

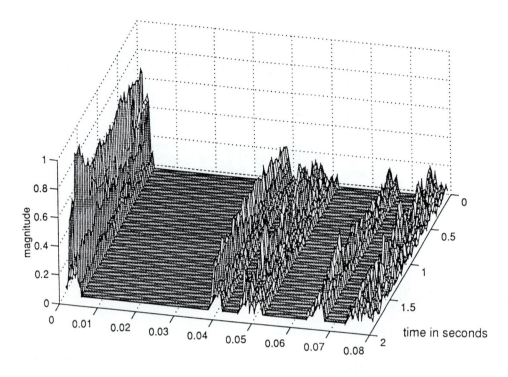

time in seconds

Figure 8.10 Sparse channel estimate history.

us examples of this coherence structure. The transmitted signal was received at 4 sensors. The sensors were located on a vertical line with an intersensor spacing of 1 m (approximately 10 wavelengths at the center frequency). For each sensor, the taps corresponding to the largest clusters of arrivals were identified. In each case, there were two significant clusters of arrivals. Let the estimate of the mth significant tap of the complex baseband, discrete-time input delay-spread function for the kth sensor at time n be denoted by $g_{k,m}[n]$. Assuming that there are M significant taps, we let

$$\mathbf{g}_k[n] = \begin{bmatrix} g_{k,1}[n] \\ \vdots \\ g_{k,M}[n] \end{bmatrix}.$$

The time-averaged auto- and cross-correlation coefficient matrices for $\mathbf{g}_1[n]$ through $\mathbf{g}_4[n]$ are shown in Figure 8.11. The top row contains the auto-correlation coefficient matrices. The second row shows the cross-correlation matrices between $\mathbf{g}_1[n]$ and $\mathbf{g}_2[n]$ through $\mathbf{g}_4[n]$. The third row shows the cross-correlation matrices

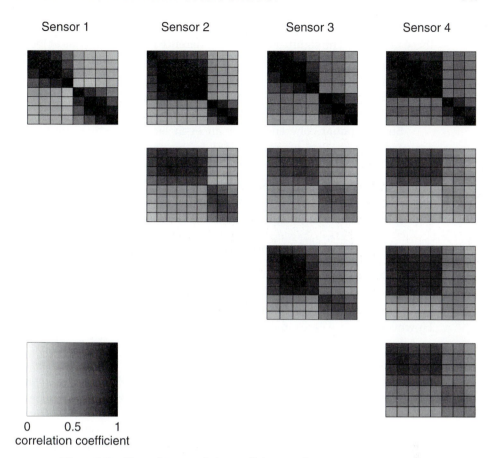

Figure 8.11 Channel tap correlation coefficient matrices.

between $\mathbf{g}_2[n]$ and $\mathbf{g}_3[n]$ and between $\mathbf{g}_2[n]$ and $\mathbf{g}_3[n]$. Finally, the fourth row shows the cross-correlation matrix between between $\mathbf{g}_3[n]$ and $\mathbf{g}_4[n]$. In all cases, the coherent averaging window was 1 s.

The block diagonal correlation structure of the matrices in the first row clearly shows the strong correlation between the taps belonging to the same ray tube and the much weaker correlation between the taps belonging to different ray tubes. Notice that for some of the matrices, the off-diagonal elements for one of the clusters of arrivals are slightly smaller than those for the other cluster, indicating a higher level of saturation for the ray tube corresponding to the cluster with the smaller off-diagonal elements. As predicted by our analysis above, the intersensor correlation for the taps in the first cluster of arrivals at each sensor is much stronger than the interpath correlation at each sensor. However, note that the taps in the second cluster of arrivals at the fourth sensor have a relatively weak correlation with the corresponding taps at the other sensors, indicating either a higher level of saturation on the corresponding ray tubes or a greater spatial separation between these ray tubes. The significant spatial coherence between arrivals at locations separated by 10, 20, and 30 wavelengths implies strong possibilities for coherent spatial filtering with arrays of receivers. In addition, it implies that significant separations between receivers are necessary for spatial diversity techniques which rely on independent channel fading to be effective.

Strong spatial coherence over an even larger effective aperture has been reported from another experiment [18]. In this experiment, the range from source to receivers was 660 m, the water depth was 17 m, the frequency of transmission was 86 kHz, and the two receivers were vertically separated by 0.992 m (57.7 wavelengths). Even with this large receiver separation, significant coherence between the signals received at the two sensors was observed.

8.1.4.4 The Vertical Multipath Channel

The preceding discussion of the characteristics of the multipath channel is most applicable to the horizontal channel (i.e., the case where the primary spatial separation between the transmitter and receiver is in the horizontal dimension). However, in some cases, the transmitter and receiver lie on a nearly vertical line and their spatial separation is primarily vertical. The multipath characteristics of the resulting vertical channel can deviate significantly from those described above.

The difference in characteristics of the horizontal and vertical channels is primarily due to the different scales of environmental fluctuations in the horizontal and vertical dimensions. In general, the horizontal gradient of these fluctuations is much smaller than their vertical gradient. Thus, ray paths connecting a source and receiver that lie in a vertical line will have little refraction, and the cluster of arrivals corresponding to a ray tube will show little spreading. The multipath that does exist will be due to successive reflections of the sound off the sea surface and sea floor. The resulting multipath structure will show narrow clusters of arrivals separated by the round-trip propagation time from the receiver to either the sea surface or sea floor. There will be almost no arrival energy in the spaces between the clusters. The input delay-spread function of the channel will thus have a sparse structure in the delay parameter. Algorithms exploiting this sparse structure are discussed in Section 8.4.2.

8.1.5 Doppler Effects

Until now, we have not paid special attention to channel fluctuations caused by relative motion of the transmitter, receiver, or significant scattering surfaces in the environment. As discussed in earlier chapters, when any of these objects moves in a manner that causes an increase (or decrease) in the length of a propagation path between the source and receiver, it induces a corresponding expansion (or compression) of the time axis for the received signal. That is, if the speed of sound propagation is c m/s, the length of a single propagation path is increasing at a rate of v m/s, and the narrowband signal $\mathrm{Re}[s(t)e^{j\omega_0 t}]$ is transmitted, then the noiseless, undistorted received signal will be[2]

$$r(t) = \mathrm{Re}\left[s(t(1 - v/c) - \tau_o)e^{j\omega_0 t(1 - \frac{v}{c}) - j\omega_0 \tau_0}\right],$$

where τ_o is a fixed delay. If the bandwidth of $s(t)$ is small with respect to its center frequency this scaling of the time axis is modeled as a frequency shift of the signal. That is, the received signal is given by

$$r(t) = \mathrm{Re}\left[s(t - \tau_o)e^{j\omega_0 t(1 - \frac{v}{c}) - j\omega_0 \tau_0}\right].$$

The mean of the frequency shift of the signal over some window of time is referred to as the *Doppler shift* of the signal. Removing this mean, the remaining frequency fluctuations in the signal are referred to as the *Doppler spread* of the signal. A significant source of Doppler spread is the scattering of the sound by the time-varying sea surface. See [23] for a discussion of this type of Doppler spread.

If a receiver does not compensate for the Doppler shift in the received signal, there remains an apparent phase rotation in the tap weights of the channel input delay-spread function at a rate of $e^{-j\omega_0 \frac{v}{c} t}$. Assuming a center frequency of $\omega_0/2\pi = 15$ kHz, the Doppler shift associated with a velocity of 1 m/s is approximately 10 Hz. Such a phase rotation of the taps of a channel can cause severe tracking problems with many adaptive algorithms. However, an important property of this type of time variation is that it is highly correlated from tap to tap. The recognition and exploitation of this fact in [44] led to a suitable equalizer structure for phase-coherent communications and is discussed in Sections 8.3 and 8.4.1.

In some situations, the different clusters of taps associated with different ray tubes can experience different Doppler shifts. A simple example of this is the situation where two ray tubes lead from the transmitter to the receiver. The first tube goes upward from the transmitter, bounces off the sea surface, and is reflected down to the receiver. The other tube goes down from the transmitter, bounces off the sea floor, and is reflected up to the receiver. Then, if the transmitter moves in an upward direction, the channel taps associated with the first tube will experience a negative Doppler shift while the taps associated with the second tube will experience a positive Doppler shift. Recent research has indicated that the techniques

---

[2]As in previous chapters, Re[·] and Im[·] denote the real and imaginary parts, respectively, of their complex arguments.

developed in [44] and described in Section 8.4.1 are often incapable of providing reliable communications when such multiple Doppler shifts are present [19].

8.1.6 Channel Latency and Coherence Times

A final distinguishing characteristic of the underwater acoustic channel is the relationship between the channel latency (i.e., the time it takes for a signal to propagate through the channel) and the coherence time of the channel fluctuations. Data presented and referenced herein indicate that the characteristics of a channel can change significantly in one-half of a second or less. However, for a channel with a range of just 1 km, the one-way propagation time from transmitter to receiver is greater than two-thirds of a second. Therefore, in many situations, the channel can change significantly in the time it takes for the signal to propagate from the transmitter to the receiver. Such a situation makes it difficult to implement modulation or coding algorithms that require accurate knowledge of the particular realization of the channel through which communication is being attempted. Platform motion can also have a significant effect on multiple-access techniques that require precise timing of signal receptions. Motion of just a single meter can shift the arrival time of a signal by two-thirds of a millisecond. This shift can span many channel symbols in a high-rate system and can occur in much less time than it takes a signal to propagate through all but very short-range channels. Thus, the synchronization of the user's transmissions to achieve reception without multiple-access interference or the use of significant guard bands will be difficult in many channels.

8.2 PLATFORM CONSTRAINTS IN UNDERWATER ACOUSTIC COMMUNICATIONS

Many applications of underwater acoustic communications involve communication with or between autonomous underwater vehicles or remotely deployed instrumentation. In both situations, the communications system faces significant constraints on the permissible size of the communications equipment and on the power available to this equipment.

The size constraint can limit the choice of the operating frequency of the system. In general, the power efficiency of an acoustic transducer is proportional to the ratio of its size to the acoustic wavelength. So, in order to obtain an efficient transducer with which to transmit the sound, a high frequency must often be used. However, the high-frequency sound will experience greater attenuation in the water than would lower-frequency sound. Thus, the frequency must be chosen to obtain a balance between transducer efficiency and signal absorption.

The constraint on available power most obviously limits the signal energy that can be radiated. However, this constraint also places limits on the complexity of the modulation and demodulation algorithms that can be used in a system. In general, the power consumed by a processor is proportional to the computational capability provided by that processor. Thus, the power constraint also constrains the computational capability available to implement modulation and demodulation algorithms. Combining the rapidly changing multipath channels whose delay spreads approach a hundred symbols with significant power limitations on oceanographic instrumentation or autonomous underwater vehicles, the computational complexity of demodulation algorithms can quickly exceed the available computational capability. For this reason, the development of computationally efficient algorithms is an important and active area of research. Some past and current work in this area is discussed in Sections 8.4.2 and 8.4.4.

8.3 A Brief History of Underwater Acoustic Communications

There has been considerable progress in improving communications over the channels discussed in the preceding sections. Here we identify those communication techniques that have advanced the field. This section is not intended to be an exhaustive historical review, and the reader is directed to [21] for a review of underwater acoustic communications prior to 1967, to [22] and the references therein for a review of acoustic telemetry prior to 1983, to [23] for a review of recent advances in phase-coherent underwater acoustic communications, and to the articles in [24] for reviews of modulation techniques, equalization and coded modulation as applied to the UAC.

Underwater acoustic communication systems can be classified according to how the transmitter and receiver combat the effects of multipath fading. Important to this classification is the concept of *diversity*, which, as discussed in Chapter 1, is the transmission of the communication message through independently faded channels [31]. *Explicit diversity* is characterized by intentional transmissions through distinct subchannels in time, frequency, geometric space, or waveform space. Due to the independence of the subchannel fading processes, the channel error probability is exponentially decreasing in the number of retransmissions, or diversity order. Coding across these subchannels (other than repetition coding) is known to make efficient use of the channel bandwidth [31]. *Implicit* diversity can be achieved by spectrally spreading the message signal over a single transmission band having a width W much larger than the coherence bandwidth of the channel. This wideband signal can be used at the receiver to resolve and identify the individual multipath arrivals spaced in delay by more than $1/W$ [32]. If these complex amplitudes can be accurately estimated and tracked, a receiver can use implicit diversity to achieve an error-rate improvement identical to that of explicit diversity [33]. Using the concept of diversity, underwater acoustic communication systems can be grouped into three classes, as to whether the links use (1) no diversity techniques, (2) only explicit diversity reception, or (3) at least implicit diversity processing.

The first class of underwater acoustic communication systems includes those that did not employ diversity techniques. This class includes most of the early analog communication systems, which used careful hydrophone placement to compensate for multiple path propagation via distinct angles of arrival. The development of reliable underwater acoustical communications began after World War II with the Gertrude, an analog, amplitude-modulation system that permitted communication to submarines [25]. Single-sideband derivatives of the Gertrude are still used in modern diver communication systems [26] and perform well for vertical or ultrashort-range horizontal links with negligible multipath propagation.

Several digital underwater acoustic communication systems also belong to this first class. The development of digital UAC links were reported as early as 1960 [27, 28]. These systems dealt with multipath by acoustic baffling and by using very low rate channel codes [29]. A typical digital communication system of this type, described in [30], permitted 4800 bits per second (BPS). Early implementations of multiuser detectors for UACs did not employ diversity methods and minimized multipath-induced distortion by transmitting vertically through the water column [36]. Digital communication links through horizontal, shallow-water channels have also been designed without diversity techniques [39]. In this case, multipath propagation was avoided through the use of many transmit transducers.

The second class of underwater communication systems use explicit diversity reception. Most of these systems employed digital modulation. Perhaps the most carefully documented communication link of this class is the digital acoustic telemetry system (DATS) [34]. The DATS was designed for transmission of digital data in an environment that exhibited frequency-selective multipath fading and extreme phase instability. The DATS transmitted coded data using on-off, multiple-frequency shift keying modulation in the 45–55 kHz frequency band, provided a coarse word synchronization reference with a gated 30 kHz header tone, and a provided continuous 60 kHz pilot tone for Doppler tracking. In one implementation of DATS, a 400 BPS digital data stream was transmitted at a baud (symbol) rate of 100 Hz by encoding 4 bits per baud with an (8,4)-Hamming code. The Hamming codeword elements selected from eight tones spanning 2 kHz at each baud period. A slow frequency-hopping scheme translated the set of eight tones for each baud period to minimize intersymbol interference. The receiver coherently estimated the Doppler shift with a phase-locked loop (PLL) whose output was used to adapt the downconversion carrier, nominally at 50 kHz. The DATS detector tracked the hopping pattern to determine the frequency span for the current word and implemented an inverse FFT to extract the squared magnitudes of the received gated tones. Non-coherent, soft-decision detection was used to estimate the 8-bit codeword.

Since the coherence bandwidth of the UAC in [34] was 2 kHz, the above system did not utilize frequency diversity per se. However, trivial changes in the implementation would permit eightfold diversity reception. Variants of this communication system are still under investigation [35]. Several commercially produced acoustic modems follow the DATS format and permit reliable transmission through highly reverberant multipath channels at modest complexity [37]. This communication link is quite insensitive to Doppler shifts as well. In the configuration described above, the DATS system can tolerate about 600 Hz of Doppler shift if the PLL is successfully tracking the pilot tone, and about 25–50 Hz of Doppler shift when the pilot PLL loses lock on the received tone. Since 0.5 m/s corresponds to 17 Hz of Doppler shift for an underwater acoustical tone at 50 kHz, this Doppler tolerance translates to a relative platform drift between 1 m/s and 8 m/s. Other configurations of the DATS have been reported for deep water channels [43].

The third class of underwater acoustic communication systems includes those modems that utilize implicit diversity reception. Specific examples of implicit diversity reception include RAKE filtering, fractionally spaced decision-feedback equalization, wideband array processing, and echo cancellation. Explicit diversity reception can be used in these systems as well. Digital underwater acoustic links in this class have demonstrated the greatest spectral efficiency to date (measured in BPS per Hertz of transmission bandwidth) of all systems operating with the same number of sensor elements, input signal-to-noise ratio (SNR) and bit error rate [23]. Communication links that employ implicit diversity processing have shown the greatest promise to provide reliable, high-speed, and power-efficient links. As of this writing, research is most active for these types of communication systems, and digital signal processing techniques for implicit diversity systems are the focus of the remainder of this chapter.

Perhaps the earliest reference to single-sensor implicit diversity for UACs is [41]. In that work, coherent echo cancellation compensated for intersymbol interference in a phase-shift keying (PSK) communication system. References to adaptive equalization as a means to provide high-speed underwater acoustic communications appeared in the literature in the early 1990s [42]. Much work has been done on the development of demodulation algorithms for multichannel receivers, and it remains an area of active research [16, 39, 45, 47, 86, 87, 88, 89, 90].

An impetus for current signal processing research in UACs is provided by the results in [44] for single-sensor reception and [45] for multisensor reception. These works demonstrated the feasibility of high-rate phase-coherent digital communications through highly reverberant shallow-water channels, by means of adaptive channel equalization. With regard to current research in this field, the most important contribution of [44] was a decomposition of the time-varying input delay-spread function of shallow-water channels into two tandem subsystems: a complex baseband input delay-spread function whose coherence time spanned hundreds of milliseconds, and a unit-gain complex amplifier whose instantaneous phase is approximately affine over tens of milliseconds. This second subsystem accounts for the Doppler shift introduced by relative platform motion, as was discussed in Section 8.1.5. Prior to this work, it was thought that shallow-water channel taps exhibited mutually uncorrelated temporal trajectories, both in magnitude and phase. Adaptive channel identification under this hypothesis would usually not produce convergence within the coherence time of the overall channel. In hindsight, the (common) phase variation of all taps yielded a rapidly time-varying channel over the averaging window for the estimator. The work in [44] presented the first demonstrable evidence that the phase trajectories for significant input delay-spread function taps were strongly correlated and that the common phase rotation of these taps could be tracked by a single, second-order, digital PLL. Employing a common phase correction, an adaptive, fractionally spaced, decision-feedback equalizer (DFE) was shown to converge and compensate for the channel distortion. Convergence of equalizer taps is not possible in some shallow-water channels without this phase compensation. This result is quite different from equalization for twisted-pair links, whose (few) equalizer taps may converge in the absence of common phase tracking, even for similar frequency offsets. An analysis of the effects of residual phase errors on adaptive equalization is presented in [46].

8.4 SIGNAL PROCESSING IN DIGITAL UNDERWATER COMMUNICATIONS

In the remainder of this chapter, we describe some of the ways that advanced signal processing can enhance the performance of underwater communication systems.

8.4.1 Detection of Linear Digital Modulation

We begin by reviewing the detection strategy presented in [44] for coherent detection of single-sensor reception of linear digital modulation. Multisensor reception is reviewed in another section. Both the channel model and the adaptive filtering metric are described in the deterministic, weighted least squares framework. Provided a scalar $\lambda \in (0, 1)$, there is a temporal window ending at time t_0 and having an approximate duration of $T/(1 - \lambda)$ seconds, over which the baseband complex channel output is modeled as

$$r(t) = \sum_n d_n g(t, t - nT) + n(t)$$

$$\approx \sum_n d_n h(t - nT) e^{j\varphi(t)} + n(t), \quad t - t_0 \in \left[-T/(1 - \lambda), 0\right).$$

Here, $g(t, \tau)$ denotes the input delay-spread function as discussed in Section 8.1.4.2, $\{d_n\}_{n=-\infty}^{n_0-1}$ represents a known or accurately detected complex symbol sequence, $\lfloor n_0 - t_0/T \rfloor = 0$, T denotes the channel symbol (or baud) period, $\varphi(t)$ is an unknown affine function described in Section 8.1.5, $h(\tau)$ is an unknown, deterministic, time-invariant impulse response, and $n(t)$ is baseband, complex, stationary Gaussian noise with a flat power spectral density over the transmission bandwidth. It is assumed that the "forgetting factor" λ was chosen to maximize the temporal window width for which $g(t, \tau) \approx h(\tau)e^{j\varphi(t)}$. It is also assumed that coarse synchronization has provided an approximate temporal reference for the impulse response and that Doppler compression is expressed as a phase rotation. Coarse synchronization may be achieved by preceding a data packet by a brief signal whose energy spectral density is broad.

An adaptive, fractionally spaced, DFE was proposed in [44] as a receiver structure and is shown in Figure 8.12.

The feedforward filter samples the observation at rate $1/T_s$, an integer multiple of the symbol rate $1/T$, and produces outputs at times nT. After phase correction, the output at time $n_0 T$ is p_{n_0}, where[3]

$$p_n = \mathbf{a}^H \mathbf{r}(n, \tau) e^{-j(\omega_d n + \theta)},$$
$$\mathbf{r}(n, \tau) = [r(nT + N_1 T_s + \tau) \quad \cdots \quad r(nT - N_2 T_s + \tau)]^T, \tag{8.20}$$

where $\mathbf{a}$, ω_d, and θ and τ are filter coefficients. The fractionally spaced front end is especially adept at correcting for an error between τ_n and the true timing phase [48]. The filter's temporal span $T_s(N_1 + N_2 + 1)$ should be chosen to allow

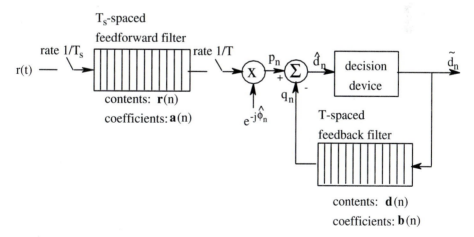

Figure 8.12   Adaptive, fractionally spaced, decision-feedback equalizer.

[3]The superscript ^{H} denotes the conjugate-transpose operator. The superscript ^{T} denotes the transpose operator.

for nonminimum phase responses, such as the one shown in Figure 8.9, and residual timing jitter due to Doppler dilation/compression.

The feedback filter in Figure 8.12 linearly combines M previous symbol decisions to suppress residual causal intersymbol interference and forms q_{n_0} at time $n_0 T$, where

$$q_n = \mathbf{b}^H \mathbf{d}(n)$$
$$\mathbf{d}(n) = [\tilde{d}_{n-1} \quad \cdots \quad \tilde{d}_{n-M}]^T. \tag{8.21}$$

The complex predecision variables $\hat{d}_n = p_n - q_n$ are quantized to form final decisions $\tilde{d}_n$.

The adaptation rules presented in [44] yield extremal solutions for the deterministic weighted squares metric

$$\mathcal{J}_n(\mathbf{a}, \mathbf{b}, \tau, \omega_d, \theta) = \sum_{k=0}^{n} \lambda^{n-k} |\tilde{d}_k - \mathbf{a}^H \mathbf{r}(k, \tau) e^{-j(\omega_d k + \theta)} + \mathbf{b}^H \mathbf{d}(n)|^2. \tag{8.22}$$

Gradients of $\mathcal{J}_n$ with respect to the receiver parameters can be obtained directly from the observations and take the form[4]

$$\frac{1}{2} \nabla_{\mathbf{a}} \mathcal{J}_n = \left(\sum_{k=0}^{n} \lambda^{n-k} \mathbf{v}(k) \mathbf{v}^H(k) \right) \mathbf{a} - \left(\sum_{k=0}^{n} \lambda^{n-k} \mathbf{v}(k) \mathbf{d}^H(k) \right) \mathbf{b} - \left(\sum_{k=0}^{n} \lambda^{n-k} \mathbf{v}(k) \tilde{d}_k^* \right),$$

$$\frac{1}{2} \nabla_{\mathbf{b}} \mathcal{J}_n = \left(-\sum_{k=0}^{n} \lambda^{n-k} \mathbf{d}(k) \mathbf{v}^H(k) \right) \mathbf{a} + \left(\sum_{k=0}^{n} \lambda^{n-k} \mathbf{d}(k) \mathbf{d}^H(k) \right) \mathbf{b} + \left(\sum_{k=0}^{n} \lambda^{n-k} \mathbf{d}(k) \tilde{d}_k^* \right),$$

$$\frac{\partial}{\partial \theta} \mathcal{J}_n = +2 \, \mathrm{Im} \left[\sum_{k=0}^{n} \lambda^{n-k} p_{k,n} e_{k,n}^* \right],$$

$$\frac{\partial}{\partial \omega_d} \mathcal{J}_n = +2 \, \mathrm{Im} \left[\sum_{k=0}^{n} \lambda^{n-k} k p_{k,n} e_{k,n}^* \right], \tag{8.23}$$

$$\frac{\partial}{\partial \tau} \mathcal{J}_n \approx +2 \, \mathrm{Re} \left[\sum_{k=0}^{n} \lambda^{n-k} \frac{p_{k,n} - p_{k-1,n}}{T_s} e_{k,n}^* \right],$$

where $p_{k,n} = \mathbf{a}^H(n) \mathbf{v}(k)$, and $e_{k,n} = p_{k,n} - \mathbf{b}^H(n) \mathbf{d}(k) - \tilde{d}_k$. In these equations, we have made the substitution $\mathbf{v}(k) = \mathbf{r}(k, \tau) e^{-j(\omega_{d,n} k + \theta_n)}$, and we have suppressed the notational dependence of $\mathbf{v}$ on the receiver's phase and synchronization parameters. We have denoted extremal values of filter parameters for cost $\mathcal{J}_n$ with the index n in the defining expressions for $p_{k,n}$ and $e_{k,n}$. A backward difference approximates the for-

---

[4]The gradient of $\mathcal{J}$ with respect to a complex vector $\mathbf{v} = \mathbf{x} + j\mathbf{y}$ is defined as $\nabla_{\mathbf{v}} \mathcal{J} = \nabla_{\mathbf{x}} \mathcal{J} + j \nabla_{\mathbf{y}} \mathcal{J}$. A complex vector $\mathbf{w}$ is normal to this gradient if $\mathrm{Re}(\mathbf{w}^H \nabla_{\mathbf{v}} \mathcal{J}) = 0$.

mal derivative in the last expression. The gradients for **a** and **b** are usually decoupled from the unknowns θ, ω_d, and τ by use of precomputed estimates. The extremal equations for **a** and **b** can be rewritten to reveal the recursive structure of their solutions,

$$\mathbf{R}(n)\begin{bmatrix} \mathbf{a} \\ \mathbf{b} \end{bmatrix} = \mathbf{y}(n),$$

$$\mathbf{y}(n) = \lambda\mathbf{y}(n-1) + \tilde{d}^*(n)\begin{bmatrix} \mathbf{v}(n) \\ -\mathbf{d}(n) \end{bmatrix}, \quad \mathbf{y}(-1) = \mathbf{0}, \tag{8.24}$$

$$\mathbf{R}(n) = \lambda\mathbf{R}(n-1) + \begin{bmatrix} \mathbf{v}(n)\mathbf{v}^H(n) & -\mathbf{v}(n)\mathbf{d}^H(n) \\ -\mathbf{d}(n)\mathbf{v}^H(n) & \mathbf{d}(n)\mathbf{d}^H(n) \end{bmatrix}, \quad \mathbf{R}(-1) = \mathbf{0}.$$

The sequence of extremal equalizer parameters $\{\mathbf{a}(n), \mathbf{b}(n)\}$ can be found by any version of the recursive least-squares (RLS) algorithm, as this structure shows. A numerically stable implementation of an order $10N$ (per update) RLS algorithm was suggested for the shallow-water acoustic channel [44], and a development of this algorithm can be seen in [50]. Here, N is the number of real coefficients in the recursion $2(1 + N_1 + N_2 + M)$. Estimation for the synchronization and phase can proceed in a variety of ways. For modest Doppler shifts, the phase estimate can be obtained via a second-order phase-locked loop (to track the affine trajectory). The synchronization update can be achieved by an early delay-lock loop, as suggested by the gradient $\partial\mathcal{J}_n/\partial\hat{\tau}_n$. An extension of these results can be found in [45] for the multiple-sensor case and in [51] for high Doppler environments.

Adaptive linear equalization has also been applied successfully to the vertical acoustic channel, which usually exhibits longer coherence times. Since the direction of propagation is tangential to the thermal gradient, multipath due to refraction is minimal. However, transmission from mid-depth to the surface is known to produce an extremely high multipath spread due to bottom reflections, as discussed in Section 8.1.4.4 and in [52]. In at least three instances, coherent or differentially coherent phase modulation was achieved in the vertical acoustic channel by adapting a linear equalizer according to a least mean squares (LMS) algorithm for well-conditioned input data [52, 53, 54]. The extremal conditions for linear equalization may be seen by forcing $\mathbf{b} = \mathbf{0}$ and ignoring the gradients for $\mathbf{b}_n$ in (8.23). Several approaches suggest self-adaptive versions of the LMS algorithm, which varies the step sizes for each gradient in order to reduce both the convergence time and tracking error, e.g., [52]. The reader is referred to [55] for a review of self-adaptive LMS algorithms and to [56] for a review of the convergence and tracking properties of the LMS family of algorithms.

The work in [44] has suggested that RLS-based adaptive equalization is required for rapidly varying shallow-water acoustical channels. Once the channel taps are derotated by the phase compensator, as shown in Figure 8.12, then the coherence time of the residual channel is lengthened considerably and permits accurate channel tracking. Despite several well-known reports of the superior convergence properties of RLS algorithms over LMS algorithms [31], there exists a preference for the latter in the underwater acoustic communications literature. Most authors prefer LMS because of the order $O(N)$ complexity (floating-point operations (flops) per update) of LMS to the order $O(N^2)$ complexity of the (stable) conventional RLS algorithm. These authors cite improvements in convergence time of newer members of the LMS family, as reviewed in [56] and references therein. RLS proponents counter with the $O(N)$ complexity of the numerically stable, fast transversal RLS (FTRLS) algorithm [49]. At least one work has employed an RLS-based update for the feedforward taps, for which the input data is not well conditioned, and an LMS-based update for the feedback taps, for which the input data is usually white [79]. Regardless of the vantage point of these works, two attributes are considered crucial for an adaptive receiver in shallow water: fast convergence time and low computational complexity.

In the next section, we summarize research efforts to combine these features in detection algorithms. Rather than focusing on the usual attempts to reduce complexity by minimizing computational redundancies, we concentrate on work that exploits the natural sparseness of the impulse response magnitude of the UAC, as well as cases for which the baud rate dominates the Doppler spread (reciprocal coherence time) of the channel.

8.4.2 Complexity Reduction in Adaptive Detection

Implicit diversity, which is usually attained via wideband signaling, permits the resolution of paths spaced by no less than the reciprocal of the transmission bandwidth. As the channel symbol rate has increased in underwater acoustic applications during the last decade, both the channel resolution and the channel memory have increased as well. The *memory* of the communication channel refers to the multipath delay spread, normalized by the channel symbol period. The channel memory is an important characterization in partially coherent reception of linear digital modulation because the number of parameters in an adaptive receiver is typically proportional to the worst-case channel memory. For example, a fractionally spaced DFE requires enough feedforward parameters to span the group of delay taps corresponding to the principal arrival, including uncorrected tap drift, and the symbol-spaced feedback taps should span the memory of the residual impulse response. Consider a medium-range horizontal acoustic link with a worst-case multipath spread of 40 ms and a baud rate of 2,500. The channel memory for this link is 100 channel symbols. If the $T/2$-spaced feedforward filter spans 16 channel symbols, then the dimension of the feedback filter might be as high as 120. Doubling the channel symbol rate would double the number of receiver parameters. Since it has been demonstrated that higher baud rates improve the phase tracking properties of the adaptive receiver [23], the baud rate and channel memory will tend to increase in future deployments of phase coherent systems.

We begin by establishing a complexity benchmark for our subsequent development in this section. To this end, we shall assume single-sensor acquisition and that $O(\mu)$ complex parameters will be adapted at the baud rate of R_B symbols/s (SPS) for a channel of memory $\mu = T_m R_B$, where T_m seconds is the temporal multipath spread. Regardless of whether LMS or FTRLS is used to update the receiver parameters, the computational requirements for adaptation will be proportional to $O(T_m R_B^2)$ flops per second. Thus, the computational burden imposed on battery-powered acoustic modems operating in a channel with the above dimensions is roughly 10 times that of modems in an average telephony channel operating at the same baud rate (see [31], p. 537). In addition, the demodulation complexity for a DFE is also of order $O(T_m R_B^2)$ flops per second, excluding decoding complexity. Real-time demodulation of these signals on board an autonomous undersea vehicle (AUV) requires careful design of the hardware and algorithms for onboard modems, which are usually constrained to use less than 20 W for both hardware and acoustic power.

There are two general approaches for reducing the $O(\mu R_B)$ complexity of the adaptive receiver described earlier. One class attempts to reduce the complexity through the parameter update rate, which coincides with the channel baud rate (R_B) in the above scenario. If the channel baud rate would exceed substantially the reciprocal coherence time of the channel (often called the *Doppler spread*), then the channel taps at one update time are strongly correlated with those of the previous update time, and the channel estimator (or equalizer parameters) may be updated less frequently. We describe this class as *reduced updating* techniques. While these techniques play a vital role in reducing the computational complexity in some adaptive receivers, space does not permit a thorough review of current approaches. Instead, we focus on approaches that find special application in underwater systems. This second, broader class of algorithms attempts to reduce the number of receiver parameters relative to the scenario described earlier. Through these techniques, the parameter dimension in the complexity product is considerably less than the channel memory μ. Regardless of whether LMS- or RLS-type updates are used, we also expect the convergence time to be smaller for this reduced set of receiver coefficients if the system is not underparameterized. We refer to algorithms of this type as *reduced parameterization* techniques. In the remainder of this section, we review algorithms in this class.

8.4.2.1 Reduced Parameterization Techniques

There are three well-documented approaches to the reduction of the number of receiver parameters: *waveform shaping, indirect adaptive equalization,* and *memory truncation.*

   Waveform shaping attempts to reduce the memory of the overall input delay-spread function through signal design at the transmitter. In effect, the transmitter introduces an additional filter tandem to the communication channel, which yields an overall input delay-spread function with less memory. While this approach has been most often employed in time-invariant communication channels, it may be employed for time-varying systems, using accurate channel information supplied by the far-end receiver. In underwater systems, this enhancement may be achieved through a reverse link with a propagation delay that is less than the coherence time of the phase-compensated channel. Waveform shaping may be performed with single transducer transmission in this environment but is considerably more effective when long transmitter arrays are used, as described in [39]. As was noted in this work, this approach works best in the shallow-water channel with stationary platforms and short-range links. Latency and coherence issues for horizontal channels was presented in Section 8.1.6. For single-transducer transmission, waveform shaping may also be achieved by designing a receiver filter to produce an output with a prescribed and controlled amount of interference. With respect to overall computational burden, the trade-off in this approach, sometimes called *partial-response equalization,* balances the complexities of the front-end filter and the residual interference suppression. This distortion can be mitigated in slowly time-varying, 2-way links by *channel precoding* at the transmitter or by subsequent equalization at the receiver. As an example of the latter approach for long memory channels, a linear equalizer followed by a maximum likelihood sequence detector has been considered. Complexity was reduced in part by the design of the linear equalizer and by a reduction of memory in the subsequent dynamic programming (Viterbi) algorithm [57]. Linear partial-response equalization followed by Viterbi-type equalization has been used in a (time-invariant) magnetic recording channel to reduce complexity.

   Indirect adaptive equalization has also demonstrated an ability to reduce the computational complexity in some time-varying multipath fading channels. Rather than (directly) adapting the equalizer coefficients in a receiver, this approach first identifies the unknown channel taps (using a tapped-delay-line model) and noise statistics and then sets the equalizer to compensate for the channel model. It has been demonstrated recently that the method of indirect channel equalization is more robust to the time variation of some multipath fading channels [62]. Other work has suggested that indirect adaptive DFEs may be implemented at a lower computational complexity than an equalizer with directly adapted coefficients [63]. In effect, the indirect adaptation method yields fewer degrees of freedom in these circumstances, which reduces the convergence time, tracking error, and misadjustment of an adaptive algorithm. As we discuss shortly, there may be other, *channel-specific* ways to reduce the degrees of freedom of this channel identification even further.

Memory truncation is a deliberate mismatch of the receiver structure with respect to the channel model, to provide fewer adapted parameters than with an unconstrained structure. Memory truncation has traditionally been achieved by ignoring scattered paths beyond a fixed delay and advance from the principal arrival and is usually determined with a knowledge of the performance sensitivity to the receiver mismatch. Many studies of performance sensitivity to mismatch exist and quantify tolerable levels of memory truncation. The sensitivity to channel mismatch was investigated for maximum-likelihood-type receivers in intersymbol interference (ISI) channels [58, 59] and to linear equalizers and DFEs in [60, 89] and references therein.

8.4.2.2 Sparse Channel Identification

We now address current work on channel-specific approaches to parameter reduction that are especially suited for underwater acoustic applications. Of relevance are channel identification techniques that reduce the computational complexity according to the *sparseness* of slowly time-varying linear systems. For purposes of this presentation, sparseness applies to a slowly time-varying linear system if its multipath intensity profile (MIP), maximally truncated in the delay axis without loss, shows a concentration of power in relatively few taps. Intuitively, a sparse system exhibits clean echoes spaced by large temporal gaps. These gaps may incur an increase in the computational complexity for both conventional channel tracking and demodulation of linear, digital data passed through this sparse system, relative to a system possessing a similar MIP, but without the delay gaps. A typical example of a sparse system is the medium-range shallow-water UAC, and a history of input delay-spread function magnitude estimates can be found in Figure 8.10. Tap magnitudes below a very small threshold were suppressed in this figure in order to more clearly exhibit the sparse nature of the channel. Note in the figure that while the multipath delay spread for this channel exceeds 80 ms, about 12% of the channel taps contribute significantly to the output. Channels such as the UAC are especially suited for sparse channel identification techniques.

The UAC has exhibited sparseness in a variety of wideband transmission experiments. Sparseness has been observed from a surface-deployed receiver in the vertical acoustic channel, in which echoes were observed due to reflections of the transmitted signal off the sea floor [52]. The medium-range, horizontal UAC has also exhibited sparseness for shallow-water transmissions near Fort Lauderdale, Florida [66], New England harbor sites [64] and for deep-water transmissions in the Arctic circle [65]. A presentation of sparse UACs can be found in [67]. While an underwater channel is not always sparse, it can be approximated as sparse with enough frequency to motivate an incorporation of sparse techniques into wideband acoustic modem designs [66]. An ideal algorithm for sparse channel identification would include full-order identification as a special case and would permit low convergence times for all system orders. In this section, we shall refer to a channel identification technique as a *sparse channel identification* algorithm if it exploits the sparseness in a system to reduce the computational complexity, convergence time, or estimation error. Sparse adaptive equalization will refer to similarly tailored techniques for coherent demodulation of linear, digital modulation through sparse dispersive systems.

There is a significant amount of prior work on rapidly converging or low complexity channel identification for sparse systems. In some of these cases, prior knowledge about the channel is presumed, such as the number of significant input delay-spread function taps (the system order), their location in delay, or the system multipath spread. In short, all of these (or their approximations) may be provided without major changes of several current modem designs [66]. These modems will initiate communication by preceding a phase-modulated data packet with a packet consisting of a short carrier-modulated Barker sequence followed by a silent period. While the purpose of the Barker sequence is to remove the receive modem from standby mode, a temporal window of the observation (or its correlation magnitude with the Barker sequence) can be stored for future processing. This magnitude sequence provides an estimate of system order, approximate locations of significant taps, and delay spread.

Reference [70] proposed a method of sparse channel identification that increments the order of the channel estimator until a prescribed squared-error criterion is satisfied. At each increment, the best delay for a new input delay-spread function tap is determined, using a criterion that is independent of this tap's magnitude and phase. Once the best tap location is determined, the magnitude and phase are estimated. This approach is suited for very sparse systems, and its order-recursive approach may impose a lengthy convergence for full-order systems. However, this technique is quite effective in rapidly identifying the location of a new tap in other sparsing methods. At least three independently derived efforts in sparse adaptive equalization appeared during the same year, suggesting that the solution had strong relevance [64, 76, 77]. In [77] and in subsequent work [52], a sparse implementation of an adaptive linear equalizer was considered for suppression of intersymbol interference in a vertical acoustic channel. Direct adaptive equalization was implemented, and the coefficients of the symbol-spaced transversal filter were adjusted with the LMS-based stochastic-gradient algorithm with time-varying step sizes. Time was partitioned into epochs lasting hundreds of symbols, over which the channel is presumed to be statistically stationary. The approach in [77] tailored its complexity for sparse systems by selecting a different subset of the transversal filter taps to adapt during each of these epochs. The two largest tap magnitudes are selected from an estimation of the transversal filter taps at the beginning of an epoch; from the relation $\frac{1}{1-z^{-D}} = 1 + z^{-D} + z^{-2D} + \cdots$, $|z| < 1$, the tap delay spacing D between these two taps estimates the multipath delay spread of a 2-tap system and, hence, the spacing of significant taps in a suitable zero-forcing equalizer (high SNR is presumed). These two taps and those at delays $2D$, $3D$,… are updated during this epoch, and the remaining taps are fixed. The work demonstrated successful, low-complexity demodulation of 4-PSK data in both 2- and 3-path channels using this approach, and comparisons between full-order filter updates and the sparse approach showed little performance degradation.

Identification of sparse systems is also an active topic of research in the controls area, where it is motivated from a model-reduction perspective [71, 72, 73, 74, 75]. A model-reduction approach was used to develop the sparse adaptive equalizer in [76]. Both the system order and the set of significant delays were presumed to be known in this work. As stated earlier, this knowledge may be approximated from Barker probes. The result is similar to a concurrent paper [64] which did not require such knowledge and which we summarize below. Details beyond this summary and experimental results can be found in [78].

8.4.2.3 Sparse Adaptive Equalization

We begin our presentation of sparse adaptive equalization by creating a framework for the sparse channel identification problem. To this end, we focus on the T_s-spaced, phase-compensated channel and ignore the common tap-phase trajectory. We will refer to Figure 8.13, which illustrates the estimation of a D-dimensional vector $\mathbf{h}$, using the known input scalar sequence $\{d_m\}$ and the observed output sequence $\{y_m\}$ satisfying

$$y_m = \mathbf{h}^H \mathbf{d}(m) + v_m,$$

where $\mathbf{d}(n)$ denotes D consecutive elements of the sequence $\{d_m\}$. To determine the best estimated output sequence of the form $\hat{y}_{k,n} = \hat{\mathbf{h}}(n)^H \mathbf{d}(k)$, $\forall k \le n$, we shall consider the deterministic least squares criterion $C_n(\hat{\mathbf{h}})$,

$$C_n(\hat{\mathbf{h}}) = \sum_{k=0}^{n} \lambda^{n-k} |y_k - \hat{y}_{k,n}|^2. \tag{8.25}$$

As shown in Figure 8.13, we are interested in representing the channel estimate in a two-step process: through a $D_s \times D$ *selection matrix* $\mathbf{S}$ ($D_s < D$) and a D_s-dimensional vector $\hat{\mathbf{h}}_\mathbf{S}(n)$ such that $\hat{\mathbf{h}}(n) = \mathbf{S}^H \hat{\mathbf{h}}_\mathbf{S}(n)$. The matrix $\mathbf{S}$ is so named since its row vectors form a subsequence of D_s rows from the $D \times D$ identity matrix, $\mathbf{I}$. Note that $\mathbf{SS}^H$

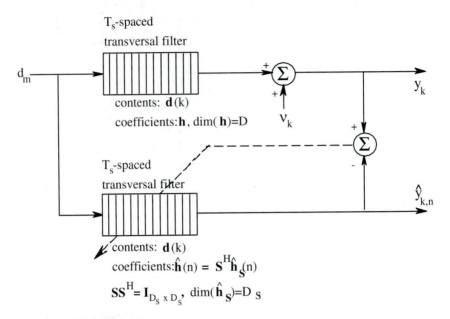

Figure 8.13 Sparse channel estimation.

is the $D_s \times D_s$ identity matrix, and that $\mathbf{S}$ allows the D_s components of $\hat{\mathbf{h}}_{\mathbf{S}}(n)$ to estimate a channel having a delay spread much greater than $T_s D_s$ seconds. We are interested in selecting D_s, $\mathbf{S}$, and $\hat{\mathbf{h}}_{\mathbf{S}(n)}$ at each time n to estimate the observation sequence y_k, $0 \le k \le n$ with an accuracy measured by $C_n(\hat{\mathbf{h}})$.

We begin with the special case of $\mathbf{S} = \mathbf{I}$ to develop the solution for $\hat{\mathbf{h}}_{\mathbf{S}}(n) = \hat{\mathbf{h}}_{\mathbf{I}}(n)$ and to relate this to the general case. When $\mathbf{S}$ is the identity matrix, the estimator $\hat{\mathbf{h}}_{\mathbf{I}}(n)$, which minimizes C_n, is found via weighted least squares regression,

$$\hat{\mathbf{h}}_{\mathbf{I}}(n) = \mathbf{R}_{dd}^{-1}(n)\mathbf{p}(n), \tag{8.26}$$

where

$$\mathbf{R}_{dd}(n) = \lambda \mathbf{R}_{dd}(n-1) + \mathbf{d}(n)\mathbf{d}^H(n)$$

$$\mathbf{p}(n) = \lambda \mathbf{p}(n-1) + y_n^* \mathbf{d}(n), \tag{8.27}$$

with initial conditions $\mathbf{R}_{dd}(-1) = \mathbf{0}$, $\mathbf{p}(-1) = \mathbf{0}$. Note that both y_i and d_i, which are used to form the linear estimator, are available to the receiver via observation, training sequences, or accurately detected data. The minimal cost for $\mathbf{S} = \mathbf{I}$ is

$$C_n(\hat{\mathbf{h}}_{\mathbf{I}}) = R_{yy}(n) - \hat{\mathbf{h}}_{\mathbf{I}}^H(n)\mathbf{p}(n), \tag{8.28}$$

where $R_{yy}(n) = \sum_{k=0}^{n} \lambda^{n-k} y_k y_k^*$.

An update complexity of $O(D)$ for the series $\{\hat{\mathbf{h}}_I(n)\}$ can be provided by an RLS-type algorithm. However, for a judicious choice of the selection matrix $\mathbf{S}$ and for sparse systems, the update complexity for the series $\{\mathbf{S}^H \hat{\mathbf{h}}_S(n)\}$ can be lessened by an additional order of magnitude. The added cost $C_n(\mathbf{S}^H \hat{\mathbf{h}}_S) - C_n(\hat{\mathbf{h}}_I)$ for this complexity decrease can also be precomputed. To see this, we quantify the accompanying increase in C_n when the estimator uses $\mathbf{S} \neq \mathbf{I}$, relative to the case $\mathbf{S} = \mathbf{I}$. For a fixed $\mathbf{S}$, the estimator $\hat{\mathbf{h}}_S(n)$ that minimizes $C_n(\mathbf{S}^H \hat{\mathbf{h}}_S(n))$ from (8.25) is

$$\hat{\mathbf{h}}_S(n) = \left[\mathbf{S} \mathbf{R}_{dd}(n) \mathbf{S}^H\right]^{-1} \mathbf{S} \, \mathbf{p}(n). \tag{8.29}$$

Due to the selection matrix $\mathbf{S}$, the channel estimator is constrained to have a sparse structure. Since $\mathbf{S} \mathbf{R}_{dd}(n) \mathbf{S}^H$ and $\mathbf{S} \mathbf{p}(n)$ have similar recursions as for the case $\mathbf{S} = \mathbf{I}$, the RLS algorithms can be used to update $\mathbf{S}^H \hat{\mathbf{h}}_S(n)$. Since the minimum cost associated with any selection matrix $\mathbf{S}$ is

$$C_n(\mathbf{S}^H \hat{\mathbf{h}}_S) = R_{yy}(n) - \hat{\mathbf{h}}_S^H(n) \mathbf{S} \mathbf{p}(n), \tag{8.30}$$

the incremental cost in reducing the estimator order to $D_s \leq D$ is

$$C_n(\mathbf{S}^H \hat{\mathbf{h}}_S) - C_n(\hat{\mathbf{h}}_I) = \mathbf{p}^H(n)\left[\mathbf{R}_{dd}^{-1} - \mathbf{S}^H(\mathbf{S}^H \mathbf{R}_{dd}(n) \mathbf{S})^{-1} \mathbf{S}\right] \mathbf{p}(n). \tag{8.31}$$

A critical step in an efficient sparsing algorithm is to find, in a simple fashion, the selection matrix $\mathbf{S}$ that minimizes D_s while satisfying a prescribed upper bound to the incremental cost given by the left-hand side of (8.31). Unfortunately, no simple rule exists for the general sparsing problem, and this fact will be shown to restrict the approach to sparse adaptive equalization. There are special cases of practical interest that permit efficient sparsing, however. For the case of *white* channel inputs, $E[\mathbf{R}_{dd}(n)] = a^2 \mathbf{I}$, the incremental cost associated with the sparse estimator has strong intuitive appeal. In this case, the increase in squared error is

$$C_n(\mathbf{S}^H \hat{\mathbf{h}}_S) - C_n(\hat{\mathbf{h}}_I) = \frac{1}{a^2} \mathbf{p}^H(n)\left[\mathbf{I} - \mathbf{S}^H \mathbf{S}\right] \mathbf{p}(n). \tag{8.32}$$

For the case of white inputs, the incremental error is the total energy in the components of $\mathbf{p}(n)$ corresponding to the rows of $\mathbf{I}$ not taken to form $\mathbf{S}$. As seen from (8.27), the energy in a component of $\mathbf{p}(n)$ is proportional to the energy of the channel tap at that location, and $E[\mathbf{p}(n)] = \mathbf{h}$. If the true channel, $\mathbf{h}$, has negligible energy at the ignored taps, then the increased cost due to a substantially simpler estimator is also negligible.

The squared error $C_n(\cdot)$ is directly related to the performance of the equalizer. For this reason, a careful selection of $\mathbf{S}$ is needed to maximally reduce the computational burden throughout a data packet without increasing $C_n(\mathbf{S}^H \hat{\mathbf{h}}_S)$ beyond a prescribed limit. There are at least two ways in which to proceed. The first approach, suggested in [66] and [79], uses the magnitude sequence output from the modem's Barker correlator as an MIP estimator. This estimator provides important information for sparse systems, such as the number of *tap groups* (a group is a set of adjacent, significant taps), their centroids and spans. Adjacent or nearly adjacent tap groups are usually combined so as to minimize the total number of groups in the response, g. This decomposition relates any sparse approximation to the superposition of g subsystems and suggests an application of the *fast modular* RLS algorithm [69], whose update complexity is $O(D_s g + g^2)$. Provided $D_s \ll D$ and the group count g is small, then $D_s g + g^2 \ll D$, and a fast modular RLS algorithm is significantly simpler than FTRLS.

The second approach does not require a priori knowledge of the input delay-spread function structure, and was suggested in [64]. In this case, it is assumed that a data packet consists of an initial training sequence followed by an information sequence, and that full-order channel estimation ($\mathbf{S} = \mathbf{I}$) occurs for a brief period at the beginning of the packet. Full-order estimation provides $C_n(\hat{\mathbf{h}}_1)$ and $\mathbf{p}(n)$, which are required for the selection of D_s, $\mathbf{S}$ and $\hat{\mathbf{h}}_\mathbf{S}$. Experimental results have shown that full-order estimation must be implemented for only a short period (about 2D symbols) at the beginning of the packet, usually within the training period. While this application of full-order channel estimation does preserve the peak processing load, it does not substantially affect the average computational burden. The average computational burden is determined strongly by the choice of $\mathbf{S}$.

We now address the connection between reduced-complexity channel identification and channel equalization. We focus in particular on fractionally spaced DFEs. Does an accurate, low-order channel estimate $\hat{\mathbf{h}}_s$, with $D_s \ll D$, guarantee a reduced number of computations for equalizer tap adjustment? Not in general, as can be seen by the condition for the extremal equalizer coefficients for a minimum-MSE DFE. For jointly stationary vectors $\mathbf{r}(n)$, $\mathbf{d}(n)$ and for a zero-mean, unit-variance white sequence $\{d_m\}$, the vectors $\mathbf{a}$, $\mathbf{b}$ must satisfy

$$\begin{bmatrix} \mathcal{R}_{rr} & -\mathcal{R}_{rd} \\ -\mathcal{R}_{rd}^H & \mathcal{R}_{dd} \end{bmatrix} \begin{bmatrix} \mathbf{a} \\ \mathbf{b} \end{bmatrix} = \begin{bmatrix} \mathbf{f} \\ \mathbf{0} \end{bmatrix}, \tag{8.33}$$

where

$$\mathcal{R}_{vw} = E\left[\mathbf{v}(n)\mathbf{w}^H(n)\right]$$

$$\mathbf{f} = [h(N_1 T_s) \ h((N_1 - 1)T_s) \ \dots \ h(-N_2 T_s)]^T.$$

The statistical cross-correlations take the form

$$[\mathcal{R}_{rd}]_{ij} = h(jT + N_1 T_s - (i-1)T_s)$$

$$[\mathcal{R}_{rr}]_{ij} = \sigma^2 \delta_{ij} + \sum_\alpha h(\alpha T)h^*(\alpha T + (j-i)T_s),$$

where σ^2 is the additive noise variance. While a sparse, fractionally spaced channel **h** yields a sparse form for **f** and $\mathcal{R}_{rd}$, the reduction in complexity for the solution of **a** and **b** is not readily apparent. One approach proposed in [78] is to apply the channel sparsing matrix **S** to the feedforward coefficient vector **a**. This method preserves the adaptive matched-filter property of the feedforward filter and reduces the complexity of the solution for the equalizer coefficients. Sparse equalization of this form has been shown to reduce the average computational complexity per update by more than an order of magnitude in shallow-water acoustic channels.

8.4.3 Multiuser Detection

One of the first applications of multiuser detection (see Chapters 2 and 3) to the UAC is described in [80] and [36]. Vertical channel transmission with negligible multipath was considered. An adaptive K-user detector was used to resolve the temporal overlap of at most K short packets in a common request channel for a dynamic time division multiple access (TDMA) system. This receiver structure was a soft-decision two-stage detector and was provided information about the signature waveforms from packet headers. Using soft decisions to reconstruct and suppress cochannel interference has been shown to decrease the symbol error rate for low thermal noise conditions [82]. The parameter K was much smaller than the modem population size, and the sensitivity of request-channel throughput with K was explored [81]. This receiver was implemented and tested at sea for $K = 2$.

Multiple access techniques for shallow-water networks usually do not include time and frequency division because of the frequency-selective fading, limited system bandwidth, and the increased difficulties with global time slotting associated with platform mobility, as discussed in Section 8.1. Code-division multiple access (CDMA) is the preferred method for resource sharing in this environment. Until the early 1990s, however, little work was available on multiuser detection in time-varying dispersive channels, other than a bank of independently operating RAKE filters. In [83], two multiuser detectors were presented for a fixed frequency-selective channel which removed the bit-error-rate floor (versus average SNR) of the conventional RAKE detector. The floor was attributed to dominating cochannel interference, which was compensated for by the multiuser receivers. Fully adaptive, single-sensor multiuser receivers tailored for the shallow-water acoustic channel were presented first in [84] and most recently in [85] and are summarized below.

Figure 8.14 presents a block diagram description of this detector, which we refer to as a *multiuser DFE*, or MDFE, and which can be compared with the multiuser DFE structures described in Chapters 3 and 4.

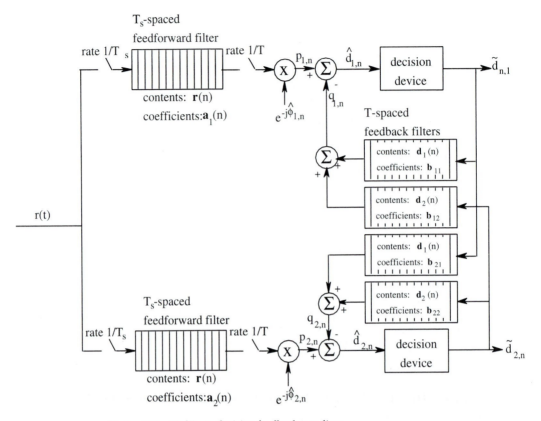

Figure 8.14 Multiuser decision feedback equalizer.

Note the strong similarity of this receiver to a bank of K conventional, fractionally spaced DFEs, with the exception of the "crossover" feedback filters $\mathbf{b}_{ji}$, $i \neq j$ in Figure 8.14. This single-sensor multiuser receiver consists of one fractionally spaced feedforward filter and K symbol-spaced feedback filters for each of the K users in the system. The sampled input signal in every feedforward section at time nT is given by

$$\mathbf{r}(n) = [r(nT + N_1 T_s) \ \ r(nT + (N_1 - 1)T_s) \ \ \cdots \ \ r(nT - N_2 T_s)]^T. \qquad (8.34)$$

The scalar output of the feedforward transversal filter for the kth user at time nT is given by $\mathbf{a}_k^H \mathbf{r}(n)$, where the combining coefficients are

$$\mathbf{a}_k = [a_{k,-N_1} \ \ \cdots \ \ a_{k,N_2}]^T.$$

We also define the output of the kth phase correction circuit at time nT as $p_{k,n} = \mathbf{a}_k^H \mathbf{r}(n)e^{-j\hat{\varphi}_{k,n}}$. Similarly, the symbol-rate feedback filter for the ith receiver containing decisions from user j has an output at time nT given by $\mathbf{b}_{ij}^H \mathbf{d}_j(n)$, where

$$\mathbf{d}_j(n) = [\tilde{d}_{j,n-1} \ \ \cdots \ \ \tilde{d}_{j,n-M}]^T$$

and

$$\mathbf{b}_{ij} = [b_{ij1} \cdots b_{ijM}]^T.$$

As suggested by the notation, $\tilde{d}_{k,n}$ denotes the final decision for the nth symbol for user k, $d_{k,n}$. The sum of all K feedback transversal filters for user j at time nT is denoted as $q_{j,n}$. For the two-user case, the MDFE receiver produces for user 1 the feedback aggregate $q_{1,n} = \mathbf{b}_{11}^H \mathbf{d}_1(n) + \mathbf{b}_{21}^H \mathbf{d}_2(n)$, while the corresponding feedback aggregate for the conventional DFE would be $q_{1,n} = \mathbf{b}_{11}^H \mathbf{d}_1(n)$.

As shown in Figure 8.14, the decision variable for user k at time n is $\hat{d}_{k,n} = p_{k,n} - q_{k,n}$. It is convenient to represent this variable as the inner product

$$\hat{d}_{k,n} = \mathbf{c}_k^H \mathbf{u}_k(n),$$

where

$$\mathbf{c}_1 = [\mathbf{a}_1^T - \mathbf{b}_{11}^T - \mathbf{b}_{21}^T]^T,$$

$$\mathbf{u}_1(n) = [\mathbf{y}_1^T(n) \mathbf{d}_2^T(n)]^T, \tag{8.35}$$

and

$$\mathbf{y}_1(n) = [\mathbf{r}^T(n) e^{-j\hat{\varphi}_{1,n}} \mathbf{d}_1^T(n)]^T.$$

Final decisions for the data $d_{k,n}$ are denoted as $\tilde{d}_1(n)$ and are given by quantization (slicing) of the decision variable. The performance metric is a weighted sum of squared errors $e_{k,n} = d_{k,n} - \hat{d}_{k,n}$ when the symbol is known by the receiver in a training mode and $e_{k,n} = \tilde{d}_{k,n} - \hat{d}_{k,n}$ in a decision-directed mode. The receiver parameters at time nT are adjusted to minimize the weighted squared error $\mathcal{E}(n) = \sum_{k=0}^{n} \lambda^{n-k} \{|e_{1,n}|^2 + |e_{2,n}|^2\}$. Defining the cross-correlation matrix between the column vector sequences $\mathbf{x}(n)$ and $\mathbf{w}(n)$ as $\mathbf{R}_{xw}(n) = \lambda \mathbf{R}_{xw}(n-1) + \mathbf{x}(n)\mathbf{w}^H(n)$, the receiver parameters $\{\mathbf{c}_{1,\text{opt}}, \hat{\varphi}_{1,n}\}$ for user 1 which minimize $\mathcal{E}(n)$ are given by

$$\mathbf{c}_{1,\text{opt}} = \mathbf{R}_{\mathbf{u}_1,\mathbf{u}_1}^{-1}(n) \mathbf{R}_{\mathbf{u}_1,\mathbf{b}_1}(n)$$

and

$$\frac{\delta \mathcal{E}(n)}{\delta \hat{\varphi}_{1,n}} = -2 \operatorname{Im}[p_{1,n} e_{1,n}^*] = 0.$$

Since $\mathbf{u}_1$ depends on $\hat{\varphi}_{1,n}$, and $e_{1,n}$ depends on $\mathbf{u}_1(n)$, an approximate and computationally efficient solution to these equations can be achieved through the decoupling approach discussed earlier.

In this section, we have reviewed an adaptive receiver structure for the MDFE for K asynchronous users. A bank of K conventional DFEs is a special case of this presentation and is given by the substitution of $\mathbf{y}_1(n)$ for $\mathbf{u}_1(n)$ in this section. In [85], the improvement due to the crossover filters $\mathbf{b}_{ij}$ is quantified. It is shown there that

$$E\{\mathcal{E}_{\text{MDFE}}\} \approx 1 - \mathcal{R}_{\mathbf{d}_1\mathbf{u}_1} \mathcal{R}_{\mathbf{u}_1\mathbf{u}_1}^{-1} \mathcal{R}_{\mathbf{u}_1\mathbf{d}_1},$$

$$E\{\mathcal{E}_{\text{DFE}}\} \approx 1 - \mathcal{R}_{\mathbf{d}_1\mathbf{y}_1} \mathcal{R}_{\mathbf{y}_1\mathbf{y}_1}^{-1} \mathcal{R}_{\mathbf{y}_1\mathbf{d}_1},$$

where $\mathcal{R}_{wx}$ denotes a statistical cross-correlation matrix. These expressions are valid under the additional assumption that the data symbols $\{d_{k,n}\}$ are uncorrelated with zero mean and unit variance and can be used to show that the MDFE outperforms the DFE in a multiple-access setting [85]. The two detectors have comparable performance in the unusual case when $\mathbf{d}_2$ and $\mathbf{y}_1$ are statistically uncorrelated.

8.4.4 Wideband Array Signal Processing

As discussed in Chapters 1, 2, and 4, the use of a spatial arrays of receivers, which we refer to as multichannel receivers, can improve the demodulation capabilities of communications systems. If the receivers are widely separated so that the fluctuations in the channels from the transmitter to one receiver are independent of those to another receiver, explicit diversity techniques can be exploited to enable the overall system to overcome fading in one or more of the independent channels. If the fluctuations in the channels to different receivers are not independent, then the coherent spatial structure of the received signal can be exploited. One of two different approaches is usually taken in coherently combining the signals at the receivers. The first approach is to attempt to capture all of the signal energy that has propagated from the transmitter to the receivers. The second is to attenuate the signals that have propagated through some ray tubes while accentuating signals that have propagated through other ray tubes. The latter approach corresponds to traditional beamforming. In the multiuser environment, the multichannel receiver can use the differences in the spatial structure of the signals transmitted by different users to distinguish between interfering arrivals [16, 90].

We begin with the approach developed in [45], which is an extension of the adaptive single-channel equalizer discussed in Section 8.4.1 and shown in Figure 8.12. The single input channel is replaced by K input channels, K feedforward filters, and K PLLs. The input to the kth channel is $r_k(t)$, the coefficient vector of the kth feedforward filter is $\mathbf{a}_k(n)$, and the phase of the kth PLL is $\varphi_{n,k} = \omega_{d,n,k} n + \theta_{n,k}$. The outputs of the individual channels are summed to form

$$p_n = \sum_{k=1}^{K} \mathbf{a}_k^H(n)\mathbf{r}_k(n)e^{-j\varphi_{n,k}}. \tag{8.36}$$

The tap weight vector given by

$$\mathbf{u}(n) = \begin{bmatrix} \mathbf{a}_1(n) \\ \vdots \\ \mathbf{a}_K(n) \\ \mathbf{b}(n) \end{bmatrix} \tag{8.37}$$

is determined with a least squares metric, as in Section 8.4.1 for the single-channel equalizer. The optimal phase correction terms $\varphi_{n,1}$ through $\varphi_{n,K}$ are individually estimated using a second-order digital PLL.

The computational complexity required to compute the optimal tap weight vector grows with K^2, yielding an often unacceptable level of complexity. Approaches to reducing the complexity associated with multichannel algorithms in the underwater environment have been developed [16, 89]. Like the sparsing approach to complexity reduction described in Section 8.4.2, these multichannel algorithms achieve their complexity reduction by exploiting particular features of the propagation environment.

The approach in [89] is built around an essentially deterministic ray propagation model. Let P denote the number of rays from the transmitter to the receiver array, and assume that $P < K$. Then, assuming that the ray paths are known, it is shown that there is no loss in performance if a K-input, P-output beamformer is used to transform the outputs of the K sensors into P-channel signal, with the pth channel carrying the signal that propagated along the pth ray. This P-channel beamformer output is followed by a P-channel equalizer. A complexity reduction is realized in calculating the weights of the multichannel equalizer since the dimension of the receiver parameter vector $\mathbf{u}(n)$ is proportional to P. When the characteristics of the P ray paths are not known or are time varying, the front-end beamformer is required to adapt. The beamformer adaptation algorithm in [89] is based on three assumptions. The first is that the signals can be reasonably modeled as narrowband signals. The second is that the transmitter is far enough from the receiver array and the environment is sufficiently homogeneous that the propagating signal is a plane wave across the aperture of the array. The final assumption is that the array is linear with uniform spacing between sensors. When these assumptions are fulfilled, a memoryless beamformer with complex-valued, unit-norm weights is sufficient to achieve the desired signal separation and a relatively simple PLL algorithm can be used to calculate the needed beamformer weights. An ad hoc approach to updating the unconstrained weights of a memoryless beamformer is described for the case where the above assumptions are not valid.

The reduced complexity multichannel beamformers developed in [16] are based upon the micro-/macro-multipath propagation model developed in Section 8.1.4. The multichannel demodulator consists of a K input, single-output tapped-delay-line beamformer followed by a single-channel equalizer such as the one described in Section 8.4.1. This structure is based upon several assumptions. The first is that the intersensor correlation of the clusters of arrivals for the dominant ray tube is reasonably strong. The second assumption is that the macro-multipath provides sufficiently different spatial/temporal structure to signals that propagate through different ray tubes so that the front-end beamformer can distinguish between these different signals by using only that structure. Given these assumptions, the front-end beamformer can eliminate most of the intersymbol interference from multipath arrivals from other than the dominant ray tube and the multiple-access interference from other transmitters without having to adapt to the fluctuations in the micro-multipath structure of the environment. Complexity reduction is achieved because the macro-multipath structure of the channel to which the beamformer must adapt changes fairly slowly and therefore the weights of the beamformer can be updated infrequently. This method of complexity reduction therefore belongs to the *reduced update* class of techniques described in Section 8.4.2. The single-channel output of the beamformer contains only a modest amount of intersymbol interference due to the micro-multipath spread of the dominant ray tube and a greatly reduced amount of multiple-access interference. A small single-channel equalizer is used to complete the demodulation process.

8.5 CONCLUDING REMARKS

In this chapter, we have presented a first-principles model for the UAC, and we have discussed many of the characteristics that distinguish this channel from the RF channels discussed in earlier chapters. Our model was developed using the concept of ray propagation through random nonhomogeneous media, by introducing the notion of ray tubes. The effect of environmental fluctuations on the communication channel were decomposed into macro- and micro-multipath input delay-spread functions. This model was used to relate channel saturation to the temporal and spatial coherence of the input delay-spread functions. We have focused on signal processing techniques for situations when the input delay-spread function exhibited a separable decomposition $g(t, \tau) \approx h(\tau)e^{j(\omega_d t + \theta)}$. This explicit modeling of the Doppler shift permits reliable phase-coherent demodulation in time-varying shallow-water UACs. We have presented techniques to reduce the computational complexity for both single-sensor and multisensor receivers; these techniques exploited features germane to the underwater channel. Multiuser detectors find a natural application in this channel and were presented in this chapter as extensions of the single-channel equalizers.

   We have also presented current research on several open problems in phase-coherent communications through UACs. Current and future research topics include adaptive detection in multipath channels that exhibit path-dependent Doppler shifts, wideband coherent detection for severe Doppler dilation, and coherent demodulation algorithms with reduced computational complexity. We conclude this chapter with an extensive bibliography detailing past and recent work in the field.

REFERENCES

[1] T. G. Bell, "Sonar and Submarine Detection," *U.S. Navy Underwater Sound Lab. Rep.* 545, 1962.

[2] L. M. Brekhovskikh, Y. P. Lysonov, *Fundamentals of Ocean Acoustics*, 2nd Edition, Springer-Verlag, Berlin, 1991.

[3] C. S. Clay, H. Medwin, *Acoustical Oceanography: Principles and Applications*, John Wiley and Sons, New York, 1977.

[4] D. Kilfoyle, J. Catipovic, "Dynamic Viterbi Decoding in Underwater Acoustic Channels under Burst Noise Conditions," in *Proc. OCEANS'96*, Fort Lauderdale, FL, 1996, pp. 832–838.

[5] M. Buckingham, C. Epifanio, M. Readhead, "Passive Imaging of Targets with Ambient Noise: Experimental Results," *J. Acoust. Soc. Am.*, Vol. 100, No. 4, Pt. 2, October 1996, pp. 27–36.

[6] J. Bellingham, H. Schmidt, M. Deffenbaugh, "Acoustically Focused Oceanographic Sampling in the Haro Strait Experiment," *J. Acoust. Soc. Am.*, Vol. 100, No. 4, Pt. 2, October 1996, p. 2612.

[7] S. Flatte, ed., *Sound Transmission Through a Fluctuating Ocean*, Cambridge University Press, Cambridge, 1979.

[8] S. Flatte, "Wave Propagation Through Random Media: Contributions from Ocean Acoustics," *Proc. IEEE*, Vol. 71, No. 11, November 1983, pp. 1267–1294.

[9] T. Duda, S. Flatte, D. Creamer, "Modelling Meter-Scale Acoustic Intensity Fluctuations from Oceanic Fine Structure and Microstructure," *Journal of Geophysical Research*, Vol. 93, No. C5, May 1988, pp. 5130–5142.

[10] T. Duda, "Modeling Weak Fluctuations of Undersea Telemetry Signals," *IEEE J. Oceanic Eng.*, Vol. 16, No. 1, January 1991, pp. 3–11.

[11] L. Ziomek, "Generalized Kirchhoff Approach to the Ocean Surface Scatter Communication Channel. Part I. Transfer Function of the Ocean Surface," *J. Acoust. Soc. Am.*, Vol. 71, No. 1, January 1982, pp. 116–126.

[12] L. Ziomek, "Generalized Kirchhoff Approach to the Ocean Surface Scatter Communication Channel. Part II. Second-order Functions," *J. Acoust. Soc. Am.*, Vol. 71, No. 6, June 1982, pp. 1487–1495.

[13] D. Dowling, D. Jackson, "Coherence of Acoustic Scattering from a Dynamic Rough Surface," *J. Acoust. Soc. Am.*, Vol. 93, No. 6, June 1993, pp. 3149–3157.

[14] R. Owen, B. Smith, R. Coates, "An Experimental Study of Rough Surface Scattering and Its Effects on Communication Coherence," in *Proc. OCEANS'94*, Brest, France, Vol. III, 1994, pp. 483–488.

[15] H. Schmidt, Personal communications regarding results of 1996 Haro Strait experiment.

[16] S. Gray, J. Preisig, D. Brady, "Multi-user Detection in a Horizontal Underwater Acoustic Channel Using Array Observations," *IEEE Trans. Signal Process.* Special Issue on Signal Processing for Advanced Communications, Vol. 45, No. 1, January 1997, pp. 148–160.

[17] P. Bello, "Characterization of Randomly Time-Variant Linear Channels," *IEEE Trans. Commun. Systems*, CS-11, December 1963, pp. 360–393.

[18] D. Farmer, S. Clifford, J. Verral, "Scintillation Structure of a Turbulent Tidal Flow," *Journal of Geophysical Research*, Vol. 92, No. C5, May 1987, pp. 5369–5382.

[19] T. Eggen, "Underwater Acoustic Communications over Doppler Spread Channels," Ph.D. Thesis, Massachusetts Institute of Technology and Woods Hole Oceanographic Institution, Cambridge, MA, June 1997.

[20] B. Woodward, H. Sari, "Digital Underwater Acoustic Voice Communications," *IEEE J. Oceanic Eng.*, Vol. 21, No. 2, April, 1996, pp. 181–192.

[21] H. O. Berktay, B. Gasey, and C. A. Teer, "Underwater Communication: Past, Present and Future," *J. Sound Vib.*, Vol. 7, pp. 62–70, 1968.

[22] A. Baggeroer, "Acoustic Telemetry—An Overview," *IEEE J. Oceanic Eng.*, Vol. 9, No. 4, October, 1984, pp. 229–235.

[23] M. Stojanovic, "Recent Advances in High-Speed Underwater Acoustic Communications," *IEEE J. Oceanic Eng.*, Vol. 21, No. 2, April, 1996 pp. 125–136.

[24] *IEEE J. Oceanic Eng.* Special issue on Ocean Acoustic Data Telemetry, Vol. 16, No. 1, January, 1991, pp. 1–177.

[25] J. A. Catipovic, *Design and Performance Analysis of a Digital Acoustic Telemetry System*, Doctoral Dissertation, Woods Hole Oceanographic Institution/MIT, May, 1988.

[26] A. Clark, "Diver Communications—The Case for Single Sideband," *Underwater System Design*, 1989, pp. 16–18.

[27] W. Dow, "A Telemetering Hydrophone," *Deep-Sea Research*, Vol. 7, 1960, pp. 142–147.

[28] P. Hearn, "Underwater Acoustic Telemetry," *IEEE Trans. Commun. Tech.*, Vol. CT-14, Dec. 1966, pp. 839–843.

[29] G. M. Walsh, A. P. Alair, A. S. Westneat, "Establishing Reliability and Security in an Offshore Command Link," *Proc. Offshore Tech. Conference*, 1969.

[30] F. R. Mackelburg, S. J. Watson, A. Gordon, "Benthic 4800 bits/second Acoustic Telemetry," *Proc. OCEANS, 1981*, p. 78.

[31] J. G. Proakis, *Digital Communications*, 3rd edition, New York, McGraw-Hill, 1995.

[32] R. Price, P. E. Green, Jr., "A Communication Technique for Multipath Channels," *Proc. IRE*, Vol. 46, March, 1958, pp. 555–570.

[33] P. Monsen, "Theoretical and Measured Performance of a DFE Modem on a Fading Multipath Channel," *IEEE Trans. Commun.*, Vol. COM-25, Oct. 1977, pp. 1144–1153.

[34] J. Catipovic, A. B. Baggeroer, K. Von Der Heydt, D. Koelsch, "Design and Performance Analysis of Digital Acoustic Telemetry Systems for the Short Range Underwater Channel," *IEEE J. Oceanic Eng.*, Vol. OE-9, No. 4, October, 1984, pp. 242–252.

[35] S. Coatelan, A. Glavieux, "Design and Test of a Multicarrier Transmission System on the Shallow Water Acoustic Channel," *Proc. OCEANS '94*, Brest, France, Vol. 3, 1994, pp. 472–477.

[36] D. Brady, J. Catipovic, "Adaptive Multiuser Detection for Underwater Acoustic Channels," *IEEE J. Oceanic Eng.*, Vol. 19, April 1994, pp. 158–165.

[37] http://www.datasonics.com/products/modems/atm850.htm

[38] J. Catipovic, "Performance Limitations in Underwater Acoustic Telemetry," *IEEE J. Oceanic Eng.*, Vol. 15, July, 1993, pp. 205–216.

[39] R. Galvin, R. F. W. Coates, "Analysis of the Performance of an Underwater Acoustic Communication System and Comparison with a Stochastic Model," in *Proc. OCEANS'94*, Brest, France, 1994, pp. 478–482.

[40] A. Essebbar, F. Loubet, F. Vial, "Underwater Acoustic Channel Simulations for Communication," in *Proc. OCEANS'94*, Brest, France, 1994, pp. 495–500.

[41] S. J. Roberts, "An Echo Canceller Technique Applied to an Underwater Acoustic Data Link," Ph.D. Thesis, Department of Electrical and Electronic Engineering, Herriot Watt University, Edinburgh, Scotland, September, 1983.

[42] G. Sandsmark, *The Feasibility of Adaptive Equalization in High Speed Underwater Acoustic Data Transmission*, Ph.D. thesis, NTH, Trondheim, Norway, 1990.

[43] J. Catipovic, M. Deffenbaugh, L. Freitag, D. Frye, "An Acoustic Telemetry System for Deep Ocean Mooring Data Acquisition and Control," *Proc. OCEANS'89*, Seattle, WA, 1989, pp. 887–892.

[44] M. Stojanovic, J. Catipovic, J. G. Proakis, "Phase-Coherent Digital Communications for Underwater Acoustic Channels," *IEEE J. Oceanic Eng.*, Vol. 19, No. 1, January, 1994, pp. 100–111.

[45] M. Stojanovic, J. Catipovic, and J. Proakis, "Adaptive Multichannel Combining and Equalization for Underwater Acoustic Communications," *J. Acoust. Soc. Am.*, Vol. 94, No. 3, Pt. 1, Sept. 1993, pp. 1621–1631.

[46] D. Falconer, "Jointly Adaptive Equalization and Carrier Phase Recovery in Two Dimensional Digital Communication Systems," *Bell Syst. Techn. J.*, Vol. 55, March, 1976, pp. 317–334.

[47] R. Coates, "Underwater Acoustic Communications," *Proc. OCEANS'93*, Victoria, BC, Canada, Vol. 3, October, 1993, pp. 420–425.

[48] R. Gitlin, S. Weinstein, "Fractionally Spaced Equalization: An Improved Digital Transversal Equalizer," *Bell Syst. Techn. J.*, Vol. 60, Feb. 1981, pp. 275–296.

[49] D. T. M. Slock, T. Kailath, "Fast Transversal RLS Algorithms," in *Adaptive System Identification and Signal Processing Algorithms*, N. Kalouptsidis, S. Theodoridis, Editors, Englewood Cliffs, NJ: Prentice Hall, 1993, pp. 123–190.

[50] D. T. M. Slock, T. Kailath, "Numerically Stable Fast Transversal Filters for Recursive Least Squares Adaptive Filtering," *IEEE Trans. Signal Process.*, Vol. 39, Jan., 1991, pp. 92–114.

[51] M. Johnson, L. Freitag, M. Stojanovic, "Improved Doppler Tracking and Correction for Underwater Acoustic Communications," *Proc. ICASSP, 1996* Atlanta, Ga., Vol. 1, pp. 575–578.

[52] B. Geller, V. Capellano, J.-M. Brossier, A. Essebbar, G. Jourdain, "Equalizer for Video Rate Transmission in Multipath Underwater Communications," *IEEE J. Oceanic Eng.*, Vol. 21, No. 2, April, 1996, pp. 150–155.

[53] A. Kaya, S. Yauchi, "An Acoustic Communication System for Subsea Robot," *Proc. OCEANS'89*, Seattle, WA, Oct. 1989, pp. 765–770.

[54] M. Suzuki, T. Sasaki, "Digital Acoustic Image Transmission System for Deep Sea Research Submersible," *Proc. OCEANS'92*, Newport, RI, Oct. 1992, pp. 567–570.

[55] A. Benveniste, M. Metivier, P. Priouret, *Adaptive Algorithms and Stochastic Approximations*, New York: Springer-Verlag, 1990.

[56] W. A. Sethares, "The Least Mean Square Family," in *Adaptive System Identification and Signal Processing Algorithms*, N. Kalouptsidis and S. Theodoridis, editors, Englewood Cliffs, NJ: Prentice Hall, 1993, pp. 84–122.

[57] L. Barbosa, "Maximum Likelihood Sequence Estimators: A Geometric View," *IEEE Trans. Inform. Theory*, Vol. 35, No. 2, March 1989, pp. 419–427.

[58] D. Divsalar, "Performance of Mismatched Receivers on Bandlimited Channels," Ph.D. thesis, UCLA, 1978.

[59] J. Habermann, D. Dzung, "Performance of Coherent Data Transmission with Imperfect Channel Estimation in Frequency-Selective Rayleigh Fading Channels: Cochannel and Adjacent Channel Interference," *AEU*, Vol. 44, No. 5, Sept./Oct., 1990, pp. 1680–1686.

[60] F. Gozzo, "Robust Sequence Estimation in the Presence of Channel Mismatch," *IBM Federal Systems Report 92-OTP-050*.

[61] M. Stojanovic, J. Proakis, J. Catipovic, "Analysis of the Impact of Channel Estimation Errors on the Performance of a Decision-Feedback Equalizer in Fading Multipath Channels," *IEEE Trans. Commun.*, Vol. 43, Feb./Mar./Apr., 1995, pp. 877–886.

[62] S. Fechtel, H. Meyr, "An Investigation of Channel Estimation and Equalization Techniques for Moderately Rapid Fading HF-Channels," *Proc. ICC'91*, Denver, CO, 1991, pp. 768–772.

[63] P. D. Shukla, L. F. Turner, "Channel-Estimation-Based Adaptive DFE for Fading Multipath Radio Channels," *IEE Proceedings, Part I*, Vol. 138, No. 6, Dec. 1991, pp. 525–543.

[64] M. Kocic, D. Brady, "Complexity-Constrained RLS Algorithm for Sparse Channels," *Proc. Conf. Inform. Science Sys.'94*, Princeton, NJ, March, 1994, pp. 420–425.

[65] M. Kocic, D. Brady, M. Stojanovic, "Sparse Equalization for Real-Time Digital Underwater Acoustic Communications," *Proc. OCEANS'95*, San Diego, CA, 1995, pp. 1417–1422.

[66] L. Freitag, M. Johnson, "A Robust and Efficient Receiver for Coherent Acoustic Communications," submitted to *IEEE J. Oceanic Eng.*, 1997.

[67] *Haro Straits data* at http://telem.whoi.edu/.

[68] P. Anderson, "Adaptive Forgetting in Recursive Identification Through Multiple Models," *Int. J. Control*, 42(5), 1984, pp. 1175–1193.

[69] D. Slock, L. Chisci, H. Lev-Ari, T. Kailath, "Modular and Numerically Stable Fast Transversal Filters for Multichannel and Multiexperiment RLS," *IEEE Tran. Acoust. Speech Sig. Process.*, vol. ASSP-40, April 1992, pp. 784–802.

[70] Y. F. Cheng and D. M. Etter, "Analysis of an Adaptive Technique for Modeling Sparse Systems," *IEEE Trans. Acoust. Speech Sig. Process.*, Vol. ASSP-37, No. 2, Feb. 1989, pp. 254–264.

[71] D. Wilson, "Optimum Solution of Model-Reduction Problem," *Proc. IEE*, Vol. 117, No. 6, June 1970, pp. 1161–1165.

[72] K. Nagpal, R. Hemlick, C. Sims, "Reduced-order Estimation: Part 1. Filtering," *International Journal of Control*, Vol. 45, No. 6, 1987, pp. 1867–1888.

[73] B. Moore, "Principal Component Analysis in Linear Systems: Controllability, Observability and Model Reduction," *IEEE Trans. Autom. Control*, Vol. AC-26, No. 1, February 1981, pp. 17–32.

[74] H. Kim, C. Sims, K. Nagpal, "Reduced Order Filtering in H-Infinity Setting," *Proceedings of American Control Conference*, 1992, pp. 1876–1877.

[75] N. Siddiqui, C. Sims: "Filter Order Reduction Using Mean Value and Covariance Matching Technique," *Proceedings of American Control Conference*, 1992, pp. 1789–1793.

[76] D. Reynolds, C. Sims, L. Tong, "Adaptive Equalization of a Digital Communications Channel with a Reduced-Order Equalizer," *Proc. 1994 Asilomar Conf*, Monterey, CA, November 1994, pp. 1428–1432.

[77] B. Geller, V. Capellano, J. M. Brossier, "Equalizer for High Data Rate Underwater Communications," *Proc. OCEANS'94*, Brest, France, 1994.

[78] M. Kocic, D. Brady, M. Stojanovic, "Reduced Complexity Equalization for Acoustic Telemetry in Shallow Water," submitted to *IEEE J. Oceanic Eng.*

[79] M. Johnson, D. Brady, M. Grund, "Reducing the Computational Requirements of Adaptive Equalization in Underwater Acoustic Communications," *Proc. OCEAN'95*, San Diego, CA, Vol. 3, 1995, pp. 1405–1410.

[80] D. Brady, J. Catipovic, "Adaptive Soft-Decision Multiuser Receiver for Underwater Acoustical Channels," *Proc. 1992 Asilomar Conf.*, Pacific Grove, CA, Vol. 2, Oct. 1992, pp. 1137–1141.

[81] D. Brady, L. Merakos, "Throughput Performance of Multiuser Detection in Unslotted Contention Channels," *Proc. INFOCOM*, Toronto, Canada, Vol. 2, June, 1994, pp. 610–617.

[82] X. Zhang, D. Brady, "Asymptotic Multiuser Efficiencies for Decision-Directed Multiuser Detectors," *IEEE Trans. Info. Theory*, Vol. 44, No. 2, March 1998, pp. 502–515.

[83] Z. Zvonar, D. Brady, "Coherent and Differentially Coherent Multiuser Detectors for Asynchronous CDMA Frequency-Selective Channels," *Proc. MILCOM, 1992*, pp. 17.6.1–17.6.5.

[84] Z. Zvonar, D. Brady, "Adaptive Multiuser Receiver for Fading CDMA Channels With Severe ISI," *Proc. Conf. Inform. Sc. Sys., 1993*, pp. 324–329.

[85] Z. Zvonar, D. Brady, J. Catipovic, "Adaptive Detection for Shallow-Water Acoustic Telemetry with Cochannel Interference," *IEEE Trans. Oceanic Eng.*, Vol. 21, No. 4, 1996, pp. 528–536.

[86] J. Catipovic, L. Freitag, "Spatial Diversity Processing for Underwater Acoustic Telemetry," *IEEE J. Oceanic Eng.*, Vol. OE-15, 1991, pp. 205–216.

[87] Q. Wen, J. Ritcey, "Spatial Equalization for Underwater Acoustic Communications," in *Proc. 22nd Asilomar Conference on Signals, Systems and Computers*, 1992, pp. 1132–1136.

[88] O. Hinton, G. Howe, A. Adams, "An Adaptive, High Bit Rate, Sub-sea Communications System," in *Proc. of the European Conference on Underwater Acoustics*, Brussels, Belgium, editor, M. Weydert, Amsterdam, Elsevier Applied Science, 1992, pp. 75–79.

[89] M. Stojanovic, J. Catipovic, J. Proakis, "Reduced-complexity Spatial and Temporal Processing of Underwater Acoustic Communication Signals," *J. Acoust. Soc. Am.*, Vol. 98, No. 2, Pt. 1, August 1995, pp. 961–972.

[90] M. Stojanovic, Z. Zvonar, "Multichannel Processing of Broad-Band Multiuser Communication Signals in Shallow Water Acoustic Channels," *IEEE J. Oceanic Eng.*, Vol. OE-21, 1996, pp. 156–166.

ACKNOWLEDGMENT

This work was supported by Grant N0014-95-1-1316 from the Office of Naval Research.

EPILOGUE

Four Laws of Nature and Society: The Governing Principles of Digital Wireless Communication Networks

Andrew J. Viterbi

Four laws, two each from the natural sciences and the social sciences, have formed the basis for the development of digital wireless communication networks. This essay describes their interaction, as well as their logical support for spread-spectrum multiple-access techniques.

E.1 OVERVIEW

In this techno-philosophical essay, we attempt to demonstrate that the implementation and success of digital wireless communication networks depends primarily on four basic laws and their underlying theories, which are attributed respectively to:

- Maxwell and Hertz
- Shannon
- Moore
- Metcalfe

The first two laws are laws of nature, while the last two, though often mistakenly thought as such, are in reality laws of behavior. The order is in the sequence of their discovery and their importance; additionally, as the field of wireless communications has matured, the emphasis and immediate relevance has shifted gradually downward in the list. Without an appreciation for Maxwell's and Hertz's theories, there would be no controlled wireless propagation of electromagnetic waves. Without an understanding of Shannon's theories, efficient use of the spectrum through sophisticated signal processing could not be achieved. Without the consequences of Moore's law, these signal processing techniques could not be implemented in a useful and economic fashion. And finally, Metcalfe's law, which we shall explore last, helps to predict the success or failure of large new network deployments and, consequently, the wisdom of business strategies involving proportionally large capital investments.

E.2 Wireless Propagation and Its Anomalies

In a remarkable sequence of achievements in theoretical and experimental physics toward the end of the nineteenth century, the basis for electromagnetic propagation was established and proved both theoretically and experimentally. Though numerous academic researchers, residing in the musty lecture halls and laboratories of that period, shared in the success, the two that stand out are James Clerk Maxwell and Heinrich Hertz. Maxwell's equations, learned by every electrical engineering undergraduate as the elegant synthesis of all the fundamental laws of electricity and magnetism, represents the framework upon which, with the aid of a few unifying steps, the theoretical proof of electromagnetic wave propagation is readily constructed. Hertz was perhaps the first to verify this theory experimentally. Thereafter, just after the turn of the century a succession of pioneering "communication engineers" defined this new profession with gradually more convincing experimental successes, culminating in commercial deployments for which the name of Guglielmo Marconi stands out for his outstanding blend of experimental and business acumen. Our purpose, however, is not to review the scientific and historical record, far better recounted elsewhere, but to note the particular features that impact modern wireless multiple access communication embodied in digital cellular networks.

Thus, we take for granted electromagnetic propagation but, as discussed in several chapters of this volume, note that the direct path from transmitter to receiver may not be the only path of signal propagation and, in some cases, may be blocked and hence attenuated far more than other indirect paths created by reflections off terrain or buildings. Consequently, for a transmitted signal

$$x(t) = A(t) \sin [2\pi f_o t + \theta (t)], \tag{E.1}$$

where f_0 is the carrier frequency and $A(t)$ and $\theta(t)$ are, respectively, the amplitude and phase modulation which bear the information transmitted, the received signal will be of the form

$$
\begin{aligned}
y(t) = \ & B_1(t - t_1) \sin\left[2\pi f_o(t - t_1) + \theta_1(t - t_1)\right] \\
& + B_2(t - t_2) \sin\left[2\pi f_o(t - t_2) + \theta_2(t - t_2)\right] \\
& + \cdots + B_L(t - t_k) \sin\left[2\pi f_o(t - t_L) + \theta_L(t - t_L)\right] + \eta(t),
\end{aligned}
\tag{E.2}
$$

where $t_1, t_2, \ldots, t_L$ are the propagation delays of the various propagation paths and $B_k(t - t_k)$ and $\theta_k(t - t_k)$ are, respectively, the received amplitude and phase for the kth path. Finally, $\eta(t)$ represents the additive noise at the receiver, partly of thermal origin, but which may also include interference caused by other emissions and transmissions not under the control of the communicators. The amplitudes $B_k(t - t_k)$ and phases $\theta_k(t - t_k)$ may be distorted versions of the transmitted amplitude and phase, which vary with time. As discussed in several of the chapters of this book, one model for this process is that of a time-varying delay line as shown in Figure E.1, where both the delays between taps, $\Delta t_k \triangleq t_{k+1} - t_k$, and the complex tap multipliers $\alpha_k(t) = B_k(t)e^{j\theta_k(t)}$ are time-varying functions. If these values can be measured exactly, then the optimum receiver, in a minimum-mean-square-error sense and in the presence of Gaussian interference, is the matched filter, which can be implemented as the time reverse[1] of Figure E.1, but with the complex amplitude $\alpha_k(t)$ replaced by its conjugate $\alpha_k^*(t) = B_k(t)e^{-j\theta_k(t)}$.

The problem, however, is in the feasibility of the measurement and its accuracy when the parameters vary rapidly in time. A fundamental limitation on feasi-

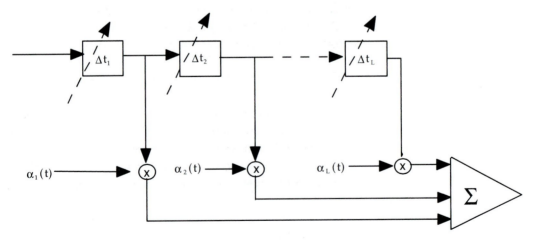

Figure E.1　Multipath propagation channel model: $[\alpha_k(t) = B_k(t)e^{j\theta_k(t)}]$.

[1]Usually referred to as a RAKE receiver.

bility, known as the uncertainty principle, dictates that measurement resolution in time is inversely proportional to the signal (and receiver) bandwidth. Thus, for example, if we wish to resolve two paths separated in time by $\Delta t = 1\ \mu s$, the bandwidth of the complex signal $A(t)e^{j\theta(t)}$ must be at least on the order of 1 MHz.

If the signal and receiver bandwidth W is less than $1/\Delta t$, then the paths will appear smeared together; for paths whose relative delays are separated by less than $1/\Delta t$, their complex amplitudes will occasionally cancel one another, thus causing deep fades in that composite component of the received signal. But measurement accuracy depends on more than bandwidth. Fundamental estimation theory (the Cramèr-Rao bound) leads to the expression for the standard deviation (inaccuracy) of the estimate for the peak-time of the received signal,

$$\sigma_{\Delta t} \geq \frac{k}{W\sqrt{E/N_0}}, \tag{E.3}$$

where W is the signal bandwidth, N_0 is the noise or interference density (assuming wideband noise of uniform density), and E is the energy of the signal component measured; k is a proportionality constant that depends in part on the definition of bandwidth but is not far from unity. Thus, the longer the measurement time, the more accurate the result, *provided* the signal parameters remain nearly constant over this duration. This phenomenon implies that for rapidly varying environments, such as with speeding vehicles, the measurement time must be shortened and the accuracy reduced. Though it would appear that wider bandwidths are always preferable, this too has its limitations. For while the far field propagation effects produce distinct multipath components (such as reflections from different buildings or hills), the near field effects may create many components, closely spaced in time. With a very large bandwidth, we may resolve these close components, but each will produce very little received energy, jeopardizing its measurement accuracy. Thus, keeping the bandwidth large enough to resolve a reasonable number of moderately spaced multipath components is important. Beyond this, it may be better to allow very closely spaced components (closer than the inverse bandwidth) to combine, occasionally causing fading of the combination through cancellation, but on the average with enough energy for accurate measurement. The key is to have a sufficiently wide bandwidth to isolate enough components (or composite components) and thus provide sufficient path diversity to guarantee an overall adequate signal strength for the optimally combined paths at the output of the matched filter. This strategy is usually effective for outdoor propagation, where the propagation time dispersion is on the order of several microseconds. Indoors, where propagation time dispersion is much smaller, other techniques for temporal and spatial diversity must be employed.

We move on now to the second set of laws.

E.3 SHANNON THEORY: LIMITATIONS ON SIGNAL PROCESSING

Claude Shannon, beginning with a remarkable series of papers in 1948, established the theoretical basis of digital communication with two well-known theorems: for source coding and channel coding. The first established the minimum bit rate required to reproduce a source signal within a given degree of accuracy; the second established the maximum rate at which transmission is achievable with arbitrarily high accuracy, in terms of channel parameters such as bandwidth and signal-to-noise ratio.[2] Underlying the proof of the channel coding theorem was the concept of signal randomness, which is closely related to wide bandwidth spread-spectrum signaling. We shall concentrate on the latter as it applies to multiple-access communication, which we have previously labeled "Shannon's Third Theorem." It is best expressed in terms of a game between a communicator and an interferer (or jammer). If both are restricted to transmit at power levels S and J, respectively, the communicator's channel capacity is bounded according to the Shannon-theoretic limits,

$$\frac{1}{2} \log \left[2\pi e \left(S + J \right) \right] - H(J) \geq C$$

$$\geq \frac{1}{2} \log \left(1 + S/J \right) \text{ bits/channel symbol,} \tag{E.4}$$

where C is channel capacity, $H(J)$ is the entropy of the jammer's signal of power J, and all logarithms are to the base 2. (This formula can be converted to bits/s by multiplying all terms by $2W$, where W is the communicator's bandwidth, assuming Nyquist-pulse signal modulation.) The minimax solution to this game, meaning the joint selection of communicator and jammer signals that maximizes capacity for the worst case jamming or that minimizes capacity for the best case communicator signal, is for both the communicator and the jammer to employ random signals whose first-order distributions are Gaussian and for which successive symbols are independent. The per symbol entropy of the jammer signal is thus $H(J) = (1/2) \log (2\pi e J)$.

Consequently, from (E.4) we obtain

$$C = \begin{cases} \dfrac{1}{2} \log \left(1 + S/J \right) \text{ bits/channel symbol} \\[2ex] W \log \left(1 + S/J \right) \text{ bits/s,} \end{cases} \tag{E.5}$$

[2]This maximum limit, known as channel capacity, has occupied legions of channel coding specialists for the intervening half-century. While simple techniques for reaching within about one-third to one-half of channel capacity (for a wideband Gaussian channel) have been known and employed for at least 30 years, only within the last few years has a composite technique involving iterative soft decoding of parallel or serial concatenated codes, known as "turbo" decoding, shown that efficiencies above 80% of channel capacity are practically achievable, provided sufficiently long decoding delays can be tolerated.

which is the usual capacity expression for a Gaussian channel. Gaussian signals can be approximated by spread-spectrum signals that are implemented by modulating the digital information onto a carrier already modulated by a random (or pseudorandom) sequence of symbols generated at a much higher rate than the information, approximately equal to the spreading bandwidth W. The ratio of the random sequence rate to the digital information rate is the spreading factor. Actually, provided the random carrier bandwidth is much larger than the information rate (the condition for spread spectrum), the symbols of the random sequence need not be Gaussian distributed. It suffices that the random sequence consist of independent, equiprobable binary symbols and thus be Bernoulli distributed; given the large spreading factor, the aggregate (sum) of the independent spreading symbols over the duration of one data symbol will approach a Gaussian distribution, according to the central limit theorem.

Returning to the communicator-jammer scenario, since the spread spectrum jamming is approximately uniformly distributed over the bandwidth W, we can define its spectral density as

$$N_0 \triangleq J/W. \tag{E.6}$$

Let the communicator's signaling bit rate be R_b bits/s, which is bounded by the capacity formula (E.5). Its bit energy is the ratio of power S to bit rate R_b:

$$E_b = S/R_b. \tag{E.7}$$

Combining (E.6) and (E.7), we have the ratio of tolerable jamming-to-signal powers,

$$\frac{J}{S} = \frac{W/R_b}{E_b/N_0}, \tag{E.8}$$

where E_b/N_0 in the denominator is the minimum value required at the receiver for tolerably low error probability. A lower bound on E_b/N_0 can be obtained from the capacity formula, for since $R_b \leq C$, it follows from (E.5) and (E.8) that

$$\frac{R_b}{W} \leq \log\left(1 + \frac{E_b/N_0}{W/R_b}\right), \tag{E.9}$$

whence,

$$\frac{E_b}{N_0} \geq \frac{W}{R_b}\left[\exp\left(\frac{R_b}{W}\ln 2\right) - 1\right] > \ln 2. \tag{E.10}$$

The lower bound, $\ln 2$, is approached for R_b at capacity and as the spreading factor W/R_b approaches infinity.

We turn now to a nonhostile and reasonably cooperative set of communicators in multiple access to a common receiver, such as a base station of a cellular mobile telephony system. If M communicators are all spread over the same bandwidth by independent random (or pseudorandom) carrier sequences and each has its power controlled so that they all arrive at the common receiving station with equal powers, S, then the demodulator for each user will be faced effectively with jamming power equal to the sum of the powers of all other users, $(M - 1)S$. It follows from (E.8) and (E.10) that the tolerable jamming-to-signal power and hence the tolerable number of other users

$$M - 1 = \frac{W/R_b}{E_b/N_0} < \frac{W/R_b}{\ln 2}. \tag{E.11}$$

Thus, the overall throughput aggregated over all users, normalized by the total (common) bandwidth occupied, is upper bounded by

$$\frac{MR_b}{W} < \frac{1}{\ln 2} + \frac{R_b}{W} \approx \frac{1}{\ln 2} = 1.4 \text{ bits/s/Hz}. \tag{E.12}$$

To approach this bound requires a very large spreading factor and error-correcting coding powerful enough to approach channel capacity. It also assumes that all the interference is caused by other users in the band, ignoring background noise of thermal or other origin. Including background noise and practically implementable coding techniques, throughputs of one-quarter to one-half of this value can be achieved, depending on the time-variability of the physical channel.

It can be shown that, with powerful enough error correction, the overall throughput MR_b can approach the classical channel capacity formula (E.5), with J equal to just the background noise *not* including any of the interference from the other cooperative users. Approaching this capacity requires that all users cooperate further in transmitting at specified but unequal powers, and that the common receiver optimally demodulates and decodes each user successively, subtracting off its effect from the common, overall received signal prior to decoding the next user. This latter procedure is an idealized form of the successive cancellation discussed in Chapter 3. At present, however, this optimal successive cancellation procedure remains a theoretical possibility only. Less ambitious cancellation or cooperative demodulation techniques abound, but they seem to yield only modest improvements.

We proceed now to review the practice of spread-spectrum techniques over the past half century.

E.4 Half a Century of Wireless Spread Spectrum: From Military to Commercial Applications

Spread-spectrum techniques for thwarting hostile interference or jamming date back to World War II. The sophisticated approach employing carriers whose spectrum is spread by a pseudorandom sequence generated by a maximum-length shift register (with linear feedback) date back to the fifties. In their simplest conceptual implementation, the binary sequences, which appear random and repeat only after $2^N - 1$ symbol times, where N is the length of the shift register, modulate the carrier by shifting its phase by $+\pi/2$ or $-\pi/2$ radians, corresponding to whether the symbol is a "0" or a "1," respectively.[3] The intended receiver knows the parameters to generate the same sequence and thus can demodulate by performing the inverse operation, shifting phase by $-\pi/2$ or $\pi/2$ corresponding to a "0" or a "1," respectively. Incidentally, by performing this inverse operation, the jammer spectrum at the demodulator will appear to be spread, even if the jammer's transmitted signal was originally narrowband.[4]

With the launch of Sputnik in 1957, the era of satellite communication began. Many of the early launches were of military satellites, which are "sitting ducks" to hostile interferers. Spread-spectrum techniques with very large spreading factors provided for a wide margin of advantage over a jammer, according to (E.8). Commercial satellites began operations in the mid-sixties and the first digital communication satellites were launched in the seventies. Spread-spectrum techniques were not employed commercially until the eighties, when they found their way into mobile terminals operating with very small and hence wide-aperture antennas in the presence of much stronger interference from stationary terminals transmitting through near-orbit satellites. With 250,000 mobile satellite terminals now installed in trucks worldwide, providing two-way communication and position location to their home bases, this spread-spectrum system has dramatically impacted the long-haul transportation industry. Finally, in the nineties, spread spectrum has had an even greater impact on the digital cellular communication industry. Here, it is usually referred to as *code-division multiple-access* (CDMA), to distinguish this access technology from frequency division (FDMA) and time division (TDMA) techniques.

The common thread through all these applications is tolerance to interference through digital signal processing. As discussed elsewhere in this volume, for cellular applications, the interference comes not only from the other users communicating through a given cell's base station but also from the transmissions of users in other cells, which contribute strongly to the background noise in the given base station. For FDMA and TDMA systems, it is generally necessary to allocate different bands or time slots to contiguous cells, thus reducing spectrum efficiency by an order of magnitude. With CDMA, all cells can be allocated the same common spectrum, a feature called *universal frequency reuse*. This of course increases the interference in each cell, reducing the number of users per cell, but only by a factor

---

[3]As noted elsewhere in this volume, this approach is generally called *direct-sequence spread spectrum*. An alternate approach, known as *frequency hopping*, uses multiple symbols of the pseudorandom sequence to select one of numerous carrier frequencies among which to hop.

[4]This case again implies a minimax solution to the game between communicator and jammer.

of about 1.6, a net gain over the other access techniques. As previously noted, to establish (E.11), each CDMA user must be power controlled to guarantee the maximum number of users per cell. But with tight power control, each transmitter emits the least amount of power needed to achieve reliable transmission, thus avoiding the usual power margin allocation and also reducing the interference to users in the same and other cells. Furthermore, the spreading feature of CDMA also guarantees a bandwidth wide enough to isolate multipath, using the RAKE receiver technique described in Section E.2. Also, with universal frequency reuse, transition between cells can be eased by performing "soft handoff"; as the mobile user approaches the edge of a first cell, it can begin communication with the second cell's base station without dropping the link to the first cell's base station. This artificially creates a dual diversity multipath condition. The same multipath RAKE receiver can thus be employed to handle soft handoff just as it does the natural multipath condition previously described in Section E.2.

Employing all the above techniques in implementing a digital multiple-access system, as well as error-correcting coding to reduce E_b/N_0, and variable rate transmission of digitally compressed voice, results in a spectral efficiency over ten times that of a conventional analog system. Hence, employing techniques based on Shannon laws improves the efficiency of a system based only on Maxwell-Hertz laws by more than one order of magnitude.

Their implementation, however, depends critically on digital signal processing technology whose practical and economic embodiment would be impossible without highly integrated solid state processors and memories. We discuss this technology and its underlying law in the next section.

E.5 MOORE'S LAW: THE SOCIO-ECONOMIC BASIS FOR DIGITAL WIRELESS

Gordon Moore, a founder of Intel Corporation, observed in the seventies that the number of devices per unit area that can be incorporated in a silicon integrated circuit (IC) doubles approximately every year and a half. We may state this fact as the formula

$$v(T) = v(T_0)2^{(T-T_0)/1.5},\qquad\text{(E.13)}$$

where $v(T)$ is the device density at time $T > T_0$, where time is stated in years. Although this may appear to be a physical law like those previously discussed (and in some minds it is considered such), it is in fact an empirical observation with no direct physical basis. Rather, it is explained by the fact that human ingenuity, coupled with market forces, produces exponential growth of technological capability, and eighteen months appears to be the time between product cycles in the semiconductor industry. This leads us to categorize Moore's Law as a socio-economic principle. Applying the formula and taking the initial year to be 1965, when one IC contained only one device, we find that by 1995 one IC of the same area could contain one million devices and by the year 2000 about ten million devices, both estimates being reasonably accurate. Within a few more decades, the atomic limit will be reached, but already experiments in subatomic storage would lead us to believe that the ultimate limit may be even further out. In any case, the millionfold growth in the last three decades has turned many a system theorist's dream into reality. The early military spread-spectrum systems consisting of multiple racks costing tens of millions of dollars are now implemented on a single chip (embedded in a palm-sized cellular telephone), costing only tens of dollars. In short, the progress described by Moore's Law was indispensable for the realization of the benefits of all three laws of Shannon (source, channel, and multiple-access coding). Yet, there remain the skeptics who do not understand, or merely ignore, the significance of Shannon's laws made practically implementable by Moore's Law and continue to design and attempt to justify systems that do not profit from these guiding principles.

E.6 METCALFE'S LAW: IMPLICATIONS FOR WIRELESS NETWORKS

Another socio-economic law of more recent vintage, due to Robert Metcalfe, states that the value of any communication network grows as the square of the number of users of the network. More precisely, if the number of users is N, then

$$\text{Network Value} \sim N(N-1)/2 \qquad \text{(E.14)}$$

since this is the number of connections[5] possible among N users. This law places an inordinate burden on the initiation of a network service that requires a significant capital cost to both the network provider and the consumer. Numerous failures attest to this fact. The most notable was "PicturePhone," an initiative of AT&T in the early seventies, which was discontinued after an initial trial period. As telephone data modems became ever more capable, offering progressively higher data rates at ever lower costs, a few manufacturers offered updated versions of telephone video terminals that could operate over the ordinary public switched network. Even these were not particularly successful because of low consumer demand both caused by and resulting from prices being high. The most common sale involved the purchase of a pair of terminals by grandparents who wished to view distant grandchildren. Turning to successes, we cite the Minitel data terminal operating over the French public switched network. This, in large part, was the

---

[5]Actually, since each link is bidirectional, one could argue that it should be $N(N-1)$.

result of heavy subsidization by the French government and thus removed the burden from the consumer. Another is the facsimile (fax) machine, which is practically ubiquitous in businesses and is becoming common in consumer households. For this success to occur, first the prices had to fall significantly and then the considerable saving and convenience became clear: a fax message consumes a fraction of the time of a telephone conversation, and more importantly, it gets through with greater accuracy than voice mail when the recipient is not present. This feature is particularly important for transoceanic calls, where the volume of fax messages now exceeds that of voice calls.

The most dramatic current network success is the Internet. Its usage has been growing precipitously in just the last couple of years, after a gestation period of over twenty years, during which the U.S. government through DARPA and, later, NSF, financed an ever-higher-speed data network to interconnect computer centers of universities and government facilities. Turned over to the private sector, its usage and required capacity have grown exponentially with time. The explanation is the growing presence of the personal computer at almost every desk and workstation of businesses and in the majority of homes, coupled with the rapidly falling costs of the embedded data modem. With tens of millions of users comes the opportunity to capture their attention as consumers, resulting in the creation of a multitude of information and other services through the World Wide Web.

Focusing, however, on our theme of digital wireless networks, the question arises as to whether Metcalfe's law applies in a strict sense. Even a single mobile phone has accessibility, through a base station, to all the fixed phones of the public switched wired network. Yet the mobile user wants to connect to the universe of users wherever he or she may roam. Thus, the operator must provide to this "first" user, connectivity through a multitude of base stations throughout a metropolitan area as well as in multiple areas and even in multiple countries. And this works in reverse as well. All the user's wired friends and associates want to reach her or him wherever she or he may be. Hence, an operator's capital expenditures dominate the network economics, at least initially. Thus, effectively, the value should grow as the square of the area covered (assuming, simplistically, a uniform density of users). Each unit area covered captures users linearly, for whom value increases also linearly with the size of the total area covered. Clearly, the network provider's economics are improved, the fewer the base stations required per unit area. A base station's coverage area and capacity, in numbers of users that it can serve, are the key economic parameters. With light usage, coverage is the principal factor. As usage grows, the capacity constraint dictates that more base stations must be provided. Both coverage and capacity are increased from two- to four-fold by the use of spread-spectrum techniques, enhancing economics and consequently network growth.

Case histories of the growth of two very different digital wireless networks are summarized in Table E.1.

TABLE E.1   Case histories of two digital technologies

	GSM (TDMA)	IS-95 (CDMA)
Technology Proposed	1982	1989
Technology Standardized	1988	1993
First Commercial Launch	1991	1995 (late)
1 Million Subscribers	1994	1996
5 Million Subscribers	1995	1997
50 Million Subscribers	1997	1999 (Estimated)
Spectral Limit*	Sooner	Later

*Limit set by Maxwell and Shannon, not Moore or Metcalfe

The first, GSM, is the Pan-European TDMA standard, now available also in the Americas and Asia. It was launched as a "Greenfield" service, meaning that new spectral allocations were provided for it. Also, most countries in which it was launched had either no previous service or a very inadequate and sparse analog service, and GSM was the only digital technology licensed in the countries of the European Union. Probably its most significant innovation was the introduction of "global roaming," whereby a phone purchased in any covered nation can be used equally in any other covered nation, wisely adhering to the just-stated wireless version of Metcalfe's law.

The second technology is the spread-spectrum, or CDMA, technology which has been the major theme of this essay. It was standardized in North America as the Telecommunication Industry Association's IS-95. Its purpose was to provide much higher spectral efficiency than the analog networks, already well established in North America. Thus, it was planned to ultimately serve an order of magnitude more users within the same spectral allocation, over the same base stations, which were nearing saturation with the inefficient analog access technology. A principal requirement, both economic and regulatory, was to not displace the existing analog users, unless they willingly chose digital service. The introduction strategy, therefore, contained three components:

- Provide superior service, including improved voice quality
- Convert only as much of the spectrum over to digital service as the user demand warranted, thus not disadvantaging the remaining analog users
- Provide for staged conversion of base stations so that initially not all stations needed to be converted to digital

CDMA was able to fulfill all three requirements. By its better than tenfold increase in user capacity, it required converting only one-tenth of the bandwidth to digital service to provide enough capacity to potentially serve all existing analog users on the network. By providing improved service, CDMA could attract the heaviest users; these represent typically 3% of users, consuming 30% of service. Thus, a relatively small percentage of conversions to the 10% digital bandwidth could alleviate congestion on the 90% of the band still serving analog users. Finally, by providing all converted users with a dual-mode analog/digital CDMA phone, the operator did not need to provide digital service on all base stations at once; when digital users roamed away from digitally equipped base stations, their phone was transferred automatically to analog without service interruptions. The statistics in the second column of Table E.1 attest to the success of this approach.

Standardized seven years later than GSM and launched four years later, CDMA now lags in customer adoption by two years. Significantly also, in almost all countries where it was introduced, the consumer had a choice between analog, CDMA, and one or two digital TDMA technologies. The most important conclusion, then, is that where there already exists an installed base, as well as considerable competition, it is essential, if not also mandatory, to provide backward compatibility to the predominant, previously existing technology.

The last entry in both columns deals qualitatively with the time at which spectral allocations will reach saturation. More users are supported by CDMA because of its greater efficiency. One approach for GSM operators to expand their capacity would be to gradually equip their base stations with CDMA technology compatible with their GSM network signaling. The high-usage customers would be provided with a dual-mode GSM/CDMA subscriber unit capable of communication with both types of base stations, in the same manner in which North American and Asian analog users were converted to digital.

We conclude with a look into the future. Unlike the wired network that already serves a substantial and ever-growing percentage of data users at varying speeds, the wireless network still offers primarily voice telephony, with only a minuscule percentage of low-speed data. This situation is certain to change over the next few years. The nomadic nature of business users, particularly for electronic mail and continuous connectivity to a home base, mandates the availability of wireless high-speed data (above 64K bits/s) beyond what is currently provided by ordinary wired phone connections. Several standards organizations worldwide are currently deliberating the merits of a variety of proposals for this service. Given the large number of current subscribers to digital cellular services, adherence to the principles of Metcalfe's law seems logical. In order not to start at the bottom of the quadratic curve, service providers would be well advised to choose a technology that is backward compatible to one of the existing services. The flexibility of spread-spectrum signaling facilitates such compatibility.

The complex interaction of technology and economics, involving the interplay of the four very diverse laws described in this essay, has created a vibrant wireless telecommunication industry, likely to serve an ever-widening percentage of the world's population, with both mobile and fixed service, for many decades to come.

List of Acronyms

ADPCM	adaptive differential pulse code modulation
ADSL	asymmetric digital subscriber loop
AGN	additive Gaussian noise
AMPS	Advanced Mobile Phone Service
ARQ	automatic repeat requests
ATM	asynchronous transfer mode
ATRC	Advanced Television Research Consortium
AUV	autonomous undersea vehicle
AWGN	additive white Gaussian noise
BCH	Bose-Chauduri-Hocquenghem
BER	bit-error rate
BPCM	binary pulse code modulation
BPS	bits per second
BPSK	binary phase-shift keying
BSC	binary symmetric channel
CCI	cochannel interference
CDMA	code-division multiple-access

CM	constant modulus or continuous media
CMA	constant modulus algorithm
CNR	carrier-to-noise ratio
CR	cross-relation
CRC	cyclic redundancy check
CSI	channel state information
CW	continuous-wave
DATS	digital acoustic telemetry system
DCT	discrete cosine transform
DES	Data Encryption Standard
DFE	decision-feedback equalizer
DFT	discrete Fourier transform
DMT	discrete multitone modulation
DPCM	differential pulse code modulation
DQPSK	differential quadrature phase-shift keying
DS-CDMA	direct-sequence code-division multiple-access
DSL (xDSL)	digital subscriber loop
DSP	digital signal processor
DTS/PTS	decode/presentation timestamp
ECVQ	entropy-constrained vector quantization
EZW	embedded zerotree wavelet
FA	finite alphabet
FDMA	frequency-division multiple-access
FEC	forward error-correction coding
FFT	fast Fourier transform
FIR	finite impulse response
FSM	finite state machine
FTRLS	fast transversal recursive least-squares
GII	global information infrastructure
GMSK	Gaussian minimum-shift keying
GSM	Groupe Speciale Mobile
HDTV	high-definition television
HOS	higher-order statistics
IC	integrated circuit
IID (i.i.d.)	independent, identically distributed
IIR	infinite impulse response
ILSE	iterative least-squares with enumeration
ILSP	iterative least-squares with projection
IP	Internet Protocol
IR	infrared

ISI	intersymbol interference
ISO	International Organization for Standards
JPEG	Joint Photographic Experts Group
JSCC	joint source-channel coding
LMS	least mean squares
LPTV	linear periodically time-varying
LS	least-squares
LTI	linear time-invariant
LTV	linear time-varying
MAI	multiple-access interference
MBone	multicast backbone
MDFE	multiuser decision-feedback equalizer
MF	matched filter
MIMO	multiple-input multiple-output
MIP	multipath intensity profile
ML	maximum likelihood
MLSE	maximum likelihood sequence estimator
MMSE	minimum mean-square error
MPEG	Moving Pictures Experts Group
MR-JSCC	multiresolution joint source-channel coding
MSE	mean-square error
MSK	minimum-shift keying
MO	multiple output
MU	multiple user/multiuser
NBI	narrowband interference
OPTA	optimal performance theoretically obtainable
OSI	Open System Interconnection
PAM	pulse amplitude modulation
PASTd	projection approximation subspace tracking-dilation
PCM	pulse code modulation
PCR	program clock reference
PET	priority encoding transmission
PLL	phase-locked loop
PN	pseudonoise
PSK	phase-shift keying
PSNR	peak signal-to-noise ratio
PTS	public telephone service
QAM	quadrature amplitude modulation
QoS	quality of service
QPSK	quadrature phase-shift keying

RACMA	real analytical constant modulus algorithm
RCPC	rate-compatible punctured convolutional
R-D	rate-distortion
RF	radio frequency
RKG	running key generator
RLS	recursive least squares
RTI	real-time interface
SA	successive approximation
SFQ	space-frequency quantization
SI	single input
SIMO	single-input multiple-output
SINR	signal-to-interference-plus-noise ratio
SIR	signal-to-interference ratio
SNR	signal-to-noise ratio
SOFAR	sound fixing and ranging
SOS	second-order statistics
SPS	symbols per second
SS	spectral selection or spread spectrum
S-T	space-time
SU	single user
SVD	singular-value decomposition
TCM	trellis-coded modulation
TCP	Transport Control Protocol
TCQ	trellis-coded quantization
TDL	tapped delay line
TDMA	time-division multiple-access
UAC	underwater acoustic channel
UDP	User Datagram Protocol
UEP	unequal error protection
ULA	uniform linear array
UV	underwater vehicles
VC	virtual circuits
VQ	vector quantizer
VR	variable rate
VSELP	vector sum excited linear predictor
WMF	whitened match filter
WSS	wide-sense stationary
WWW	World Wide Web
ZF	zero-forcing
ZFE	zero-forcing equalizer

Index